PLANT SCIENCE RESEARCH AND PRACTICES

A COMPREHENSIVE FIELD GUIDE TO MEDICINAL PLANT BIODIVERSITY OF MYANMAR AND MAURITIUS

PLANT SCIENCE RESEARCH AND PRACTICES

Additional books and e-books in this series can be found on Nova's website under the Series tab.

PLANT SCIENCE RESEARCH AND PRACTICES

A COMPREHENSIVE FIELD GUIDE TO MEDICINAL PLANT BIODIVERSITY OF MYANMAR AND MAURITIUS

MAYARAM UNIYAL

D. K. SEMWAL

AND

RUCHI B. SEMWAL

nova science publishers
New York

Additional color graphics may be available in the e-book version of this book.

Library of Congress Cataloging-in-Publication Data

ISBN: 978-1-53617-425-0
Library of Congress Control Number: 2020931166

Published by Nova Science Publishers, Inc. † New York

CONTENTS

PREFACE

At present 80% of the needs for healthcare are depended on rural areas where about 25% of the existing services are located. Almost 80% of the world's population meets their primary healthcare needs through traditional medicine. Medicinal plants, used in the home for self-administration, are maintained in home gardens or obtained locally. The majority of medicinal plants, used by traditional healers, is available in markets or collected from the wild. Medicinal plants are a source of alternative crops and income for local people in future rural development programs.

Traditional medicinal plants have been considered solely part of the health sector, but they are part of the agriculture and even of environmental program. The demand for medicinal plants is increasing at such a rate that natural stocks in the wild are being destroyed. Hundreds of species are over-harvested and face extinction if they are not protected or cultivated. India and China are the first countries to seriously grapple the efforts on medicinal plants. They are ensuring the long term health of the resource upon which billions of people trust in their lives. Since ancient time people have used medicinal plants for their health hazards.

A preliminary survey on medicinal plants of Mauritius was conducted during December 1996 to February 1997 whereas the survey of Myanmar was conducted during the months of October-November 1998 under the Indian Technical and Economic Cooperation (ITEC) Programme of Govt. of India. It was observed that there is tremendous scope for the development of an indigenous system of medicine in both countries. In the present study, it was observed that there is a good scope for acclimatization of the medicinal plants in these countries. It is also observed that there is an extensive possibility of attaining success in the development and cultivation of medicinal plants through proper conservation of the soil. Hence, it is suggested that the governments of Myanmar and Mauritius should increase the development of the medicinal plants gardens, herbariums and pharmacies in their respective country. There seems to be an extensive possibility of attaining the success of growth and development for the cultivation of medicinal plants through proper soil management. In addition, the governments should be given major emphasis on the purity and quality of herbal products considering both raw materials and finished products.

The present study and survey tours reveal the identification of more than six hundred medicinal plants found in the forest areas, nearby population areas and also in the botanical gardens. In addition to the compilation of the plants' species, the information about their medicinal values and other important uses has also been obtained from locals. The information gathered during the survey includes Ayurvedic plants found in these areas, possibility of propagation of the plants in the gardens, indigenous remedies and the traditional claims, guidelines for pharmaceutical assessment, product information for drug standardization, classification and preparation of drugs, manufacturing of drugs in Pharmacies and Drugs and Cosmetics Act, etc. These countries possess good technical capabilities to undertake the major initiatives in this direction.

The present textbook titled "A comprehensive field guide to the medicinal plants biodiversity of Myanmar and Mauritius" is divided into five chapters, in which the first chapter is based on the background of Myanmar including its people, tradition, agriculture, and other useful information. The second chapter covered general information about Mauritius Island with major emphasis on the traditional medicinal system of the country and its trade. The details of about 500 plants are given in chapter 3-5. The third chapter covers medicinal plants found in Myanmar but not in Mauritius whereas the fourth chapter comprised of the medicinal plants of Mauritius which are not found in Myanmar. All the common plants which are often found in both countries have been described in Chapter 5. The vernacular name, taxonomy, occurrence, Phytochemistry, traditional and biological importance of each plant is described in a concise way. In addition, photographs of selected plants are also given in separate section. This textbook can be used as a field guide during the survey of medicinal plants. We hope that this information in the form of a book will certainly be benefitted to the readers especially to the students of plant science and other relevant subjects.

Chapter 1

General Information About Myanmar

Introduction

Myanmar is the name which has long been used by its people to describe their homeland which the British called Burma, also known as the Golden Land for its rich land the wealth of its agriculture and minerals. Geographically, Myanmar is the largest in mainland South-East Asia sharing borders with Bangladesh, India, China, Laos and Thailand. With a total area of 675, 577 square kilometres, it is twice the size of Vietnam and approximately the size of Great Britain and France combined.

Myanmar enjoys a tropical climate with three general seasons. The rainy season during the South-West monsoons from mid-May to mid-October, the dry cool season from mid-October to mid-February and the hot season from mid-February to mid-May before the rains begin. The best time to visit Myanmar is from October to February.

Myanmar is a union of many nationalities as many as 135 groups, with their own languages and dialects. The ream Myanmar embraces all nationalities: Bamar, Chin, Kachin, Kayas, Kayin, Mon, Rakhine and Shan. The Bamars make up 68 percent of the total population of over 41 million. Myanmar lies in a meeting place of two of the world's great civilizations, China and India, but its culture is neither that of India nor China exclusively, but a blend of both interspersed with Myanmar social life and festive occasions that occur each month, Myanmar citizens are known for their simple hospitality and friendliness.

Myanmar is a new tourist destination in South-East Asia where one can enjoy the beautiful scenery of a peaceful and tranquil country with delightful terrain, crystal clear waterfalls and mountain ranges. Myanmar offers a variety of natural tropical beauty snow-capped mountains, silvery waterfall, magnificent, lakes, enchanting caves, unpolluted, unspoiled, alluring wide sandy beaches, ravines and the unique culture of different ethnic groups in addition to the famous thousand-year-old glittering golden pagodas from which begot the name, Myanmar, the Golden Land.

In the northern extreme of Myanmar are the snow-capped mountains of the Himalayas among which Khakaborazi with the height of 5881 meters above sea level stand as the highest mountains in South-East Asia. On the West, mountain ranges stretch, forming a massive wall terrain comprising numerous North-South ridges.

Besides the giant mountain ranges, Myanmar has a series of valleys and two coastal trips. The Ayeyarwaddy, recognized as the major lifestream of Myanmar, pursues its way southward from the confluence of its headwaters just 27 miles north of Myitkyina. Along the banks of River Ayeyarwaddy stand many ancient capitals of Myanmar King s where many existing old monuments reveal their glorious greatness and reign. Among others, Began located in the heart of Myanmar, with its thousands of ancient pagodas, recognized by world travellers as one of the challenging wonders of the world, revealing the rich cultural heritage of Myanmar and making Myanmar one of the archaeological treasure houses in Asia and the word.

Myanmar, the Golden Land is incredibly rich in natural resources. The Imperial Jade, the best pigeon blood rubies and the finest sapphires, all come from this soil of Myanmar. The fame of Myanmar Gems, Jade and Pearls have attracted many of the wealthiest jewel dealers in Asia and the world. Myanmar teak is Asia and the world's wonder, especially the sight of thousands of elephant and riders who each year extract more than a million tones of teak and other hardwoods from Myanmar's deciduous monsoon forests; showing off Myanmar's thriving elephant culture and its economic importance.

LAND

Physically, Myanmar falls into three well-marked geomorphological division: the Western Hills, the Central Belt and the Shan Plateau in the east with a southward continuation of the highland in the Tenasserim strip. The sea washes Burma on her western, southern and south-eastern coastlines, serrated and rocky for the most part, and covering a distance of about 1,200 miles from the Naaf Estuary in Arakan to Victoria Point in Tenasserim, which marges with Thailand and the Malay Peninsula.

Western Hills

The Western Hills stem from the vast mountain knot in the Tibeto-Chinese borderlands and swing southward through the Naga and Chin Hills and the ArakanYomas in a great arc some 700 miles along the sea to Cape Negrais, just west of the Irrawaddy Delta. In the

Kadu, Hpon, Maru, Lashi, Asi, Nung, Daru, Taungnyo, Taman, Yaw, Mro, Chaungtha, Myaingtha, Intha and Danu. Besides the Chins and Kachins, the second sub-group includes the Naga, Gauri and Duleng. The Lolo group includes the Lolo, Lisu or Yawyin, Lahu, Muhso, Kwi, Kaw and Ako.

The Chins moved into the western hills and the Kachins following later halted in the triangle near the headwaters of The Irrawaddy. The Lolos came down the Mekong valley and small groups settled on the eastern fringe of Burma. The principal wave of the Myanmar and proto-Myanmar moved southwards leaving small settlements as they came to form minor tribes of their stock in northern Myanmar.

Thai-Chinese

The third pressing wave of migration, composed chiefly of the Thai-Chinese race, came into Myanmar between the 13th and 14th centuries from Yunnan, where since the 7th century they had formed the kingdom of Nanchao.

At this period, fusion and intermingling between the first two waves of migration, through the medium of Buddhism, has resulted in the emergence of one culture, one religion and one history. But the Thai-Chinese group did make its presence felt by end of the 14th century. This third group includes the Shans, Karens, Shan-Myanmar, Shan-Tayok, Taungthus and the Thais.

Fusion

From the 14th century until the annexation of Myanmar by the British in the mid-19th century, a process of unification between the three main groups of migrants progressed rapidly in the country through the spreading of the civilizing factor of Buddhist culture. Until the British arrived, only the people in the outlying hills remained untouched by the civilizing media of Buddhism. The rest of the people of Myanmar embraced Buddhism, thus leading to one growth of but one culture, one written language and one literature in the country.

The term "Myanmar" may be traced through gold plaques with particular Buddhist motifs and inscription found near Prome to a period between the 6th and7th centuries A.D., although the chronicles record the city-state of Shrikitshara (near Prome) in the 8th century B.C. and the existence of two earlier city-states in the north. Today, the main races in Myanmar are the Kachins, Kayahs, Karens, Chins, and Shans.

According to the estimates of the latest census, out of a total population of about 25,246,000 inhabitants of the country, 644,000 belong to the Kachin State, 106,000 to the Kayah State, 743,000 to Kawthoolei (formerly known as the Karen State), 331,000 to the

Chin Special Division, and 2,550,000 to the Shan State, leaving 20,872,000 to Myanmar Proper. The major racial groups comprise about 15,644,000 Myanmar 2,005,000 Karens, 1,723,000 Shans, 655,000 Chins, 290,000 Kachins and 51,000 Kayah (formerly known as Karenni).

Myanmar Proper

The Myanmar Proper occupies an area of about 150,433 square miles of the country. Myanmar settled in the Dry Zone areas of Central Myanmar since the 9th century and it is here that sites of their ancient capitals of Pagan, Ava and Mandalay may be found.

It is the Dry Zone which remains the true home of Myanmar and the presence of a considerable number of Myanmar in Lower Myanmar is a development of the last hundred years, arising chiefly out of the growth of the rice industry in Myanmar. By religion, Myanmar are predominantly Buddhists. Agriculture is the primary industry of Myanmar and Myanmar are thus essentially a people with the realistic and unsophisticated outlook of those who live close to the soil.

The Mons settled in the Irrawaddy Delta and later in what are now Thaton and Amherst Districts in the Tenasserim. Like Myanmar, the Mons are also Buddhists.

Kachins

The Kachins live chiefly in the districts of Bhamo and Myitkyina which from the Kachin State. High mountains, some rising to 19,000 feet, rim the northern and eastern boundaries of the State. Excepting the wide flat Hukawng Valley, the northern part of the State is mostly mountains, while the southern part is hilly with vast stretches of flat country, particularly along The Irrawaddy.

According to ethnographers, the original name of the Kachin race is Jinghpaw belonging to the Tibeto-Myanmar subfamily. The five parental groups of Jinghpaw are generally recognized as the Lahpai, Lahtawng, Marip, Maran and N'Hkum. The majority of the Kachins are Animists (Nat-worshippers). A considerable number also embraces Christianity and Buddhism. The Kichin State, with its administrative headquarters at Myitkyna, has an area of 34,379 square miles approximately.

Kayahs

Formerly called Karenni or red Karens, the Kayahs inhabit the Kayah (formerly Karenni) State lying on the eastern slopes of the range which are dominated by Mt. Nattaung to the east of Toungoo in Central Myanmar and just south of the Shan State.

Ethnically, the Kayahs belong to the Mon-Khmer branch of Mongoloid stock. Besides the Kayahs, the Kayah State is inhabited by many other ethnic groups which comprise the padaung, Shan, Bre, Bew, Geba, Geko, Taungthu, Manu Manaw, Paku, Yintalaing, Yinbaw and Intha, each group speaking its own language or dialect. The Kayah State consists of the three sub-state of Kantarawadi, Bawla-ke and Kyai-pho-gyi and covers an area of 4,529 square miles approximately.

Although the small est in size of the constituent States of the Union, the Kayah State is rich in natural resources, such as hardwoods, wolfram, antimony, lead, zinc, tin and precious stones. The world-famous Mawchi Mines and the Baluchaung Hydro-electric Power Plant are located within the State. The administrative headquarter of the Kayah State is Loikaw, about 88 miles south of Taunggi.

Karens

The Karens, of Thai-Chinese origin, settle in large numbers in the eastern Toungoo District, in the Irrawaddy delta and the Tenasserim Division. Karens may be divided into two main classes, the Plains and the Hill Karens. The former comprising the Pwo and SgawKarens, form the bulk of the Karens and are found near Tonugoo, in the delta and in the Tenasserim Division.

Hill Karens are mostly Buddhists, while there are many Christians amongst the Pwos and Sgaws. Though Christianity has found a more receptive audience amongst the Karens than any other race in Myanmar, the Karens remain preponderantly more Buddhist than Christian. The Kawthoolei or Karen State with administrative headquarter at Pa-an, 33 miles north of Moulmein, covers an area of approximately 11,731 square miles. It comprises Papun Township, Hlaingbwe Township, Pa-an Township, Kawkareik Township, Kya-in Township and Thandaung Township.

Chins

The Chins occupy the hilly areas known as the Chin Hill and the Hill Tracts of Arakan between the Chindwin River and the borders of India and Pakistan. The Special Chin district consists of narrow valleys and steep mountains, several rising to over 9,000 feet. The area of the Division is about 13,907 square miles. The chins belong to the Tibeto-Myanmar branch of Mongoloid stock. The principal ethnic groups in the Division are Lai, Simbhrin, Kumi, Cho, Siyin and Matu. The majority of the Chins are Animists, while Christians and Buddhists number roughly about 30 per cent of the population. The administrative capital of the Division is Falam, about 86miles north of Kalewa town on the Chindwin.

Shans

The Shans inhabit the eastern part of Myanmar known as the Shan plateau which occupies about 56,000 square miles. Under the British, the Shans in Myanmar, through the system of a federation, retained their old feudal institutions through "Sawbwas" or feudal lords. But in April 1959, the Sawbwas or Saophalongs voluntarily renounced their hereditary rights.

The Shans belong to the same ethnic group as the Thais of Thailand, Viet-Nam, Laos and Cambodia.

The Shan State with an area of 60,155 square miles originally comprised 34 sub-states. The administrative capital of the State is Taunggyi. Under the Revolutionary Government's policy for complete racial equality, a new procedure was introduced for classification of the various national races in the country. Such terms as "racial minorities" or "nationalities" are removed from official usage. Also abolished was the order of precedence in the listing of the races according to population strength. With effect from the Union Day, February 12, 1964, the official list of races was proclaimed according to Myanmar alphabetical order.

Foreigners in Myanmar

Myanmar has a substantial population of foreign residents, chiefly those hailing from the two big neighbouring countries of India and China. The origins of their settlement in Myanmar date back to the earliest time. However, the vast majority of the present Indian and Chinese communities are the second or third generation of those who immigrated here in continual and swelling waves during the British regime. Occupational pursuits of the Indian and Chinese communities embrace a wide range, but their preponderance was in the fields of commerce and industry. The foreign population as of September 1966 is estimated to total 54,584 Indians, 80,057 Chinese, 16,092 Pakistanis and 11,699 other nationals.

It is known that about one-third of the per-war Indian population in Myanmar did not return from India where they evacuated during World War II. After Myanmar's independence, the progress of Burmanization in trade and industry narrowed the scope for foreign enterprise in this country. Hence, there has since been a steady annual decrease of foreigners in Myanmar. Also, their numbers have been reduced by many Indians, Pakistanis and Chinese adopting Myanmar citizenship. A further drop in the foreign population in Myanmar followed the nationalization of private business establishments in the last three years. The numbers of foreigners who have left Myanmar upto June 30th 1966 totalled 146,006 Indians, 13,855 Pakistanis, 2,934 Chinese and 5786 other nationals.

HISTORICAL BACKGROUND

The political history of Myanmar presents a clear perspective only after the first forging of the bond of political unification within the country in the 11th century. Prior to that, the history of Myanmar was largely the history of migration of different ethnic group from Central Asia, Tibet and China into Myanmar, in parts interwoven with legend and folklore.

Myanmar Monarchy

The 11th century is renowned as the golden age of Myanmar political unit and cultural progress. King Anawrahta (1044-77), founder of the Pagan Dynasty welded the former group of independent Myanmar state into one kingdom. From his capital at Pagan in Central Myanmar Anawrahta extended his control over the Irrawaddy Delta, the Thaton area and the hills east of the Sittang. Through his initiative and influence also spread throughout Myanmar the Theravada form of Buddhism to replace the Mahayana form. Pagan, visited by famous foreign travellers like Marco Polo, was in its heyday known as the city of four million (4,486,733) pagodas and is still considered today as one of the famous ruined cities of the world. It remained the capital of the Pagan Dynasty till the 13th century when the Mongols under Kublai Khan overthrew the Myanmar king and sacked the city. Myanmar then split into small principalities, governed partly by Shan and Mon chief unit in the 16th century when Kings Tabinshwehti (1531-1550) and Bayinnaung (1550-1581) again established unit which lasted till the early 18th century.

At the turn of the 17th century, there arose an interlude of political disunity caused by the Mon chiefs of the Irrawaddy Delta who regained their independence and also conquered a large part of Central Myanmar. But in the years 1752-58, Alaungpaya, an erstwhile headman of the Shwebo area, rallied Myanmar to re-establish a united Myanmar kingdom again. Alaungpaya'sAya Dynasty lasted till 1885. During this period, under Bodawpaya (1782-1819) the Maynmarempire attained its widest expansion, stretching from the hills of Assam to the coastline of Tenasserim. Alaungpaya, in his campaigns against the Mons, occupied their ancient citadel of Dagon and renamed it "Yangon", meaning the end of strife, to serve today under the modern international name of Rangoon as the capital city of Myanmar.

Colonial Period

Alaungpaya's successors soon become involved in wars with the British in India (1824 and 1885). Finally, the Third Anglo-British War of 1885 led to the capitulation of King

Thibaw and annexation of the whole of Myanmar to British India. Military defeat failed to subdue and demoralize the freedom-loving Myanmar in to tame acceptance of the accession of their country to British India. From the Outset, Upper Myanmar regarded the cradle of the Myanmar race, offered stubborn resistance to the British forces and it looks over five years for the well-equipped British army of 35,000 strong to "pacify" this region. Not to be forgotten was the early uprisings of Shans, Kachins, Chins and other ethnic groups against the British. These as well as much later uprising in Lower Myanmar, notably the Tharrawaddy Rebellion led by Saya San in 1930, demonstrated that the people could never be reconciled to any form of alien rule.

In succeeding decades, Myanmar political upsurgence assumed legal forms of expression. These activities from time to time succeeded in wresting constitutional reforms from the colonial authority. Though on a restricted scale, parliamentary democracy was introduced in Myanmar, gradually broadening in scope through the pressure of political agitation. The small legislature set up in 1897 was twice enlarged in 1909 and 1920. Myanmar was elevated from a province under a Chief Commissioner to that under a Lieutenant-Governor. In 1920, the "diarchy" system introduced in Indian under the Montagu-Chelmsford Reforms was extended to Myanmar, allowing her a new Legislative Council in which 79 out of 103 seats were filled by election (for the most part from popular constituencies). Myanmar was permitted some degree of independent matters classified as "Central subjects" remained under the control of the Government of India.

Such political concessions fell short of Myanmar objectives in the light of growing political awareness. They failed to restrain the political disaffection which resulted in a demand for separation of Myanmar from British India. Response to this demand came in the enactment of the Government of Myanmar Act of 1935 which enabled Myanmar in April 1937 to become a separate and distinct political unit. Under the new constitution provided by this Act, a bicameral legislature was set consisting of a half-elected, half-nominated 36-member Senate and a House of Representatives, the latter having 132 seats chiefly filled on the basis of a wider franchise for a five-year term. It also provided for a Council of not more than ten Ministers to advise and aid the Governor. General elections-the only one held under the new constitution-in November 1936 enabled Dr. Ba Maw, leader of the Sinyetha Party, to form the Government and become the first Premier of Myanmar. Labour unrest and political disorders such as student strikes and racial riots led to the fall of the Ba Maw Government in February 1939. It was succeeded by the Government of U Pu who resigned in 1940 and was replaced by U Saw of the Myochit Party. U Saw remained Premier till January 1942 when he was arrested and interned on grounds of connection with the Japanese. The next Premier was Sir Paw Tun.

The country was still dissatisfied with the successive constitutional reforms which did not really give the people the voice and power to settle their own affairs. Parts of Myanmar like the states were kept out of control of the Government in Rangoon, even though some of them were financed from the central budget. The Ministers Council remained purely advisory and the authority of the Governor was supreme even over the legislature. These obstacles to political advancement found resentful reaction among the people. Spurred by the renaissance of nationalism then mounting in tempo, all political parties were united in demanding full independence. Of these, the most revolutionary in character and militant in outlook was the "Thakin Party" or, to give its formal name, DobamaAsiayone. Thakin (later famous as Bogyoke, meaning General) Aung San was among the outstanding leaders of this organization which was a rallying point for patriotic youths to fight for the liberation of their motherland.

Such an uncompromising stand against colonialism and militancy of attitude towards national liberation became manifest in a nation-wide movement at the beginning of World War II. Several nationalist leaders were jailed but ThakinAung San and 29 other youths, later prominent as the "Thirty Comrades" succeeded in slipping out of the country. While the British Empire in Southeast Asia tottered under the blows of the Japanese, ThakinAung San and his comrades, after a period of military training in Japan, accompanied the Japanese army into Myanmar and quickly organized the Myanmar Independence Army to liberate the country. When the British forces evacuated from Myanmar in May 1942, the Governor, together with the premier, Finance Minister and sum senior officials accompanied them to India where they kept a Government of Myanmar in being with headquarters at Simla.

Final Stage of Freedom-Struggle

During the Japanese occupation, a Myanmar civil administration was established in March 1942 under the Japanese Military Administration. On July 24, the Myanmar Independence Army was demobilized, to be reorganized and formally installed as the Myanmar Defence Army on August 26 with ThakinAung San, now holding the rank of Major-General, as its commander. In August 1943, "Independence" was proclaimed for Myanmar and a new constitution came into force under which Dr. Ba Maw becomes the Adipadi or Head of State. But by now Myanmar opinion had turned decisively against the Japanese, whose real motives and attitudes were becoming increasingly apparent. A clandestine organization known as the Anti-Fascist Organization formed under General Aung San and Thakin May in August 1943 circulated a secret manifesto avowing its aims upon the people to sabotage the war effort and to attack the "fascist Japanese marauders" As proved by subsequent events, this clarion call received a wide response from all races of Myanmar, enabling the organization of an effective underground resistance movement. Open resistance broke out on March 27, 1945, when patriotic partisans struck the Japanese in the Prome area while British forces were swooping down on Upper Myanmar. Thereafter followed the Japanese route from Myanmar and British reoccupation or Rangoon on May 5. By this time the AFO had become a popular front embracing all political groups including the racial group and in August changed its name to the Anti-Fascist People's Freedom League.

British Governor Sir Reginald Dorman-Smith returned to Myanmar in October 1945 and military administration was replaced by a civil government. A deadlock ensued between him and the AFPFL in mid-1946 on the composition of the proposed Governor's Executive Council. Nationwide agitation followed, accentuated by a wave of strikes. He was recalled and his successor, Sir Hubert Rance, on September 27 reached an agreement with the AFPFL on the formation of the Executive Council with General Aung San as Deputy Chairman (equivalent to Prime Minister) and six out of nine members of the council nominated from the AFPFL.

The AFPFL supremacy in the new council signified the Myanmar determination and impatience for independence. Convinced of this significance, on December 20, 1946, British Premier Clement Attlee invited a Myanmar delegation to London to negotiate with the British Cabinet on details of the transfer of power. On January 1, 1947, General Aung San led the delegation to London. Its talks with the British Government led to the successful conclusion of the Aung San-Attlee Agreement on "methods by which the people of Myanmar may achieve their independence either within or without the Commonwealth as soon as possible".

Skepticism was voiced in some interested, including foreign quarters, on the willingness of frontier races to become merged in a single independent state on the attainment of the country's independence. However, thanks to the fraternal spirit prevailing among the frontier people and statesmanship displayed by General Aung San at the conference of all Myanmar nationalities held at Panglong in the Shan State on February 12, 1947, representatives of the Kachin, Kayah, Karen, Chin, Shan and other races spoke with Proper in a single state. Their reactions were also studied by a nine-member Frontier Areas Enquiry Committee headed by British Member of parliament, Mr. D.R. Rees-Williams. Election to the constituent Assembly, provided under the independence agreement, took place on April 7, 1947. Of the 255 seats filled, 210 were for Myanmar Proper (including 24 for Karens and 4 for the Anglo-Myanmar community) and 45 for the frontier areas as recommended by the Enquiry Committee. The outcome of the elections was overwhelming support for BogyokeAung San, with candidates on his AFPFL ticket winning more than 170of the non-communal seats in the Constituent Assembly.

The Constituent Assembly which began its first session on June 10, 1947, passed a historic resolution, moved by BogyokeAung San, providing for the proclamation of Myanmar as an independent republic. It next proceeded on its task of drafting the constitution. While the country stood on the threshold of freedom, BogyokeAung San, supreme of Myanmar's July 19, 1947, feel at the hands of assassins. This tragedy shocked the nation but his successor U Nu was able to form a Cabinet overnight and give the lead in finalization of arrangements for Independence.

The new Constitution of the Union of Myanmar was unanimously adopted by the Constituent Assembly on September 4, 1947. A final pact, necessitated by the change of Myanmar Government, the Nu-Attlee Treaty, was signed in London on October 17, 1947. A Bill to provide for the Independence was passed in the British Parliament, receiving royal assent on December 10, 1947. Accordingly, the long-cherished Myanmar goal was realized when on January 4, 1948, amid ceremony and international felicitations, Myanmar took her place within the comity of free and sovereign nations.

RELIGION

Buddhism, the teachings of the Buddha, is professed by the great majority of the people of Myanmar. The Buddhism that prevails in Myanmar is the Theravada Buddhism, which is to be found also flourishing in other countries like Ceylon, Thailand, Cambodia and Laos. The main tenets of the Theravada Buddhism as embodied in the Theravada Buddhist literatures of the three pitakas are - The Four Noble Truths, The Noble Eight Fold Path, The Seven Bojjhanga, the Low of Karma and the law of The Dependent Origination, The Theory of Anatta, the Four Paramattha Truths, The four Stages of Ariyas, and the realization of the different Bodhis, like Savaka-bodhi, Pacceka-bodhi and Sammasam-bodhi.

Buddhism, being the religion of the majority, is predominant and outstanding in the social life of Myanmar. Noteworthy, in this connection is the program for the promotion of Buddhism sponsored by various agencies under the Ministry of Religious Affairs. Soon after its taking over the administration, the Revolutionary Government started to reorganize the structure and to streamline the operations of the Buddha Sasana Council with the object of making it more effective in contributing to the moral, intellectual and spiritual development of the people.

The council engages itself in such activities as translating and publishing the original PaliPitaka Texts and their commentaries; compiling "TipitakaPali-Myanmar Dictionary" and "Guide to Tipitaka"; establishing Buddhist Missionary centres, especially with the aim liquidating illiteracy; holding of annual examinations at varying grades on Buddhist Texts; rendering aids to various meditation centre, and providing pre-requisites to the engaged in religious studies. Twenty-seven volumes of the 40 original Pali texts have been translated and published. One hundred and thirty-two missionary centres in remote hill tracts areas have been opened and there are altogether 336 registered meditation centres all over Myanmar.

Advanced Buddhistic Studies

The composite circular building standing in the quiet and attractive grounds of the Kaba-Aye (World Peace) Pagoda at Rangoon, houses the International Library for Advanced Buddhistic Studies. Opened in 1955, it is a centre of the post-graduate research study in Theravada Buddhism, as well as in the deeper aspects of Myanmar culture and Eastern philosophies. Its facilities are also open to interested visiting foreign scholars. Its evolution is directed towards the eventual formation of a composite World Buddhist University. Among its facilities are its rich and are museum and library collections, including historically authentic information on Myanmar religious edifices such as pagodas and monasteries.

Classes of postgraduate nature at seminar level centre around (1) Pail language, (2) Abhidhamma Studies and (3) Buddha's Basic Teachings and Kammatthana Practices, enriched further by the allied subjects like Philosophy, Psychology and Literature together with Library, Museum and Research facilities. The collections of the Institute are 8076 English volumes, 5386 Periodicals, 5070 Myanmar volumes, 8072 sets of Palm-leaf and rare manuscripts and 1542 museum objects.

Other Institutions

The Tipitakadara Selection Board annually holds scriptural examinations for the monks. The Pali University and Dhammacariya Department each year holds Pali University examination for which degrees, certificates and prizes are awarded to successful examinees. Under this Department, there are over 83 monasteries where hundreds of students are studying the doctrine of Buddhism. A tradition dating back to 1875 of holding Pali Patama by an examination is being perpetuated every year by the Pali Education Department.

While promoting the purification and elevation of the Buddha Sasana the policy of the Revolutionary Government is also to uphold the principle of freedom of worship in respect of all religions. State grants are therefore also made towards other religions professed in Myanmar, such as Christianity, Islamism and Hinduism.

Cultural Life

Myanmar's culture is intrinsically her own, developed by fusion of the cultural heritage of the indigenous races. This culture is further enriched by blending with influences in religion, philosophy, art and literature of neighbouring countries.

Mons

Moreover, a civilization which came to exercise a powerful influence on Myanmar was that coming from Myanmar themselves, the Mons of Myanmar. The Dvaravati Kingdom of the Mons in 835 A.D. was centred in southern Thailand, though many Mons inhabited the Irrawaddy Delta and the Tenasserim. As offshoots of the Khmer Empire of Cambodia, whose monuments, including the Angkor Wat, remain today as some of the noblest expressions of human artistic genius, the Mon state around the Gulf of Thailand and Martaban maintained close contact with each other and developed a civilization which eventually exerted due influence in Myanmar.

Pyus

The Pyus, the first inhabitants of Myanmar, also contributed to Myanmar culture. Inscription in their language, using archaic lettering, have been found near Prome and Shwebo. These inscriptions revealed that a Vikrama dynasty ruled from 673 to 718 A.D It is believed that this dynasty inaugurated the present Myanmar Era which began in March 6378 A.D.

Today, there exists a unity of culture among all the peoples of Myanmar. The mainsprings of this common heritage are the cultures of the Mons, the pyus, the Mramas or Myanmar and the Shans. The four cultures have blended into one, and its subdivisions of the Karens, Kayahs, and Kachins are but the expression of the same at a different stage of its development.

Cultural Unit

The Mons and the Pyus the first among the peoples of Myanmar to emerge from a comparatively primitive into an advanced stage of culture. The Mramas of Myanmar then living side by side with the Pyus succeeded to the Pyus as well as to the Mons politically and culturally. The Mramas and the Pyus first merged into one race.

The next conquered the Mons of Thaton and merged with them both racially and culturally. Thus the Pyus, Meamas and Mons amalgamated into one race and one culture. At the fall of Pagan in the 13th century, the Shans grew into ascendency, a position they maintained for some decades. During this period of Shan ascendency, Myanmar literature reached almost its full stature. The Golden Age of Myanmar history manifested during the Pagan Dynasty of Anawrahta in the 11th century, was the age of achievement mainly in architecture, but some advance was also made in sculpture, painting, music, ceramics and other branches of art. In the pursuit of these, religion was the predominant factor and the religious spirit inspired all cultural activities.

For more than two and a half centuries, Pagan flourished as the centre of Buddhist culture and civilization. In any appraisal of Myanmar culture, account must be taken of the profound depth to which Buddhism has saturated Myanmar culture, ideas, aspirations and way of life. Aside from bring the chief influence that has moulded the Myanmar way of life, Buddhism has provided the subjects and the outlet of expression in Myanmar art, architecture, sculpture and carving, as is evident from the most lavish work of art, sculpture and carving that went into construction of pagodas, temples and monasteries in Myanmar.

Music

At the pack of her achievements, Pagan was the famous centre for national music, including some adaptation from other Asian countries. Music was performed not only on all ceremonial occasion but also during daily offerings at important pagodas. At this period, the dulcimer, which is the prototype of the modern piano, found its way to Myanmar from Persia.

Myanmar music may, at first hearing, sound disconcerting to ears familiar with only Western music. But foreigners who become accustomed to it find that like most Oriental music it is a happy medley of highly contrasting sound, compared to the harmony that typifies Western music. Historical origins of Myanmar music can be traced back to over 1,500 years. It is on a heptatonic scale. Myanmar musical instrument was long regarded as the most highly perfected of their types, being praised in a Chinese chronicle of the year 802 AD. The traditional Myanmar orchestra is made up of the "saing-waing" or circle of drums, the "key-waing" or circle of gongs, the bamboo bar xylophone and the oboe and often with the inclusion of a large drum.

Drama and Dancing

Myanmar classical dramatic performances clearly fall into two sections: dancing of pure rhythmical movement and a dance drama unfolding a take, generally from scriptural versions of Lord Buddha's life on various planes of existence. Modern Burmese dancing is only an improvisation of the traditional, retaining the pristine art by its grace and symmetry of movement.

Zat, the most popular "Zatpwe" or drama as its main theme portrays epic stories from Myanmar Buddhist literature or history. Its side attractions are short modern play and dances, accompanied by plenty of clever clowning. Whereas "Zatpwes" used to be only all-night show, nowadays some are performing only up to midnight in the interests of public health and convenience.

Most popular is the "anyein" mainly a dancing show accompanied by a short play. Its duration is short, ending at or even before midnight. The "yein" or ballet with its group dancing and singing presents a colourful, graceful spectacle. The "Yokthepwe" or marionette show, with its skillfully manipulated human and animal figures, is a unique form of entertainment. These "yokthepwes" may consist of particular "Zat" sequences or the entire drama or the "Zat".

Lacquerware

Interiors of Buddhist monasteries provide excellent examples of gold lacquer-work. Prome and Pagan are the centre of Myanmar gold lacquerware, which are now displayed in vases, boxes, and small tables, etc. In coloured lacquer-work, Myanmar has attracted world attention. Mythological Buddhist motifs form the stock designs of the artist. Lacquerwares are probably the only subjects still conserving the various compositions and designs originated during the glorious Pagan period.

Lacquerware manufacture was introduced from the Yuns of northern Thailand, hundreds of years ago, and these wares are still known as "yun." Another source of pride in Myanmar art-craft is the making of exquisite bamboo-and-silk umbrellas with their colourful designs as produced in Bassein. Best examples of Air frescoes and plaques with inscriptions in pali, found in Pagan, Mandalay and Ava, illustrate the religious and secular life of Myanmar down to the end of the middle ages.

Myanmar was also found of moulding objects in bronze, especially bronze bell. The largest bell in Myanmar and probably the largest hanging one in the world is to be found at Mingun near Mandalay. It is a close rival to great bell of Moscow, for Mingun Bell possibly weighs from about 90 to 100 tons. It is 12 feet in height and 10 feet in external diameter.

Language

The Myanmar language is of the Tibeto-Chinese group. However, unlike Chinese, it is not ideographic, that is, having base characters indicated by pictures. So it has an alphabet and a script. It is chiefly mono-syllabic. It is also tonal, with abrupt shortening and extending of the sound of words to interpret their different meanings. Like all languages, it has its share of long-words of foreign origin. Modern Burmese language has again become simple and colloquial.

The origin of the Myanmar language many be gauged from the fact that there are about 12,000 words that are common to the Myanmar, Tibetan and Chinese, while about 1,000 words are traceable to Tibetan and not less than 750 to the Chinese. Today, Myanmar is the main language spoken and written in the country though it is but one of the languages of the Myanmar group. About 70 per cent of the people of Myanmar speak a language of this Myanmar group as the mother tongue. The written language is the same all over the country, although local variations occur in pronunciation.

One particular phenomenon of Myanmar is the predominance of the Myanmar language in the country. The political dominance of the Myanmar group through the ages gradually established the Myanmar language as the principal literary medium from the 16th century and it is only in the history of writing in Myanmar that adequate sources of literary history are found.

Literature

Throughout Myanmar's comparative isolated history since the basis of education has been largely forthcoming from the religious orders, Myanmar writing has been dominated by Buddhism and much of the best work has been that of religious teachers. Myanmar literature is rich, extensive and interesting. Literature in a lesser volume also exists in other indigenous languages.

Beginning from the translation of the original Pali foundations of the Pitakas, Myanmar authors produced an enormous secular literature on various useful subjects, especially law, medicine, astronomy, astrology, politics, prosody, poems, drama, fiction and music. Myanmar literary composition, both in prose and verse seem to have been based initially on the pali styles of composition, following the prosodial and rhetorical rules of pali and Sanskrit. But, whereas in pali and Sanskrit versification the bases of metrical units are short and long vowels, and in English accented and unaccented syllables, in Myanmar, they are rhymed and unrhymed words.

Myanmar prose literature is also rich in the inscription, royal instruction, friendly letters and chronicles, while the "Donchins" in prose are meant for group dancers. Myanmar literature has a wide range of style and composition, from the heavy and the serious to the light and merry. In the early days the subjects were mostly scriptural and didactic, but later on, romantic elements were introduced with pure aesthetic aims.

From the middle of the 18th to the middle of the 19th century, Myanmar literature received a great fillip from the patronage of the Alaungpaya dynasty and some influence was felt from the impact of other civilizations which came to be represented in Myanmar. Works on history, statecraft and science, and shorter verses and odes instead of long narrative poems grew into great popularity. An innovation was introduced when writings were accomplished in a style mixing rhymes of verse with cadences of prose. This style used for drama and discursive seriocomic narratives. Romances of court life, literary epistles and love letter gave Myanmar a many-sided and interesting literature.

Modern Myanmar writers are, however, drawing on some current resources. In the past, the prosaic styles have seldom been supple and adequately precise for modern tastes. It is in the direction of improved clarity and ease of expression that Myanmar writers are moving today. Western influences have had their effect on Myanmar literature, but the nationalist movement leading to Independence has arrested the sway, and Myanmar literature while retaining its pristine purity is drawing on current resources to enable it to take its place among the modern languages and literature of the world.

Cultural Renaissance

Various form of Myanmar culture, which languishes during the period of British occupation, being to gain strong national attention and interest with the attainment of independence in 1948. Due attention to this trend towards the revival of Myanmar culture was also paid by the Union Government of Myanmar, Whose realization of the role of culture as a significant aspect of nation-building led to concrete encouragement by the formation of a Ministry of Union Culture in April 1952.

As a first major step in the programme of cultural renaissance, a Cultural Institute was established in June 1952 for the development and maintenance of a National Library, National Museum, and National Gallery of Art, Academy of Art, and a National Academy of Music, Drama and Dancing.

Activities of the Institute include the Publication of cultural literature, provision of classes and lectures on subjects of culture, interest, holding of a periodical exhibition of fine arts and also displays of music, drama and dancing. Production of musical notes and recording of classical song, of books and pamphlets on folklore, custom and traditions, and microfilming of ancient Mon and Myanmar literature have also been started.

National Museum

Proud possession of the National Museum include the Lion Throne, the Royal Couch of King Mindon's Queen, the Crocodile Harp, a Bronze Cannon of 1757 AD, the Silver Reliquary embossed with figures of the Buddha's and their disciples dated about the 7th century AD and may other exhibits of antiquity and works of art.

The Lion Throne, used in the Supreme Court of King Thibaw, serves as a perfect specimen of exquisite carving. It is about 27 feet in height with its fringes and outer wings pointing in one piece or in wave-like sections, inwards towards the centre. The Silver

Reliquary bears lines of inscriptions in Pyu and Pail languages pointing to the existence of Buddhism in Myanmar in the early centuries of the Christian Era.

Mandalay Regalia

Of great value and historical importance to Myanmar, the Mandalay Regalia was returned by Britain to the Myanmar Government during the visit of General Ne Win to London in 1964. A remarkable gem-studded assortment of about 53 items, the regalia were kept in the Victoria and Albert Museum in London ever since it was taken away when Mandalay capitulated to the British in 1885. Now returned to Myanmar, it consists of a sceptre, swords, personal jewellery of the Myanmar monarchs, salvers and a variety of ornate bowls and other receptacles.

Other museums in Rangoon well worth a visit are the Defence Services Historical Research Museum in Stewart Road and the Natural History Museum at the former Union of Myanmar Club premises, both of which are open to the public. BogyokeAnug San Museum in Tower Lane, located at the house where Myanmar's foremost national hero had lived and contained his personal effects and furniture, is open to the public during the Martyrs' Day week, beginning on July 19, every year.

State Schools of Fine Arts

The system of State Schools for Fine Arts reorganized with an eye to the needs envisaged by the new socialist order in Myanmar. For this purpose, one of the first steps taken was to provide more spacious accommodation for the State School of Fine arts and the State School of Music and Drama which shared the same premises in the Jubilee Hall grounds. So the State School of Fine Arts was shifted to the former AFPFL headquarters in Rangoon.

The standard of both the teaching techniques and qualifications of new students for these schools were also improved. Foreign instructors, as well as veteran local artistes, conducted courses in Myanmar dramatic arts which have a historical affiliation with those of neighbouring countries. A State School of Fine Arts, Music and Drama also exist in Mandalay for Upper Myanmar. Research on such arts is carried on by the Music and Drama Department of the Cultural Institute. It has a fine collection of old and contemporary workers on music and drama and gramophone recorders, including those of performances by per-war famous singer and musicians.

At this department, classical songs are also being put down in staff notation. The study is made on instruments of ancient renown for preparation of a history of Myanmar musical instruments. The department's State Orchestra and Cultural Council Dramatic Troupe perform at state functions and since the advance of the Revolutionary Government at state Factories for the entertainment of workers.

Archaeological Department

This department is busy in excavation work at the latest big archaeological discovery in Myanmar at the Peikthanomyo (described under "Places of Interest" in late pages of this handbook). This excavation has unearthed the ruins of a citadel which was but vaguely known in traditional accounts and legendary history of about 2,000 years ago.

Excavations by the department are also underway at the site of the Paunggu Pagoda in the ancient city of Pagan. These excavations yielded fine relief sculptures on stone plaques and the latest discovery was a clay votive tablet bearing figures of the Buddha and two lines of writing in old Mon script recording the offering of the tablet by the donor. At Hmawza (Old City of Prome) stands a field museum housing 100 selected stone and terracotta objects collected at the site by the department.

National Library

The National Library, maintained by the Cultural Institute, is located in the City Hall of Rangoon. It houses an interesting collection of rare books and antiques. Up to the middle of 1966, its collection totalled 40,500 books, 13,335 palm leaf manuscripts 23,200 of other kinds of manuscripts, 17,000 issues of magazines and 41,700 issues of newspapers. The manuscripts include such old works as NanzinpoukSaphyayKyan (Answers to Royal Queries) and early recorded Ramayana classics.

A library which has expanded and gained public appeal in recent years is the Information Directorate Library in Pansodan Street. It is a handy centre offering official publication, periodical and newspapers. Another leading library in Rangoon is that of the SarpayBeikman Institute at the Corner of 37th and Merchant Streets. Private libraries are also extensively to be found in every town.

PAGODAS

Following a time-honoured custom of building pagodas as shrines to the memory of the Buddha, there are innumerable pagodas scattered all over the country in Myanmar. It has been truly said that dominant feature in any scene in Myanmar is that pagoda and the music of pagodas bells tinkling in the breeze has enchanted many a foreign victor. Most of the well-known pagodas in Myanmar are inevitably found at ancient capital or in place that is of historical significance.

Shwedagon

The towering Shwedagon Pagodas in the capital city of Rangoon is the most famous shrine in Myanmar and as the largest Buddhist temple of its kind in the world remains the most venerated place of worship for Buddhists all over the world. The shwedagon has a history dating back 2,500 years, and accounts of it are intermingled with the founding of the city of Dagon or Rangoon.

Accounting to legend, two merchants of the Kingdom of Okkalatravelling abroad secured hair relics of the Buddha. King Okkalapa built a golden Pagoda enshrining the sacred relice on Singuttara Hill in the town of Okkala which was later to known as Dagon in the 11th century. The importance of the town grew with that of the shwedagon Pagoda.

Originally only 27 feet high, the pagoda was raised to a height of 66 feet while as the royal tradition successive Mon and Myanmar king repaired, enlarged and regilt the pogoda. It was during the reign of famous Queen Shin-sawbu (1453-1472) that the Shwedagon first assumed its present appearance, while King Sinbyushin of Ava in 1777 raised the pagoda to its present height of 326 feet.

The perimeter at the base of the pagodas, is 1,420 feet, and this base is surrounded by 64 smaller pagodas, with four larger once in the one each side. There are also four Sphinxes, one at each of the four corners, with three "Chinthes" or Leogryphs on each side of them. On the platform, the pagoda is surrounded on all sides by several shrines buildings (tazanugs) and "Zayats" or rest houses. The platform is paved with marble slabs and most of the buildings have unique wood-carvings in floral designs and also pillars with mosaic work. Figures of the Buddha seated in the buildings are generally of alabaster or brass. There are also two huge bronze bells. The Maha Tissada Ganda Bell in the northwest corner, presented by King Tharrawaddy in 1841, weighs about 40 tons and is 8½ feet high. The Maha Ganda Bell, presented in 1778 by Singu Min, son of Sinbyushin, weights about 16 tons and is 7 feet high.

Ascent to the pagoda is by electric lifts installed on the eastern, western and southern sides and by any of the four flights of stairways, one opposite the centre of each face, with flower and candle booths lining the stairways. The principal stairway is on the south side from pagoda Road. The entrance to this stairway is flanked by two large Leogryphs, 30 feet high, with statues of ogres and mythical alchemist in the background. The pagoda itself is divided into a base of 2 feet 3 inches in height; three terraces called "pichayas" with the first terrace containing 64 small pagodas reaching a height of 15 feet 10 inches; and the other terraced respectively of 38 feet 5 inches and 25 feet 10 inches in height with a circumference of 344 feet 8 inches; the inverted "thabeik" or begging bowl and the "baungyit" or twisted turban with embossed bands standing to a height of 41 feet. Next to the "banungyit" is the "Kyalan" or ornamental lotus flower which is 31 feet 5 inches high.

Reaching the top is the "hnget-pyaw-bu" or plantain bud-shaped spire 52 feet 11inches in height with a girth of 65 feet tapering to 23½ feet at the top. Resting on the top is the "Hti" or umbrella 33 feet high with a diameter of 45 feet 6 inches at the bottom and 14½ feet at the top. On top, the Hti is a 15 feet shaft to which end is hinged the van 4 feet 8 inches long and 2½ feet wide. The "Seinbu" or Diamond Bud which crowns the apex of the vane shaft is a golden globe, 10 inches in diameter and inlaid with diamonds and other precious stones.

Sule

Another famous Pagoda in the city of Rangoon is the Sule Pagoda, which serves to make the exact centre of the city. The Pagoda was built about 2,250 years ago during Mon dynasty and enshrines Buddha's Relice brought over in ancient times by two Buddhist missionaries from India. The pagoda is about 157 feet in height and is seven feet higher than the Independence Monument in Bandoola Square.

Kaba Aye Pagoda

This pagoda, dedicated to the cause of world peace, was completed in early 1951. It is about seven miles north of the centre of Rangoon. This pagoda has a perimeter base of 118 feet and is also 118 feet high. The pagoda site also served as the focal point for the construction of a Buddhist university, known as the International Institute for Advanced Buddhistic Studies, and also the holding of the Sixth Great Buddhist Council.

Ananda

The ruins of Pagan contain the largest concentration of pagodas in the country and the monuments, which were erected during a period in the 12th-century date back to 108 A.D. The majority of these well-preserved pagodas, whose architectural beauty continues to enrich posterity, were built during the Pagan dynasty (1057-1284 A.D.) before the Mongol invasion by Kublai Khan. The most prominent of the temples and pagodas of Pagan is the Ananda, a massive structure which was built by King Kyansittha in 1091 A.D. The main features of the Ananda include a solid central block crowned by a four-sided tapering tower, and with two galleries or colonnades, one inside the other, running the whole way round the central portal portion and equipped with main entrances in the middle of each of the four sides. Standing against the central block and facing each of the entrances are four gigantic gilded images of the Buddha.

At the western entrance, and in front of the image of Buddha are the statues of Kyansittha and Shi Arahan, the latter being one the founders of Theravada Buddhism in Upper Burma. The colonnades contain stone sculptures of exquisite workmanship, representing scenes in Gotama Buddha's life, and terracotta tiles representing scenes from the Jataka stories which are previous existences of the Buddha. The projections which are previous existences of the colonnades produce an effect which makes the entire structure to resemble a Greek Cross. Close to the Ananda is the Museum, which contains exhibits illustrating the iconography, architecture and the religious history of Pagan.

Shwezigon

About five miles from the ruins of the city of Pagan is the present-day town of Nyaung-U. Situated close to this town is the Shwezigon built by King Anawrahta (1044-1077 A.D.) but completed by Kyansittha. The Shwezigon is a solid cylindrical structure and forms the prototype of other Buddhist pagodas in Myanmar, and indeed strongly resembles the Shwedagon of Rangoon.

Set in panels around the terraces of the pagoda there are enamelled plaques illustrating scenes in the previous lives of the Buddha. The pagoda enshrines the frontal bone and a tooth of the Buddha and consequently is held in great veneration by Myanmar Buddhists all over the country. There is an annual festival held in honour of the pagoda.

Thatbyinnyu

The pagoda was built by Alaungsithu, grandson and successor of Kyansittha in 1144 A.D., and is one of the premier pagodas of Pagan. That byinnyu signifies "omniscience" one of the attributes of the Buddha. This pagoda is the loftiest all buildings in Pagan (210 feet) and contains five stories; the first and second used as residence by monks in the old days, the third-the shrine proper in which is housed a large image of the Buddha, the fourth used as a library while the fifth was a pagoda embodying sacred relics.

Gawdawpalin

The Gawdawpalin was built by Narapatisithu (1173-1210 A.D.) and in general, plan resembles the Thatbyinnyu, though it is not so bulky while its spire is more graceful. The history of this pagoda id recorded on a stone inscription erected nearby.

Mahabodhi

The pagoda was built by Nandaungmya (1210-1234 A.D.) and unlike others in Pagan, is a tall, obelisk stupa. It is the only specimen of its class to be found in Myanmar. The whole structure is covered with niches bearing seated Buddhas and interspersed with ornamental panels and mouldings.

Shwesandaw

This is the first pagoda built by Anawrahta after his conquest of Thaton in 1057 A.D. The pagoda enshrines some Sacred Hairs of the Buddha obtained from the Mon Kingdom. The pagoda is also known as the Mahapeinne from the fact that each of the four corners of its lowest terrace is guarded by three Hindu gods. It is a cylindrical solid stupa with five terraces while the others have only three.

Htilominlo

Built by Nandaungmya in 1218 A.D., this massive pagoda is a double-storeyed temple surmounted by a "sikhara" The walls are filled with writings and frescoes, and instead of terracotta tiles, it is decorated with pieces of sand-stone glazed in green colour. There are four Buddhas facing the cardinal points on each of the lower and upper stories.

Shinbinthalyaung

Within the precincts of the Shwesandaw Pagoda is a colossal recumbent image of the Buddha built in the 11th century. Another recumbent image of gigantic proportion representing the Buddha in the act of entering Nirvana may be found in the Manuha Temple near Pagan.

Shwethalyaung

About 50 miles from Rangoon, on the outskirts of the ancient Mon capital of Pegu, is another colossal recumbent image of the Buddha, the Shwethalyaung, considered the largest complete image of the human figure in the world and measuring 180 feet in length and about 50 feet in height at the shoulder.

Other Pagodas

Other famous pagodas in Myanmar include the Shwemawdaw of Pegu, the well-known pagoda of Kyaiktiyo, near Thaton, built on top of a boulder balanced on the edge of a precipice, Shwesandaw of Prome, Kaung-hmudaw of Sagaing, Mahamuni or Arakan pagoda at Mandalaya Thanbokde at Monywa and many others, as most large towns in Myanmar are centred around large pagodas in the area.

National Custom

Religion and Society

Myanmar social life rests mainly on the stable foundations of Buddhism. This explains why it is imbued with the glorious ideals of self-abnegation and love. Religions teachers or "Phongyis" hold a high place in Myanmar society and their influence pervades all social strata. The "Dhamma" they preach on religious occasions is listened to with rapt attention by all devout Buddhists. Every village and hamlet has a "phongyi-kyaung" or monastery maintained by the community.

These monasteries also exert a strong influence in the educational field. On specified days men and women flock to the monasteries to listen to religious discourses by the monks, while at an early age Myanmar children are sent to monasteries to learn the fundamentals of Buddhism and thus incidentally to read and write. This accounts in large part for the high rate of literacy in the country.

Also, according to time-honoured custom, every Myanmar Buddhist male over seven years of age becomes ordained as a novice to a monastery for the period, which may be for Lent, or a year or several years, or even for as a short period as a fortnight. Through tradition, it is now incumbent on every young male Myanmar Buddhist to go through this "Shinpyu" ceremony. In the case of children of people who cannot afford it, some charitable society or a group of people or a person of some wealth would always come forward with the offer to stand the expenses. Myanmar also expresses deep respect for his religion by refraining from wearing footwear in such religious places as pagodas and "phongyi-kyaungs." This custom, as a measure of respect, has also gradually entered into social life.

Characteristics

Myanmar is a land of colour, not only through its science qualities but also due to the natural habits and characteristics of the people themselves. Proverbially noted for their highly developed sense of generosity and hospitality, Myanmar are very friendly and cheerful, witty and humorous, presenting an ever-smiling appearance but, nevertheless, conscientious in their work, duties and responsibilities. Myanmar is a land of laughter, and Myanmar usually greet with smiles.

The dress and costume of Myanmar also lend greatly to the colourful and picturesque quality of the country, attired as the people are in clothes of bright, gay and vivid colours. The "aingyi" (jacket) and the colourful "longyi" (sarong) constitute the costume of Myanmar with differences in style and composition between those of the men and women.

Festivals

Myanmar is a land of festivals. Myanmar culture centres around these festivals since Myanmar life and custom are so inextricably bound up with religion. Most of these festivals are in the nature of great social holidays, while a considerable number are the patronal festivals of pagodas. Moreover, each month of the Myanmar calendar has its own particular festival.

Two great festivals in Myanmar mark the beginning and end of the Buddhist Lent, which usually occurs during the rainy season and has duration of about three months. The full-moon of the Myanmar month of "Waso" (July) marks the beginning of "Wa" or the Lent. Waso Festival commemorates the sojourn of the Buddha in Tawadeintha Heaven, the abode of the Devas, to preach the "Abidhamma" to the celestial beings.

During Waso, marked by its solemnity and the absence of usual merry-making, alms are freely distributed and "phongyis" or monks are provided with new raiments and other temporal requirements during the Lent. For three months Buddhist monks observe a period of spiritual retreat and stricter religious duties, while laymen also observe a Lenten period of fast, prayer and meditation. Buddhists do not celebrate marriages during Waso, and shifting of abodes in this period is not encouraged.

The ending of the Lent is marked by the "Thandingyut" Festival or the Festival of Lights, which occurs on the full moon of the Myanmar month of Thadingyut (October). The festival is ushered in with rejoicing and alms-giving. This festival commemorates the Buddha's return from Tawadeintha Heaven when the Devas lined His route and illuminated the way. During this festival, lasting about three days, myriads of lights illuminate not only the Pagodas and monasteries but also all Myanmar houses.

Thingyan

Ushering in the Myanmar New Year is the three-day "Thingyan" or Water Festival, which usually falls around April 13. The date of commencement varies each year, according to the Myanmar calendar. On the morning of New Year's day images of the Buddha are washed ceremoniously in scented water and "phongyis" are invited to sumptuous meals. Traditionally, this festival id celebrated with great verve and gaiety by the people for a period of about three days, marked chiefly by the custom of throwing water on one another. The sprinkling of water during this festival is symbolic of blessing, goodwill and respect. Decorated floats, vehicles laden with merry-making water throwers, and wayside pandals offering free refreshments to passers-by, and at the same time throwing water on them are a happy feature of this festival.

The "KasonNyaung Ye" Festival is celebrated on the full moon of the Myanmar month of "Kason" (about early May). Pouring of water on the "nyaung bin" or the Bo tree commemorates the occasion when Buddha attained enlightenment while meditating under a Bo tree, and the subsequent watering of that particular tree by the followers of Buddha. The "Tazaungdaing" Festival is celebrated on the full moon of "Tazaung-moon" (mid-November). This festival is also an occasion for illuminations and the sending aloft of fire-balloons. It is the occasion for "Kahtein" ceremonies when the Buddhist monks are provided with various monastic requisites. Besides these main festivals and those celebrated monthly, there are various other local festivals including those of the larger pagodas, local pagodas, the "phongyi-byans" and the "Shinpyu" ceremonies.

Shinpyu

Shinpyu ceremonies for boys and the ear-boring ceremonies for girls have deep socio-religious significances steeped in both religion and tradition. According to religious custom in Myanmar, every Myanmar boy who reaches the ago for seven or more is initiated in to the Buddhist Order and spends a few days or weeks in the monastery. Ear-boring for the girls are performed at the same time if they happen to be sisters to the boys. No particular time of the year is set apart for these ceremonies, which are usually attended by great pomp and splendour. Phongyi-byans are cremation ceremonies of venerable monks famed for their piety and learning. The "phongyi-byan" or Return of the Great Glory signifies that the monk "returns" to the highest heaven of Devas and is, therefore, an occasion for rejoicing. The monk's body is sometimes preserved till an appropriate time for elaborate cremation ceremonies.

Holidays

Besides the festivals, there are four other holidays commemorating recent historical events. Independence Day celebrates the attainment of Myanmar's independence from the British on January 4, 1948; while Union Day is in honour of the occasion on February 12,

1947, when the late BogyokeAung San on the eve of Myanmar's independence, succeed in obtaining the unity loyalty of the different racial groups in Myanmar.

Resistance Day celebrated on March 27 marks the occasion in 1945, when Myanmar began armed and organized resistance against the Japanese in Myanmar. This occasion is observed by the Defence Service as the "Defence Service Day." The tragedy of the Cabinet assassinations, when BogyokeAung San and his Cabinet were assassinated on July 19, 1947, on the eve of Myanmar's independence, is commemorated on Martyrs' Day. Festivals and other occasions officially recognized as public holidays include the following:

Independence Day	January 4	Beginning of Buddhist Lent or Waso	Mid-July
Union Day	February 12	Ending of Buddhist Lent or Thadingyut	Mid-October
Full Moon of Tabaung	Late February	Dewali (Hindu Festival)	October
Peasants' Day	March 2	Tazaungdaing Light festival	Mid-November
Resistance Day	March 27	Bakri-Idd (Muslim Feast)	Late November
Thingyan of Myanmar New year	Mid-April, 3 days	National Day	Late November or early December
Full Moon of Kason	Late April or early May	Christmas Day	December 25
Workers' day	May 1	Karen New Year	Late December or early January
Martyrs' Say	July 19		

Myanmar Calendar

The Myanmar calendar consists of months dependent on the waxing and waning of the moon. There are 12 Myanmar months of 29 and 30 day duration alternately in the year which correspond approximately to the following months in the Gregorian calendar:-

Tag	March and April	Thadingyut	September and October
Kason	April and May	Tazaungmon	October and November
Nayon	May and June	Nadaw	November and December
Waso	June and July	Pyatho	December and January
Wagaung	July and August	Tabodwe	January and February
Tawthalin	August and September	Tabaung	February and March

Names

Myanmar admits all peoples of all classes on a footing of equality. This fact comes out in bold relief during festival and ceremonies. Magnanimity and spirit of co-operation have thus infused life, animation and grave into Myanmar social life. One of the main causes for a classless society has been the Myanmar system, always in vogue, of individuals gearing only personal names without the use of any surname. This lake of surname identification has thus cheeked stray aristocratic tendencies.

Myanmar names are usually with as honorific, a prefix according to the age or status of the person, e.g., Maung (brother); Ko (elder brother); Ma (sister); Daw (aunt); and U (pronounced Oo, Uncle). Thus a man will be known as MaungSein as a young boy, KoSein as a lad and U Sein as he advanced in a year or acquired a high social status. In the case of women, a younger woman Ma Sein will be addressed as DawSein as she becomes matronly or reaches a higher social level.

Women in Myanmar

The women of Myanmar, as has been frequently observed enjoy great freedom of movement and activity. Yet despite their freedom and equal rights by law and by the custom in property, ownership, divorce, business and inheritance, Myanmar women keep very largely to the traditional place of women in the home. Industry and thrift are characteristic virtues of Myanmar women and these virtues are developed by custom through various business activities. Again though the home has been chosen as the best expression of their personality, Myanmar women do not dislike what is called men's professions.

Many Myanmar women have already taken up the medical and teaching professions, while quite a number of them are qualifying for the law and journalistic fields. In recent years many are also taken an active part in the politics of the country. Marriage is a simple, unostentatious in Myanmar based on love, though the due parental agreement is always sought for. There is no financial consideration or such as the dowry. The married girl does not change her name on marriage. The marriage is a civil contract, with both husband and wife having equal claim to the property. Widows are free to marry by law and by usage, and divorce is allowed and granted by the elders of the community, though divorces are rare.

TRADE

The highest executive authority for state trading activities in Myanmar is the Trade Council, constituted in 1965 under the Law for Construction of a Socialist Economic System. It is a nine-member body headed by the Trade Minister and including other Cabinet Ministers as required for the achievement of inter-departmental coordination and assistance.

Under the council are big state trading corporations which according to their allotted lines of business handle the commerce of the whole country. Their business covers a wide range from export and import to distribution of both imported and local commodities through the Peoples' Shops-System. For regulation of internal trade, the Council maintains trade offices in every city, district and township.

Exports

Rice exports for 1965-66 amounted to about 1,115,164 tons compared with 1,308,688 tons for the previous fiscal year. The annual average of rice exports is approximately 1½million tons. Countries which are leading regular customers for Myanmar rice are India, Pakistan, Japan, Malaysia, Ceylon, Indonesia and the USSR.

Other exports from Myanmar annually are generally cotton, pulses, rubber, oilcake, beans, ores and metals, paraffin wax and salt. Of timber exports between October 1965 and September 1966, that of teak came to 135,455 cu. Tons valued at K 162,965,487 and that of other hardwoods like pyinkado to 708 cu. Tons valued at K 275,670. Since 1959-60, some quantity of veneer and plywood was exported from Myanmar. Such exports in 1965-66 were 208 tons of veneer valued at K 1,773,677 and 63 tons of plywood valued at K 52,125. Of particular significance is the achievement of a favourable trade balance since the advent of the Revolutionary Government. The total value of exports from Myanmar for 1965-66 on F.O.B. basis totalled K 925.3 million and that of imports on c.i.f. the basis for the same period totalled K 808.3 million, leaving a surplus of K 117 million in the balance of trade.

Industry

Rice mils and saw-mills, a familiar sight throughout Myanmar display their dominant role in the industrial life of the country, employing the largest percentage of its industrial labour. Yet no less striking and significant is the meteoric rise in a variety of industrial projects and enterprises in the country during the post-independence years.

Some major enterprises like the oil industry and mines have been in existence almost from the beginning of British rule in Myanmar. But prior to independence, practically, only extractive industries operated in Myanmar. They extracted natural resources of the country which were merely processed into semi-finished material for export.

In independent Myanmar, the urge became strong for industrialization to ensure self-sufficiency of manufactured goods which had to be annually imported in large value and at a disadvantage to the country's balance of payments. The largest among industries, requiring a high initial outlay, like the Steel Mill and Myanmar Pharmaceutical Industry, were established by the government, a logical step in view of the limited private capital potential in a small country. Nationalization of key industries became important for

Myanmar as they were nearly all foreign-owned and enjoying rights granted under colonialism tantamount to monopoly. Thus, soon after independence, the Government took over foreign-owned sawmills and timber leases.

Petroleum

Petroleum is the most valuable of Myanmar's mineral assets. Prior to the British denial action in the oil industry during World War II, Petroleum production in Myanmar averaged 275 million gallons annually. The richest oil belts in Myanmar lie in the ancient geological Myanmar Gulf and are notable for the remarkable continuity of yield over a long period of exploitation. At the height of oil production in Myanmar, there were about 4,400 wells functioning. The crude oil was brought down, either by tankers or by a 300-mile pipeline, to be refined at Syriam near Rangoon. After the war, a refinery was also constructed at Chauk in the oilfield area itself. The mining industry has been completely brought under state ownership and operation by the Revolutionary Government.

The BOC was the first industry to be taken over the Revolutionary Government. The joint venture operation of the industry, though making a positive contribution to the rehabilitation of the industry, had still much leeway to cover in order to obviate Myanmar's annual imports of 65 million gallons of petroleum and other petroleum products, valued at about K 33 million. The Revolutionary Government purchased the 49 percent British share in the joint venture as well as the fixed assets of two subsidiary companies, the Myanmar Oil Company (Myanmar Trading) Ltd. and Indo-Myanmar Petroleum Company. Thus, on January 1, 1963, the entire oil industry passed into Myanmar ownership and was renamed the People's Oil Industry. Success came early in operation of the oil industry as a nationalized enterprise. For the first time since its postwar rehabilitation, the oil industry showed a profit, which was also unprecedented in the initial years of any nationalized concerns in Myanmar.

The profit amounted to K 40,816,000 due to a 13 percent increase in sales during the first nine months following nationalization. Crude oil production also rose to a postwar peak in 1962-63 of 161,847,049 gallons. During the same period, the POI also exported 10,266 tons of wax, valued at K 6,311,752. By the second half of 1963, Myanmar became self-sufficient in major petroleum products. Since the nationalization of the oil industry, oil exploration was intensified in Myanmar. Hence, new oilfields were discovered and some of these are now in actual production in the Prome and Myanaung areas.

Minerals

Mineral resources, another important segment of Myanmar's economy, are chiefly non-ferrous base metals. Myanmar has substantial deposits of these and other commercially valuable minerals, principally silver, lead, gold, tin, tungsten (wolfram), copper, precious stones, limestone, zinc concentrates and antimony. Damages suffered during World War II reduced Myanmar's annual exports of 160,000 tons of metal and ores valued then at K 57 million to about 32,000 tons valued at over K 49 million today.

The famous Bawdwin Mines, near Namtu and Lashio in the Shan State, was nationalized in 1965 and taken over from the former joint venture known as the Myanmar Corporation (1967) Limited. It annually produces the following:- metallic zinc concentrate 13,280 tons, copper matte 289 ton, nickel speiss 255 tons, refined led 15,229 tons, lead slag 293 tons. At pre-war rates, the Bawdwin Mine's productions are valued at over K 21 million.

Agriculture and Forests

Agriculture is the economic mainstay of Myanmar, employing about 80 per cent of the working population and representing about 70 per cent total value exports in the post-war era. The total cultivated area in the whole of Myanmar up to the 1965-66 agricultural year was 21,684,194 acres of which the largest portion of 12,390,058 acres is devoted to paddy which is milled into the rice. Rice is the staple food of the people as well as their biggest cash crop and export item. Other crops grown extensively include groundnut, sesamum, pulses and beans, millet, jute, sugar cane, cotton, rubber and maize.

Prior to World War II, Myanmar held the position as the biggest exporter of rice in the world, with as annual rice export of 3,000,000 tons valued at about K 240, million (£ 18 million or $ 48 million). Myanmar also stood fourth on the list of the world's rice producers, following China, India and Japan with a total rice production of 7,500,000 tons annually. Her rice production in 1963-64 stood at about 7,720,000 tons. Post-war Myanmar is striving her utmost to recover her pre-war rice production and export levels. She is in various ways rehabilitating her agriculture which has suffered from wartime neglect and subsequent insurrection.

Agricultural Development

Agricultural development was accorded top priority in state planning in Myanmar after she regained independence. Hence, the Revolutionary Government in 1962 launched a gigantic land reclamation scheme to restore to cultivation about two million acres of formerly fertile land which in wartime had turned in to wasteland so as to recover the pre-war total of 12½ million acres under paddy cultivation in Myanmar Proper alone. Such wasteland lies mainly in the Pegu and Irrawaddy Divisions.

Implementation of other schemes to heighten agricultural productivity is also being intensified in the past few years. These schemes include major irrigation projects which had been held in abeyance during previous governments. The Mu Valley Project, considered as one of the biggest irrigation projects in Southeast Asia, will, when completed, provide water in the Lower Chindwin District for 700,000 new acres and more adequate irrigation to some existing 300,000 under-irrigated acres. Construction of this project is financed partly by the Union Nations Special Fund (K 4.8 million) and partly by Myanmar (K 2,722,000).

Another big irrigation project is the Kyetmauktaung Dam in Central Myanmar which was completed in 1967 at a cost of K 31,200,000 met out of the aid programme of the Soviet Union. It irrigates 52,000 acres and is the only means of providing employment for inhabitants of 18 villages of the area.

Exploring all possibilities to boost agricultural productivity, the Revolutionary Government in 1964 launched the K 5,400,000 hydro-meteorological project. This project will receive the United Nations grant of K 2,600,000, with the balance to be borne by Myanmar. It will expand the meteorological and hydrological services in Myanmar, and by the scientific study of rivers and streams as well as water levels at major riverine towns, aid soil research as well as give timely warnings to farmers of oncoming rains, storms and floods.

A vigorous Crop Expansion drive is also proceeding to step up the sown acreage of crops. Some of the crops cultivated under this drive like groundnut, maize, beans and pulses have already exceeded the set output targets. Further stimulus to agricultural progress has been in the forms of farm mechanization, increased cultivation of selected crops — especially by double cropping and side cropping — development of animal husbandry, greater use of fertilizer and higher appropriations for agricultural loans which in 1963 rose to an all-time peak of K 70 crores, double the figure granted by the previous Government for 1961-62.

Forest Wealth

The forests of Myanmar, covering about 149,898 square miles or 57.28 per cent of its total area, possess about 76 per cent of the world's supply of teak, the most valuable and versatile species among hardwoods and a major source of foreign income to the country. Nationalization of timber export was imposed in 1949 and timber extraction was out by the State Timber Board mainly, though some were allowed for this activity to licensed lessees.

Chapter 2

General Information about Mauritius and Its Traditional Medical System

Introduction

Mauritius is an island of approximately 1865 square km in which 330 km of coastline almost entirely surrounded by coral reefs, a central plateau and impressive mountain. It is situated in the South-West Indian Ocean (20S/57.5E) approximately 2,000 km from Durban, 1,800 km from Mombasa, 6,000 km from Perth and 4,700 km from Mumbai.

The central plateau rises to a level of some 600 meters marked by a deep extinct volcanic crater, crisscrossed by rivers, stream and waterfalls. Bordering this tableland is the volcanic origin of the island. From these mountains, several peaks emerge such as Point de la Riviere Noire (823 m) and Le Pouce (812 m).

Forest Vegetation of the Island

In the central part of the island at about 600 meters above sea level, the average maximum daytime temperature varies from about 20-30°C. South Western regions are warmer and relatively drier than the East and South. A green cover of vegetation is maintained over much of the country throughout the year. The plain and mountain areas are dense evergreen forests. The vegetation comprises mainly tree associated with the number of shrub and herbaceous. Soil mostly consists of sand with clay and silt. In different places, soil sandy loam, and clay are found with occasional mixture of gravel. Following types of forests are found in the country.

1. Tropical evergreen mountain forest
2. Low land forest

3. Less dense forest
4. Scrub forest
5. Tropical ravine forest.

The vegetation in these areas is mainly of trees, shrub and herbs like *Pemphis acidula* (Lythraceae), *Scaevola taccada* (Goodeniaceae), *Cynodon dactylon* (Poaceae), *Cyperus stoloniferus* (Cyperaceae), *Sporobolus virginicus* (Poaceae), *Canavalia rosea* (Fabaceae), *Tournefortia argentea* (Boraginaceae), *Pongamia pinnata* (Fabaceae), *Terminalia catappa* (Combretaceae), *Casuarina equisetifolia* (Casuarinaceae), *Rhizophora mucronata* (Rhizophoraceae), *Thespesia populnea* (Malvaceae), *Diospyros egrettarum* (Ebenaceae), *Tarenna borbonica* (Rubiaceae), *Clerodendrum heterophyllum* (Lamiaceae), *Indigofera tinctoria* (Fabaceae), *Furcraea foetida* (Asparagaceae), *Vitextrifolia* (Lamiaceae), *Albizia lebbeck* (Fabaceae), *Acacia nilotica* (Fabaceae), *Bidens pilosa* (Asteraceae), *Tamarindus indica* (Fabaceae), *Lantana camara* (Verbenaceae), *Flacourtia indica* (Salicaceae), *Syzygium cumini* (Myrtaceae), *Litsea glutinosa* (Lauraceae), *Wikstroemia indica* (Thymelaeaceae), *Cocos nucifera* (Arecaceae), etc.

Climatic Conditions

In the central part of the island, at about 600 meters above, the average maximum day-time temperature varies from about 20°C in August to above 26°C in February. The minimum temperature is of the order of 14°C in August and rises to 20°C in February. Along with coastal areas, the temperature is generally 3-5 degrees higher; the western and northern regions are warmer and relatively drier than the East and South. A green of vegetation is maintained over much of the country throughout the year. April-June and September-November are usually the most pleasant months.

Climate statistics – central region

Month	Daily high (°C)	Daily low (°C)	Daily hours of sunshine	Average humidity (%)
January	27	20	7	82
February	27	21	7	83
March	27	20	7	82
April	26	19	7	81
May	25	17	7	81
June	23	16	7	81
July	22	15	7	80
August	22	15	7	80
September	23	15	7	79
October	24	16	7	78
November	26	18	7	77
December	27	19	7	78

The hours of daylight are approximately 0530 hours to 1900 hours in summer (November-April) and 0645 hours to 1740 in winter (May-October).

Climate statistics – west and north

Month	Daily high (°C)	Daily low (°C)	Daily hours of sunshine	Average humidity %
January	31	24	8	80
February	31	24	8	83
March	31	24	7	83
April	30	23	7	82
May	29	21	7	81
June	27	20	7	81
July	26	19	7	79
August	26	19	7	78
September	27	19	7	77
October	28	20	8	78
November	30	22	8	78
December	31	23	8	79

Climate statistics – east and south

Month	Daily high (°C)	Daily low (°C)	Daily hours of sunshine	Average humidity %
January	29	23	7	83
February	29	23	7	84
March	29	23	7	84
April	28	22	6	84
May	26	20	6	82
June	25	18	6	81
July	24	18	6	80
August	24	17	6	80
September	25	18	6	80
October	26	19	7	80
November	28	20	7	80
December	29	22	7	82

Rainfall

Month	Central		West		North		East		South	
	Days >5mm	Total (mm)	Days >5mm	Total (mm)	Days >5mm	Total (mm)	Days >5mm	Total (mm)	Days >5mm	Total (mm)
January	9	233	4	93	5	91	7	114	7	132
February	12	383	4	118	6	172	9	257	9	207
March	11	337	4	118	6	118	9	238	10	247
April	10	257	3	56	7	141	9	184	10	191
May	5	90	1	11	3	46	5	64	5	65
June	6	94	1	18	3	44	3	47	5	73
July	6	92	1	13	2	32	4	59	5	74
August	7	105	1	11	3	38	3	41	5	67
September	5	101	1	31	2	43	3	53	4	71
October	3	59	1	11	2	17	2	36	3	45
November	4	66	2	26	2	34	3	49	5	75
December	8	186	2	70	4	100	6	143	7	153

Important Places of Island

Port-Louis

Capital and main port of Mauritius, Port-Louis founded by the French Governor, Mahe de Labourdomma is in 1735. The harbour lies sheltered in a semi-circle of mountains. The town has plenty of character, and show in certain quarters signs of its past elegance. Off the main square, palm-lined place d' Armes, there are some particularly fine French colonial buildings, especially the Government House (eighteenth century) and the Municipal Theater, built around the same time. There are two Cathedrals, Anglican and Catholic, a Mosque, a Chinese, a fine Supreme Court, some 18th century Barracks, a Natural History Museum. To see a fascinating cross-section of Mauritian life visit the lively covered market. Champ de Mars, originally laid out by the French for military parades is now a racecourse. The best high-up views of the racecourse, town and harbour, are from a splendid boulevard called Casino, a fine Supreme Court, some 18th century Barracke, a Natural History Museum. To see a fascinating cross-section of Mauritian life visit the lively covered market. Champ de Mars, originally laid out by the French for military parades is now a racecourse. The best high-up views of the racecourse, town and harbour, are from a splendid in the time of William IV. South of Port-Louis is Le Reduit, French colonial residence of the former Governor-General of Mauritius, set in magnificent gardens. From Port-Louis you can drive through the residential townships of Beau Bassin, Rose Hill, QuatreBornes and Vacoas.

Domaine Les Pailles

At a stone's throw from Port- Louis the Domaine Les Paolles, a mature park spreading over 300 acres at the foot of a mountain range where the visitor will be able to discover some old ways of living in an enchanting environment. Domaine Le Pailles offers a wide range of facilities: a gastronomic restaurant "Lc Clos St. Louis", a less formal restaurant "La Canelle Rouge", an Indian restaurant 'Indra', an exclusive Chinese Cuisine 'Fu Xiao', a pizzeria, a swimming pool, a mini-golf, a Casino. There are also possibilities of driving in London Rovers in the mountain where deer, monkey and rare birds live harmoniously, riding in horse-drawn carriages of in a train. In the garden is found a replica of an ox-drawn sugar mail, an alembic to distil the domain's own rum, a riding club, a spice garden and a natural spring.

Currepip

It is the island's main urban centre, a pleasant town which lacks the atmosphere of Port-Louis but has good shops and restaurant. Between Curepipe and Floreal is the spectacular Trou aux Cerf, an extinct crater 85 m deep and more than 200 m wide from the rim of which one enjoys an extensive view of the island.

Mahebourg

It is one of the main fishing centres. It lies in the bay of Grand Port and has a Naval Museum, housed in the French colonial mansion wherein 1810 the England and French naval commanders, both wounded in battle, were brought to be given medical care. A handicraft village built in typical island architecture is found on the premises of the museum.

Pamplemousses Gardens

The Royal Botanic Gardens are situated in the village area of Pamplemousses which lies about seven miles North East of Port Loius (Latitude 20°06'and Longitude 57°34'). The village might be taken its name from a variety of Citrus plant commonly called the "Pamplemoucier" (*Citrus maxima*), a tree which grows in the vicinity and which may have been introduced by the Dutch from Java. The fruit which resembles a large grapefruit, is thick-skinned and slightly bitter, it is called "bambolmas" in Tamil. This, it is through, is the origin of the word "Pamplemousse" given by the French.

These beautiful garden, which were once described as "one of the marvels of the world", and had for a long time been "ranked third among all the garden that could be admired over the surface of the globe" had successively been known "Jardim de Mon Plaisir", "Jardin de Mont Plaisir", "Jardin des Plantes", "Le Jardin National de L'Isle de France", "Jardin Royal", "Jardin Botanique des Pamplemousses", "The Royal Botanical Gardens, Pamplemousses", and finally "The Royal Botanical Gardens, Pamplemousses", the name by which it is known now. The gardens are known to naturalists throughout the world for their collection of indigenous and exotic plants, including the giant Victoria regain water lilies and many species of palm trees. Of particular interest is the talipot palm, which is said to flower once every sixty years and thereafter dies.

Aquarium

Facing the calm water of the lagoon between Pointe aux Piments and Tour aux Biches is the Aquarium populated by 200 species of fish, invertebrates, live coral and sponges, all coming from the water around the island. An open-circuit seawater cycle of I million litres run through the 36 tanks every day. The Aquarium offers a unique opportunity of admiring the colourful treasurers of the Indian Ocean.

Eureka Creole Museum

Eureka is a unique Creole house built in 1830 in a magnificent garden surrounded by the waterfalls of MokaRivre and the Moka Range. A visit to Eureka is a must to understand fully the charms of the tropic.

Casela Bird Park

This park is in the district of Black River, Stretches over 20 acres of land and contain more than 140 varieties amounting to 2,500 birds Specimens from the continents may be there, but the main attraction is the Mauritian Pink Pigeon which is one of the rarest birds in the world. Other attraction are the tigers, fish pounds, tortoises, monkeys, orchids (seasonal), and the overall green scenery and peaceful atmosphere created by the trees, streams and small cascades.

Grand Bassin, within a short distance of Mare aux Vacoas, is one of the island's two natural lakes. It rests in the crater of an extinct volcano and is a place of pilgrimage for a large number of Mauritius of the Hindu faith, especially on the occasion of MahaShivaratree.

Black River Gorges National Park

This 6574 hectare park, proclaimed in 1994, protects much of the remaining native forests of Mauritius and provides opportunities for the visitor to enjoy the spectacular natural scenery and some of the unique endemic plants and birdlife. The National Park is easily reached from Vacoas or Curepipe, via La Mare aux Vaocas. At Le Petrin, a visiter information centre is being developed with picnic facilities. A boardwalk provides access to the typical dwarf forest fo the region and a fenced conservation management area (from which invasive non-native plants have been removed) protect a sample of the remaining health land. From here visitors can walk to Macchabee Forest (7 km return) or down through the Gorges to Black River (15 km).

From the road across Plaine Champagne towards Chamarel hear is easy access to viewpoints at Alexandra Falls, over BelOmber and over the spectacular black River Gorges. There are also a number of longer walking trails, including one to the island's highest point, Black River Pack (828) details of which are contained in a brochure available free from the National Park and Conservation Service, Ministry of Agriculture and Natural Resources, Reduit. Road access into the Black River Gorge from the coast is being upgraded and a Visitor Centre and camping area will be developed there.

Chamarel (Coloured Earth)

A twisting metalled road from Case Noyale village to the coloured earth at Achmarel. This is a mound of undulating land stretching in contrasting layers of colour, and the patches of blue, green, red and yellow earth are believed to be the result of weathering. The nearby chamarel waterfall emerges from the moors and primaeval vegetation and is startlingly beautiful.

La Vanille Crocodile Park

Near Riviere des Anguilles, nestled in the wild South, is this farm breeding Nile crocodiles imported from Madagascar. The site offers a vast park with a nature walk through luxuriant forest studded with freshwater springs. A small zoo of animals found in the wind in Mauritius adds to the attraction.

Souillac

It is a seaside on the rugged coast in the Savanne district. The appeal of Souillac lies in its scenery, for it is not a bathing spot. In striking contrast to the soft curves of the northern coastline, the high cliffs here drop abruptly to the sea and the bracing winds of the Indian Ocean blow all the year-round. A charming feature is a garden overlooking the sea, named after Dr. Charles Telfair. At the southern end of the village, a favourite viewpoint is Gris-Gris on the clifftop. Facing it, the sea engulfs itself in dark rocks which have hollowed out by the constant crashing of the waves.

Rochester Falls

It is near to Souillac and well worth to visit. The road leading to the falls crosses a sugar estate. Spectacular joints have been formed by the contraction of the Lave due to sudden cooling.

Ile aux Cerfs

Off the east coast, 20 minutes by both from Pointe Mauritius is the island resort, Ile Aux Cerfs. Tickets for boat transfers are available on the spot. The island has two restaurants, a boathouse and miles of beautiful beaches.

Domain du Chasseur (Domaine des Grands Bois)

Hidden in the hill at AnseJonchee near Mahebourg on the South-Eastern side of the island is a magnificent park covering 2000 acres. On forest-clad slopes at an altitude of 300 to 500 metres live stage and deer, monkeys and wild boars among luxurious vegetation of ebony, eucalyptus, ravenala, palm trees and wild orchid. Rare species of birds such as the kestrel can be seen in the woods. The domain also boasts 10 thatch-roofed loge-type bungalows and an open restaurant. Overlooking the hills, a wide choice of venison and seafood is available.

Le Domain de L' Ylang

Ylang rests at the foot of AnseJonchee Mountain, Vieux Grand Port. Its main resource is the plantation of Ylang-Ylang from the fragrant yellow flowers, of which a perfume is distilled. The domain commands a panoramic view over the bay of Vieux Grand Port and neighbouring islets. Hiking possibilities exist for nature lovers and food sampling of from products can be organized on request.

Le Val Nature Park is situated in the South East of the island at Cluny in the Grand Port District. Le Val derives its name from the valley as it is surrounded by luxuriant mountains. It offers the following attraction: spring water with natural aquatic life like shrimps, eels and carps; Anthuriums and Andreanum greenhouses, Prawn ponds, watercress ponds deer park, monkeys, birds, endemic plants and animals.

MAURITIUS TRADITIONAL SYSTEM OF MEDICINE

Thousands of years ago, several therapeutic substances have come into light such as Silajatu, a remedy for diabetes and rheumatism. In addition, bones of octopus, leaves of the neem tree (*Azadirachtaindica*), horns of the red deer, and skulls on which cranial surgery had been performed are the high antiquity of the medical sciences in India. In the Vedic period (from 5000 B.C. to 2000 B.C.), medicine was already an established profession. Rigveda 9.112 speaks of the healing of fractures whereas 10.97 is dedicated to healing herbs. The Ashvins were famed physicians and wonder-working surgeons. They were adepts in treating blindness, paralysis, and rejuvenation. They replaced the lost leg of the soldier by an iron one. The Atharva Veda is considered to be the source of all Ayurveda classical medicine of India. It speaks of the use of an unidentified plant kustha which fights against takman or malarial fever. It speaks of prosthetic limbs, artificial eye and newly set dentures. During the time of Budha, Jivaka, the 'thrice-crowned king' of physician surgeons, was expert in paediatrics and even exceeded in brain surgery. He successfully performed an intricate abdominal operation. He was a disciple of the renowned Atreya of Texil, a pioneer in India's history of medical sciences. During his region, Emperor Asoka provided hospitals throughout his dominions and even beyond in the lands on its frontiers, like the Greek king named Antiochus and of those kings who are neighbours of the Antiochus.

Medicinal herbs, whether useful to man or to the animal, have been brought and planted wherever they did not grow. Megasthenes has been the Greek Ambassador to the court of the Mauryan Emperor speaks of the high level of India's attainments in obstetrics, preventive medicine and dietetics. The oldest available medical textbook is the Samhita of Charaka, who according to Tradion, was the Royal Physician to Kaniska. His work is divided into eight parts: duties of a doctor and diet, the eight chief diseases, general pathology, anatomy and embryology, diagnosis and prognosis, special therapy and general therapy. His woke enjoyed renown far and wide: it was translated into Persian at an early date and into Arabic in the 8th century A.D.

Susruta, on the other hand, is famous for his contributions to surgery. The Mahabharata represents him as the son of Visvamitra. The treaties of Susruta, inter Asia, deals with pathology, anatomy, embryology, therapeutics and toxicology. Almost all the important operations are mentioned by him including a description of 121 surgical instruments. Like Charaka, he won fame beyond his motherland India. In the 9th century, he was renowned in Cambodia in the east and Arabia in the west.

Bhela was another eminent medical scientist whose Samhita refers to Susruta. Navanitaka is available in the oldest known medical manuscript of Sanskrit written birch bark. It was discovered from Kashgar. Paleographically, it belongs to the 4th century A.D. The age of her manuscript itself has unique importance for tracing the evolution of the medical science in India.

The Bhojaprabandha gives an account of the cranial operations performed by the Asvinis on king Bhoja who reigned from 1010 to 1056 A.D. Its special interest was the fact that it performed anaesthesia to cure him of cholesteatoma which is a brightly shining pearly tumour like the Saphara to which it is aptly compared. It also mentions the Sanjivani drug to revive the patient from anaesthesia or coma to consciousness. Medical science has yet to discover such a drug.

Indian medicine travelled to the capital Caliphate, where Hindu physicians were in charge of the Imperial Hospital. The great Harun al-Rashid was cured of a severe illness by the Hindu physician Mankh. This enhanced the reputation of Hindu medicine throughout the Saracen Empire. Mankh translated a Sanskrit work on MateriaMedica into Arabic. Thereafter, the Muslim scholars of the Caliphate read the standard Sanskrit text of Charaka, Susruta and Vagbhata. They got their knowledge of the internal administration of iron, oxides of arsenic, mercury and other metals from the Hindu. From the Saracens, the Hindu discoveries were assimilated by Europe. The considerable extent to the Hindu achievements in medicine and mathematics, Greece had, even earlier, borrowed the use of several medicinal plants from India. In modern times Hindu medicine has given the plant *Rauvolfiaserpentina*, which is widely used for the treatment of blood pressure.

More than two thousand years ago, Indian culture and civilization reached Indonesia. Indian medicine was also found congenial in this complex of islands. To this very first Day, Bail has preserved a rich Ayurvedic theory and practice. Over 250 medical texts are known in palm leaf manuscripts, e.g., the Tibetan medicine is based on that of India. Several Sanskrit works were translated into Tibetan, including the ‘Four Tantras’ whose original Sanskrit it now lost. Tibetan fastness, untouched by war, preserved the rich heritage of Indian medicine until the Chinese occupation. The Chakpori Medical College at Lhasa was the oldest institution of its kind. From Tibet, Indian medicine travelled to the Khalkha, inner Mongol and the distant Buryats of Siberia.

HISTORY OF MAURITIUS

The abolition of slavery in 1833 throughout the British Empire brought a new lease of life to many who had lived under the yoke of suppression and exploitation for centuries without knowing that they too were the human beings. But this freedom did not prove to be an occasion to rejoice as another set of people were tripped to fill in the gap caused by loss of labour force on plantation and manufacturing industries all over ‘under the sun’. It was this time when India came into the picture as the potent source of labour supply wherever the large scale migration of labour to Mauritius to work on the sugarcane plantation.

Mauritius is an island nation in the Indian Ocean about 1,200 miles off the southeast coast of the African continent. The country includes the island of Mauritius, Rodrigues (350 miles east), and the outer islands (Agaléga, St. Brandon and two disputed territories). The islands of Mauritius and Rodrigues (107 miles southwest) form part of the Mascarene Islands, along with nearby Réunion, a French overseas department. The area of the country is 2,040 km^2. The capital as well as the largest city is Port Louis. The island of Mauritius was visited during the Middle Ages by the Arabs and then by the Portuguese, who named it Dina Arobi and Cirne, respectively. The island was uninhabited until the Dutch Republic established a colony in 1638, with the Dutch naming the island after Prince Maurice van Nassau. The Dutch colony was abandoned in 1710, and, five years later, the island became a French colony and was named Isle de France in 1715. Dutch could not visualize the importance of Indian labour in developing the Island and the French followed by the English did so successfully. Later in 1721, it was ceded permanently to the French East India Company, and it was only then when the real settlement began on the island. Slaves were imported from Madagascar, Mozambique and West Africa to the island and in 1810, this island was passed into the hands of Great Britain.

Sugarcane had become its main crop while sugar is the only major industry in the country. The mode of production showed its inherent incapacitates to manage with their demands of the industrializing societies, which needed mobile and skilled labour. Slaves had do incentive or necessary training to comply with their needs. Also important was the Anti-slavery Society which questioned the validity of slavery throughout 'under the sun' in 1833. This created an acute labour shortage in their colonies like Mauritius, Fuji and Trinidad. The answer came in the form of Indian Labour which was also under the firm grip of the imperial bloodsuckers. Thought the Indians had been shipped out to Mauritius before also as convicts or domestic servants. It was only after 1834 that they were shipped out in bulk and this 'labour drain' became very systematic exploitation machinery supported by the Imperial government of both the countries. In 1835, the immigration started 1160 men, 61 women, 22 boys and 11 girls landed on the alien and hostile soil of Mauritius. The planters who bed received point-blank refusals from the freed slaves sighed with relief. They kept on asking for more and more people to cultivate their estate and also to expand the area under cultivation. To meet the increasing demand of the planters, a vicious chain of agents and suppliers came into operation which used all fair and foul means like coaxing, threatening and kidnapping. Like the Europeans in 14th and 15th centuries who were shipped to the Americas, these ignorant labourers had also been showing daydreams about the wealth and prospects in the new land. But this was not all, much their enthusiasm and zeal crushed in the journey. The agent used to load them in a ship like cattle never provided them with palatable or sufficient food and almost no sanitary facilities were available. As many as 3820 out of 25468 who were shipped to Mauritius between 1834 and 1839 died on the way. There was no consolation ever when they reached their destination. The living condition of these helpless labourers was very pitiable in the estates. Other than in the Act passed, the condition of these immigrants was on limited working hours. Though the residence, ration and free medical treatment were promised besides the monthly wage of Rs 5 for the 5 indentured years, the rate of mortality was miserably high. As many as 4035 persons died between 1834 and 39, not being able to stand the adverse surroundings.

As a typical capitalist, the Mauritian planters wanted the maximum profit with no inclination for the improvement of technology or searching new stable market. They either depended upon the suppression of labour to push the down the cost of production. They were many Acts and laws passed to get the uninterrupted and maximum possible labour out of the immigrant Indian and all combines together, brought him very close to the slave. There was the 'double cut' which meant for one day's absence from work would mean loss of two days wages, 'corvee' under which Indians had to work for public constructions etc. On their supposed free day Sunday and most grievous of all laws was that a labour should remain confined within the fence of the estate and should take to be a 'Vagrant'. Every step was taken to ensure that no Indian hand should go unused. In only one year 1869-70 about 30,824 Indians were arrested on the charge of vagrancy.

The new of suffering inflicted upon the helpless Indians hundreds of miles offshore travelled, throughout the length and width of India fast and effectively. There were brief periods of suspension of immigration, but the imperialist forces in Mauritius kept it in practice unabated till 1907. Even after that somehow or the other outflow of labour kept ongoing. It was only with the rise of strong nationalist movements of India that immigration was completely banned moved by his personal experience of the condition of the condition of the Indian laborers, Gandhi wrote in "India opinion" on 26th of Nov. 1909. "Those who read about these suffering will be convinced of the need to put atop on the system of indenture. The all too numerous incidents of the kind prove again and again that this system is indistinguishable from slavery". In 1925, Sir K. M. Singh, who was sent by Imperial Government to evaluate the labour conditions in Mauritius, writes: "It is evident, therefore, that the wages given to the recently recruited immigrant did not enable him to earn a sum sufficient to support himself, wife and three children and leave a saving as anticipated by the Govt. of India." No attempt was made by the protector of immigrants or anyone else to see that wages earned came up to the scale laid down.

The overall contribution made by the Indians in the development of Mauritius economy was so distinct that very soon the white planters had no regrets of the abolition of slavery, for the latter were Idle, lazy and drunkards while the Indians were extremely tractable under proper treatment. Most of the Indians were peace-loving and hardworking and wanted to still down in Mauritius forever. This certainly gave Mauritius the labour when the economy was in dire need, but it was only in 1853 that immigrants were granted citizenship. Indians by their efficient and honest labour made Mauritius prosperous in real terms. By 1884 the acreage of cane fields increased by 42% more than that in 1853 and officially encouraged to make the Island their adopted motherland. During the course to their struggle for survival, they adjusted to the climate and demands of the island and thus evolved an Indo-Mauritian community which was to become the most formidable section of the country. By 1846 only the Indo- Mauritians contributed 1/3rd of the island population and by 1861 it rose to 1/3rd a percentage which is still maintained.

For a better understanding of the Indo-Mauritius Relationship, it can be divided into four broad phases, (1). 1834-1906; (2).1907-1947; (3).1948-1968; (4).1968-1984.

The first period was dominated by the continued labour from India. Since India like Mauritius was under foreign control there was almost no attempt made to establish government or people level relationship. The old batch came back after the indenture period with horrific stories and the new batch would leave with hopes and fears for the future. Except a few people like Gandhi who visited the island in 1901 or Mahilall Doctor in 1907, there was no major interaction besides the genre's concern of the Indian political leaders and the people alike for the condition of Indian Labour.

During the second period, Indian was too busy in its militant struggle against the yoke of foreign rule and hence, this period proved to be a lean period for the Indo-Mauritian relationship with the island. Nehru called it a 'little India beyond the seas. In 1948, a High Commissioner was sent to Port-Louis and that was the first real step in the direction of healthy interaction between the two countries. Welcoming Mr. M.K. Kidwai, the new Commissioner to Mauritius, the Mauritius time looked back in retrospection. "Every commissioner for the Government of India has gone out of the way to unreservedly help all our people. Each one has been an ambassador of peace and goodwill and has largely contributed to our welfare.

India contributed to Mauritius's overall development, richly and steadily. Financial aids, loans, scholarship to students and exchange of cultural groups and Commissioners were the major themes of Indo-Mauritius Co-operation. India always supported the cause of Mauritius's Independence and voiced its concern at international forums. This period was marked by two visits by P.M. Sir S. Ramgoolam to India and the much-awaited independence of Mauritius. The fact that the first premier of Mauritius was an Indo-Mauritian facilitated the growth of good relation. This period was marked by two major drawbacks Mauritius was not yet independent and India was still struggling to emerge on its own feet.

It was only with Mauritius's independence (12th March 1968) that the relationship between the two countries took a major turn. In December 1967, Mauritian P.M. Sir SewoosagarRamgoolam came to India on a seven-day visit. He had come to India to renew the Indo-Mauritius relationship and to discuss plans for the Island Independence. The opportunity was also utilized by India to propose the launching of a Textile Mill in

Mauritius for which expert's report has been submitted to the Mauritius government in 1965. The Mill would be owned by the Birlas.

Henceforth, Mauritius always looked to India for help and guidance and India has always tried to come up to their expectation despite its own gigantic problems. The relationship between India and Mauritius is not just a two-country affair. The majority (67%) of the population is of Indian descent and despite and despite all alterations by missionaries and economic suffering, they maintained their cultural heritage and social system. India feels a father like a duty to help Mauritius. After Mauritius's independence, the co-operation between the two countries have touched ever possible field and the two governments are still exploring many new vistas.

Chapter 3

TRADITIONAL MEDICINAL FLORA OF MYANMAR

INTRODUCTION

Since traditional medicine has been shown to have intrinsic utility, it should be promoted and its potential developed for the wider use and benefit of mankind. It needs to be evaluated, given due recognition and developed so as to improve its efficacy, safety, availability, and wider application at low cost. It is already the people's own health care system and is well accepted by them. It has certain advantages over imported systems of medicine in any setting because, as an integral part of the people's culture, it is particularly effective in solving certain cultural health problems. It can and does freely contribute to scientific and universal medicine. Its recognition, promotion, and development would secure due respect for a people's culture and heritage.

Traditional medicine has a holistic approach, i.e., that of viewing man in his totality within a wide ecological spectrum, and of emphasizing the viewpoint that ill health or disease is brought about by an imbalance, or disequilibrium, of man in his total ecological system and not only by the causative agent and pathogenic evolution. These are some of the main reasons why traditional medicine needs to be promoted and developed. Perhaps, from the operational point of view, the most cogent reason for radical development and promotion.

- Traditional healers of the tropical rainforest have an immense knowledge finding students to pass on their knowledge before they die.
- Traditional health care is essential for 80% of the world population, which has little access to modern medicine. Plants are vital ingredients of traditional healing, Local birth attendants, bonesetters and some herbal healers are well respected, also by other medical systems.

The advantages of traditional health care are:

- It is embedded in the local culture, plant medicines are cheap and the healer is a member of the community. Moreover, it provides more attention and spiritual care to the patient than Western medicine.
- Plant medicines are usually ignored by graduates from medical schools. However, it is often not realized many effective drugs are based on plants and indigenous knowledge. They are provided as native plant products or semi-synthetic modifications and require reliable supplies of plant material.

Conservation

- These forests influence the climate, both locally and worldwide.
- They regulate rainfall and water supply and prevent drought and floods.
- They protect against erosion of the soil and dramatic landslides.
- They harbour an immense flora and fauna and an invaluable genetic diversity.
- They provide a whole range of useful products, in addition to timber and firewood.

Myanmar System of Traditional Medicine

The system of medicine that is being practiced throughout the country is the Myanmar system of medicine which has a glorious tradition, dating back to the 8th century. This system has its origin in Ayurveda and has been improved and modified over the years drawing inspiration from the philosophy of Buddhism. It was enriched by Myanmar traditions and adaptations throughout the centuries. It has become a system of medicine which can truly be called the Myanmar system of Traditional Medicine.

The Government first become interested in Traditional Medicine around 1928, realizing the need for improvement, a committee was constituted to explore the possibilities of the development of traditional medicine. Although the committee recommended the implementation of various measures, no action had been taken in a meaningful manner. Even after Independence, although an office of Traditional Medicine was established in the year 1953, only limited Government activities were possible in this field until the Revolutionary Government came into power in 1962. Since 1962, the following measures have been taken up by the Government for the development of traditional medicine Myanmar:-

1. Establishment of the Traditional Medicine Council.
2. Amendment of the Traditional Medicine Practitioners Act.
3. Compilation of the list of Traditional medicine practitioners including those qualified under the standards prescribed by the Government.
4. Increasing the number of Traditional Medicine dispensaries.
5. Extension of training course in Traditional medicine.
6. Deputation of trainees abroad for training in Traditional Medicine.
7. Opening of Institute of Traditional Medicine and Traditional Medicine hospitals.
8. Instiation of operational studies/research in traditional medicine.

Based on the above measures the Government formulated a national programme to change the development of Traditional Medicine. The national programme envisaged the following activities.

1. Establishment of Traditional Medicine Institute at Mandalay.
2. In-service training of Traditional Medicine Practitioners.
3. Operational studies/research in Traditional Medicine Practices.
4. Reorientation training of the registered private traditional medicine practitioners.
5. Establishment of a division for the manufactures of traditional medicine in the Department of Traditional Medicine.
6. Survey, cultivation and collection of medicinal plant.
7. Publication of textbooks in traditional medicine.
8. Survey and publication of ancient manuscripts.
9. Establishment of traditional medicine dispensaries.
10. Publication of training manuals and handbooks on traditional medicine.

Present Scenario

The traditional medicine in the country is deep-rooted among the community and this form the main-stay rural health services with or without state patronage. Many of the T.M.Ps. have started their lives as monks and there are different revels of training skill. They may be broadly categorized as follows:-

1. Semi-trained/TMPs and healers (unregistered).
2. Self-trained TMPs who have registered.
3. Graduate of recognised Institutes/Universities of Traditional Medicine e.g., Myanmar, Chine and India.

At present, category (1) and (2) are mainly utilized in the BTM dispensaries and the third category is mostly reinforced in the teaching institutions at Mandalay. In 1972, under the new organization of the Government, the Department of Traditional Medicine was opened under the Ministry of Health. This is the focal point for all activities concerning Traditional Medicine. Traditional Medicine practitioners are required to register with the Department of Traditional Medicine to enable them to practice.

The present book describes the most popular plant species which are common medicinal plants of the country. These species are including both wild and cultivated in Myanmar. The detailed description of these plant species including Phytochemistry, taxonomy and medicinal uses are given below.

Acanthaceae

Andrographis paniculata (Burm. f.) Nees

Vernacular Name: Say-ga-gyi (Burmese); Creat, kariyat (English); Kalmegh, kirayat, mahatita (Hindi); Bhunimba, kirata (Sanskrit).

Description: Erect annual herb, 0.4-1 m high; stems quadrangular, much-branched; leaves opposite, short-petioled, oblong-attenuate at both ends, glabrous, pale beneath; inflorescence is axillary and terminal raceme or panicle; flowers white with pink spots; capsule linear-oblong, acute at both ends, pubescent; seeds numerous, oblong, and glabrous.

Distribution: Widely cultivated in Southern and Southeastern Asia.

Part Used: Whole plant.

Phytochemistry: Plant contais andrographolide, neoandrographolide, panaculoside A-C, flavonoids, andrographin, panicalin, and apigenin 7-4-dimethyl ether.

Uses: Plant used in common cold and respiratory tract infections.

Hygrophila auriculata (Schumach.) Heine

Vernacular Name: Le-pa-du, supa-daung (Burmese); Gokulakanta (English); Kokilaksha (Sanskrit).

Description: Thorny, red-stemmed annual growing to 60 cm; bright blue flowers and small, flat, dark red seeds.

Distribution: Native to tropical Asia and Africa; widely distributed throughout tropical regions.

Part Used: Root.

Phytochemistry: Seeds contain uronic, palmitic, stearic, oleic, and linoleic acids; flowers contain apigenin-7-O-glucuronide and apigenin-7-oglucoside; the whole plant contains lupeol, betulin, and stigmasterol; leaves contain alkaloids, carbohydrates, proteins, steroids, glycosides, flavonoids, tannins, phenolic compounds, fats, and oils.

Uses: Seeds used in the treatment of blood disorders, biliousness, gonorrhoea, spermatorrhea, and fever; seeds are ground into a paste and given in buttermilk for diarrhoea.

Justicia gendarussa Burm. f.

Vernacular Name: Bur-bawanet (Burmese); Nilinirgundi, Udisambhalu, Kaka adusa (Hindi); Vataghni (Sanskrit).

Description: An erect undershrub, 0.6 to 1.2 m in height with profuse subterete branches, stems and branches dark violet; leaves simple, opposite, lanceolate, or linear-lanceolate, 7.5 to 12.5 cm long, short-petioled, glabrous, dark violet-green above and pale green beneath, main nerves about 8 pairs, midrib and main nerves prominent, on the under surface; flower white, spotted purple within, clustered in the interrupted spikes; fruits glabrous capsules.

Distribution: Found in tropical Africa and Asia.

Parts Used: Whole plant.

Phytochemistry: Plant contains beta-sitosterol, alkaloids, lupeol, friedelin and aromatic amines.

Uses: Plant is useful in asthma, rheumatism and colics of children; roots and leaves are useful in chronic rheumatism, cephalalgia, hemiplegia, facial paralysis, otalgia, hemicranias, cough, bronchitis, dysmenor-rhoea, amenorrhoea, internal haemorrhages, intermittent fivers, ascites and debility.

Achariaceae

Hydnocarpus anthelminthicus Pierre ex Laness.

Vernacular Name: Kalaw (Burmese); Sponge-berry, Common chaulmongra tree (English).

Description: Tall tree, 10 m high or more; bark whitish-grey; leaves alternate, coriaceous, shining greenl entirely pinkish when young; flowers unisexual or polygamous, in axillary raceme, both sexes on the same plant; fruit large, globose; seeds numerous, angular, embedded in the pulp.

Distribution: Grown in Southeast Asia, China, India, Myanmar, Thailand and Vietnam.

Parts Used: Seeds.

Phytochemistry: Seeds contain lipids; fatty acids, glycerides of chaulmoogric, hydnocarpic and gorlic acid.

Uses: Seeds are used in treating leprosy, scabies, impetigo and some other dermatitides; applied externally in the form of medicinal oil or ointment.

Anacardiaceae

Lannea coromandelica (Houtt.) Merr.

Vernacular Name: Bur-na-be (Burmese); Wodier Jhingam (English); Jhingan, Mohin (Hindi); Jhingini, Jingini (Sanskrit).

Description: A medium-sized to large tree 12-28 m in height with grayish bark rough, exfoliating in thin irregular flakes; leaves imparipinnately compound, crowded at the ends of branches, leaflets membranous, 7-9, oblong-ovate, green above, brown beneath, glabrous, base acute or rounded, often oblique, main nerves 6-8 pairs; flowers small, yellowish or purplish, the male racemes compound, the female simple; fruits reniform, compressed, 1-seeded red drupes.

Distribution: Found in East Asia - Southern China, Indian subcontinent, Myanmar, Thailand, Cambodia, Laos and Vietnam.

Parts Used: Bark, leaves.

Phytochemistry: Plant contains tannins, L-arabinose, D-galactose, jeolic acid, phlobatannin and phenolics.

Uses: Bark is useful in cuts, wounds, ulcers, ophthalmia, gout, ulcerative stomatitis, odontalgia, sprains, diarrhoea, and dysentery; leaves are useful in elephantiasis, inflammations, neuralgia, sprains and bruises.

Semecarpus anacardium L.f.

Vernacular Name: Bur-thitsiebo, che-thee (Burmese); Marking nut tree, Oriental cashew (English); Bhela, Bhilava (Hindi); Bhallatakah, Arusharah (Sanskrit).

Description: A medium-sized to large tree, 15-25 m in height, with grey bark exfoliating in small irregular flakes; leaves simple, alternate, obovate-oblong, round at the apex, coriaceous, glabrous above and more or less pubescent beneath, main nerves 15-25 pairs; flowers greenish-white, fascicled in pubescent panicles; fruits obliquely ovoid or oblong drupes, 2.5 cm long, black when ripe, seated on a fleshy receptacle which is yellow when ripe.

Distribution: Native to India; found in the outer Himalayas to Coromandel Coast.

Parts Used: Fruits.

Phytochemistry: Fruits contain anacardic acid, cardol, catechol, anacardol, fatty acids, semicarpol and bhilawanol.

Uses: Various parts of the plant are used in the alimentary tract and certain dermatological conditions; useful in illnesses related to the heart, blood pressure, respiration, cancer and neurological disorders; used for improving sexual power and increasing sperm count, curing diseases related to the digestive system.

Apocynaceae

Ichnocarpus frutescens (L.) W.T.Aiton

Vernacular Name: Bur-tawsaba (Burmese); Kalidudhi, Dudhi, (Hindi); Ulpalasariba, Palindi, Shyamlata, Sariva, (Sanskrit).

Description: A much-branched extensively climbing rusty-villous evergreen, laticiferous woody climber; leaves simple, opposite, elliptic-oblong to broadly lanceolate, acute or acuminate, main nerves 5-7 pairs; flowers greenish-white, fragrant, numerous in axillary or terminal panicles of cymose clusters; fruits straight or slightly curved, cylindrical follicles, usually 2, divaricate, seeds black, comose, white.

Distribution: It is native to China, India, Southeast Asia, Myanmar and northern Australia.

Parts Used: Roots.

Phytochemistry: Phytosterol, triterpenes, flavonoids and various phenolic compounds.

Uses: The plant has a large number of traditional medicinal uses, including treatment for rheumatism, asthma, cholera, and fever. Extracts of the plant inhibit tumours, protect liver cells from damage in acetaminophen overdose, and reduces complications of hyperlipidemia.

Pergularia daemia (Forssk.) Chiov.

Vernacular Name: Phala-kantaka (Burmese); Utaran, Sagovani, Jutak (Hindi); Kurutakah, (Sanskrit); Trellis-vine (English).

Description: A slender foul-smelling perennial milky twining herb with hispid stems; leaves simple, opposite, suborbicular, cordate, acuminate, velvety pubescent beneath, margins ciliate; flowers greenish-yellow or dull white tinged with purple, in axillary long peduncled umbellate or corymbose clusters; fruits reflexed follicles with long beak and soft apines, seeds many, ovate, truncate at the apex, densely velvety-pubescent on both sides.

Distribution: Throughout tropics and subtropics of Africa, Europe, and Asia.

Parts Used: Whole plant.

Phytochemistry: Plant contains betaine, hentriacontane, lupeol, α-and β-amyrins, β-sitosterol, 5β-stigmast-7en-3α-ol, 5β-stigmast-8-en-3α-ol, calctin, calotropin, calotropagenin, uzarigenin and coroglaucigenin.

Uses: Plant used as an elmintic, laxative, antipyretic and expectorant, besides treatment of infantile diarrhoea, malarial intermittent fevers, toothaches and colds; the plant has hepatoprotective, antifertility, anti-diabetic, analgesic, antipyretic and anti-inflammatory properties.

Wrightia tinctoria R.Br.

Vernacular Name: Let-htok-phyu (Burmese); Pala indigo plant, (English); Mitaindrajau, Dudhi, (Hindi); Svetakutajah, (Sanskrit).

Description: A small deciduous tree, upto 10 m in height with pale grey, smooth, thin bark abounding in yellow milky with opposite divaricate branches; leaves simple, opposite, elliptic-ovate or obovate oblong, acuminate, glabrous, main nerves 6-12 pairs; flowers white, fragrant, in lax terminal cymes; fruits follicles in pairs, pendulous, cylindrical, tips adhering at fruit; seeds linear, pointed at the apex with a deciduous coma.

Distribution: Found in India, Southeast Asia and Australia.

Parts Used: Leaves, bark and seeds.

Phytochemistry: The plant contains wrightial, a triterpenoid chemical, along with cycloartenone, cycloeucalenol, β-amyrin, and β-sitosterol.

Uses: Bark is useful in treatment of diarrhoea, piles, ringworm and other skin diseases; seeds also have anti-dysenteric properties; used as astringents, antihelminthics, aphrodisiacs and to reduce fever; bark and seeds are used to treat flatulence; root bark extract is used orally as an antidote for snakebite; powder of the bark is also used for treatment of kidney stones.

Asparagaceae

Drimia indica (Roxb.) Jessop (Syn. Urginea indica (Roxb.) Kunth)

Vernacular Name: Padaing-Kyet-thun (Indian squial) (Burmese); Indian squill, white squill, Sea-onoin, (English); Jungli-piyaz, Ban-piyazi, Ban-palandu, (Hindi); Vanapalandu, (Sanskrit).

Description: A glabrous herb; bulb dull white or pale, ovoid, 2.5-5 cm diameter or more; leaves radicle, flat, narrow pointed; flowers appearing before the leaves, drooping or spreading, distant in a terminal raceme, 15-50 cm long; perianth 10.8 cm long, bell-shaped. 6-parted; segments white, with 3 green ribs in the centre, tapering downwards; ovules several valved; seeds many, flat and black.

Distribution: Distributed in tropical and southern Africa, the Indian subcontinent and eastwards to Vietnam.

Parts Used: Bulbs.

Phytochemistry: Fresh squills contain glycosides, scillarin-A, and scillarin-B.

Uses: Bulb has anthelmintic, cardiac, stimulant, deobstruent, digestive, diuretic, emmenagogue, and expectorant properties; useful in asthma, ailments of hearts, calculous affections, ailments of hearts, calculous affections, cough and bronchitis, paralytic affection, promotes urination, in rheumatism and skin diseases; externally used for removing corns and warts.

Bambusaceae

Bambusa arundinacea (Retz.) Wilid.

Vernacular Name: Kyakat-wa (Burmese); Thorny bamboo, (English); Bans, Kantabans, (Hindi); Vamsah, Venuhh, (Sanskrit).

Description: A tall thorny bamboo upto 30 m in height with many stems, tufted on stout rootstock, sodes prominent, the lowest rooting, internodes upto 45 cm long, stem sheath coriaceous, orange-yellow, streaked, glabrous or puberulous beneath, base rounded, ciliate, tip stiff, midrib narrow, leaf sheaths with a short bristly auricle, ligule short; spikelets glabrous, yellow or yellowish-green, in very long panicles, often occupying the whole stem, floral glumes, 3-7 in number, the uppermost 1-3 male or neuter, lodicules 3, hyaline, 1-3 nerved, ciliate; fruits oblong grains, beaked by the style base, grooved on one side.

Distribution: Wild in most parts of tropical India and Pakistan, growing up to 1000 m altitudes in the Nilgiris and hills of southern India; north into China and Myanmar.

Parts Used: Roots, leaves sprouts, grains and bamboo manna.

Phytochemistry: Stem consists of cellulose and hemicellulose (xylans, arabans, polyuronides) and lignins; seeds contain protein, carbohydrate, β-carotene, thiamine, riboflavin, niacin, and ascorbic acid.

Uses: Roots are useful in leprosy, skin diseases, burning sensation, discolourations, strangury, ringworm, otorrhea, arthralgia and general debility; leaves are useful in ophthalmopathy, lumbago, haemorrhoids, diarrhoea, gonorrhoea, amenorrhoea, dysmenorrhoea, wounds, skin diseases and fever; spourts are laxative, thermogenic, anti-inflammatory, digestive, carminative, anthelmintic and diuretic, and useful in inflammations, ulcer and wounds, dyspepsia, nausea, intestinal worms, flatulence and strangury; grains are useful in intestinal worms and general debility; bamboo manna is useful in vomiting, diarrhoea, hyperdipsia, burning sensation, leprosy, jaundice, cardiac diseases, haemoptysis, haematemesis, haemorrhages, bronchitis, cough, asthma, tuberculosis, strangury, stomatitis, syphilis, fever, ophthalmia and general debility.

Betulaceae

Betula utilis D.Don

Vernacular Name: Buza-Pit-pin (Burmese); Himalayan Silver Birch, Indian Paper Birch (English); Bhojpatra (Hindi); Bhurjan, Bahulavalkalah (Sanskrit).

Description: A medium-sized deciduous tree upto 20 m in height with bark having horizontal lenticels and pink inner layers, peeling off in large papery layers; young shoots, petioles and leaves silk; leaves simple, ovate, acute, sharply irregularly serrate, base cuneate or rounded, sticky when young with yellow resinous scale; flowers in catkins, male catkins at the tip of the long shoots, female spikes solitary on the top of the dwarf shoot; fruits narrower than the bracts, the wings narrower than the nut.

Distribution: Native to Northern India; distributed to nearby areas in the Indian subcontinent.

Parts Used: Bark.

Phytochemistry: Plant contains betulin, lupeol, oleanolic acid, betulic acid, lupenone, and karachic acid.

Uses: Bark is useful in convulsions, hysteria, epilepsy, bacterial infections, wounds, diarrhoea, dysentery, haemorrhages, haemoptysis, haematemesis, cough, bronchitis, dyspepsia, otalgia, otorrhea, jaundice, bilious fever, leprosy, skin diseases and general debility.

Boraginaceae

Trichodesma indicum (L.) Lehm.

Vernacular Name: Kul-fa-thein (Burmese); Andhahuli, Chotta-kulpha, (Hindi); Adhahpuspi, Andhapuspi, (Sanskrit).

Description: A hispid, much branched, erect annual herb, upto 45 cm in height; leaves simple, opposite, very variable in shape, sessile, ovate, oblong or lanceolate-oblong, obtuse or subacute, clothed with stiff hair, arising from tubercles; flowers pale blue, changing to pink or white on drooping axillary stalks, sometimes in terminal few-flowered cymes; cone of anthers large and very pointed clothed on the back with dense white hairs; fruits pyramidal 4-ribbed nutlets, smooth on the outer, rugose, regularly pitted on the inner face.

Distribution: Throughout Afghanistan, subtropical Himalaya, India, Burma, Philippines and Mauritius.

Parts Used: Whole plant.

Phytochemistry: Steroidal, beta-sitosterol and phenolics, catechin and gallic acid.

Uses: Plant is considered anodyne, anti-inflammatory, carminative, depurative, diuretic, emollient, ophthalmic and pectoral; used in the treatment of arthralgia, inflammations, dyspepsia, diarrhoea, dysentery, strangury, skin diseases and dysmenorrhea; cold water infusion of the leaves is depurative; leaves and roots are used as remedy for snake bites; flowers are pectoral and sudorific; root is pounded, mixed with water and given to children as a treatment for dysentery; root is pounded and made into a paste which is applied as a poultice to reduce swellings, especially of the joints.

Burseraceae

Boswellia serrata Roxb. ex Colebr.

Vernacular Name: Loban, Arabi (Burmese); Indian Olibanum tree (English); Salai, Kundur, Luban (Hindi); Sallaki, Kunduruskah (Sanskrit).

Description: A medium-sized deciduous tree with ash-coloured papery bark peeling off in thin flakes, young shoots oubescent; leaves alternate, imparipinnate, crowded at the ends of branches, leaflets opposite, pubescent, crenate-serrate; flowers small, white in axillary racemes; fruits trigonus drupes, three-valved; seeds compressed, pendulous.

Distribution: Native to India and Pakistan; found in Myanmar and Vietnam.

Parts Used: Bark, gum-resin.

Phytochemistry: Bark contains boswellic acid, tannins, pentosans, lignin, holocellulose, and β-sitosterol.

Uses: Bark is good for arthritis, asthma, dysentery, ulcers, haemorrhoids and skin diseases; gum-resin is useful in fevers, diaphoresis, convulsions, dysentery, urthrorrhea, orchiopathy, bronchitis, asthma, cough, stomatitis, syphilitic diseases, chronic laryngitis, jaundice and arthritis.

Caprifoliaceae

Valeriana jatamansi Jones

Vernacular Name: Kantabalu (Burmese); Indian valerian, (English); Tagar, (Hindi); Tagarah, Hatah, (Sanskrit).

Description: The tufted, hairy, herbaceous perennial upto 45 cm in height with thick horizontal root-stock basal radical leaves long-stalked, deeply cordate-ovate, usually toothed or sinuate, cauline leaves only a few, much smaller, entire or sometimes pinnate; flowers white or tinged with pink in terminal corymbs, often unisexual, the male and female on different plant; fruits oblong, compressed, hairy or nearly hairless.

Distribution: Throughout Asia: Afghanistan, Bhutan, China, India, Myanmar and Pakistan.

Parts Used: Roots.

Phytochemistry: Actinidine, carotene, calarene, calarenol, elemol, jatamols A and B, jatamansic acid, jatamansone, nardol, nardostachonol, norseychelanone, seychellane, seychellene, spirojatomol, valeranal, valeranone, virolin; angelicin, jatamansin, jatamansinol, oroselol.

Uses: Roots are useful in arthralgia, ulcers, wounds epilepsy, convulsions, dyspepsia, flatulence, colic, ascites, constipation, jaundice, cardiac debility, dry cough, asthma, strangury, amenorrhoea, dysmenorrhoea, seminal weakness, chronic and intermittent fever, ophthalmopathy, skin diseases, falling or hair, splenopathy, nephropathy, gleet, pharyngitis, leprosy, erysipelas and general debility.

Celastraceae

Celastrus paniculatus Willd.

Vernacular Name: New-ni, myin-gon-doong (Burmese); Climbing staff tree, Balck oil plant, Intellect tree (English); Malkangani, Malkunki (Hindi); Jyotismati, Pitataila (Sanskrit).

Description: A large climbing unarmed shrub with long slender elongating branches which are reddish-brown and covered with elongate white lenticles; leaves simple, alternate, ovate or obovate, crenulate, coriaceous, glabrous, lateral nerves arching; flowers greenish-white in terminal drooping panicles; fruits capsules, depressed-globose, 3-lobed, bright yellow when ripe, opening to expose the brown seeds covered with orange-red aril.

Distribution: Native to throughout the Indian subcontinent.

Parts Used: Bark, leaves, seeds.

Phytochemistry: Seeds contain sesquiterpene alkaloids, celapagin, celapanigin, celastrine, paniculatine and celapanin.

Uses: Plant is found to improve memory and cognitive functions; beneficial in neurological diseases and pain disorders including muscle cramps, backache, sciatica, osteoarthritis, facial paralysis and paralysis.

Colchicaceae

Gloriosa superba L.

Vernacular Name: Si-mee-dauk-new (Burmese); Supper Lilly, Tiger Lilly, (English); Karihari, languli, (Hindi); Langali, Agnisikha, (Sanskrit).

Description: Herbaceous climber; rootstock tuberous, cylindrical, 15-25 x 2.5-4 cm, pointed at both ends, 'V' or 'L'-shaped, fleshy, white; roots fibrous; branchlets hairless, cylindrical, weak; leaves alternate, rarely opposite or in whorls, egg-shaped-lanceolate, stalkless, about 5-15 x 2-4 cm, base heart-shaped, apex produced into a coiled tendril, margin entire, hairless, papery; lateral nerves parallel; flowers bisexual, axillary, solitary or somewhat sub-corymbose at the ends of branches, 8-10 cm across; floral stalk 6-14 cm long with reflexed apex; perianth lobes 6, oblong-lanceolate, about 6 x 1 cm, with crisply wavy margins, greenish at first, then yellow, passing through orange and scarlet to crimson colour; capsules ellipsoid-oblong, 3-5 x 1-2.5 cm splitting by 3-valves; seeds numerous, globose, dorsally compressed, with warty projections, straw-coloured.

Distribution: Native to Africa and Asia, but it is known worldwide as an ornamental plant, a medicine, a poison, and a noxious weed.

Part Used: Seeds and tubers.

Phytochemistry: Tubers contain colchicines, benzoic and salicylic acid, sterols and resinous substances; 3-demethylcolchicine, 1,2-didemethylcolchicine, 2,3-didemethylcolchicine, N-formyl-N-deacetylcolchicine and colchicocide; seeds also contain gloriosine.

Uses: Tuberous roots are useful in curing inflammations, ulcer, scrofula, bleeding piles, white discharge, skin diseases, leprosy, indigestion, helminthiasis, snake bites, baldness, intermittent fever and debility; roots are given internally an antidote for snake poison; considered useful in promoting labour and expulsion of the placenta; seeds are used for relieving rheumatic pain and as a muscle relaxant.

Combretaceae

Terminalia chebula Retz.

Vernacular Name: Bur-phaunga (Burmese); Chebulic myrobalan (English); Hara, Harad (Hindi); Haritaki, Pathya, Abhaya (Sanskrit).

Description: A moderate sized to large deciduous tree with a cylindrical bole, rounded crown and spreading branches; leaves ovate, elliptic or obvate, glabrous to tawny-villous beneath with a pair of large glands at the top of the petiole; fruits glabrous, shining, ellipsoidal, obovoid or ovoid drupes, yellow to orange-brown in colour, faintly angled, upto 3.75 cm long; seeds hard, pale yellow.

Distribution: Native to South Asia from India and Nepal east to southwest China, and south to Sri Lanka, Malaysia, and Vietnam.

Parts Used: Fruits.

Phytochemistry: Triterpenes arjunglucoside I, arjungenin, and the chebulosides I and II; chebulin, ellagic acid, 2,4-chebulyl-β-D-glucopyranose, chebulinic acid, gallic acid, ethyl gallate, punicalagin, terflavin A, terchebin, luteolin, and tannic acid; chebulic acid, luteic acid and terflavin B.

Uses: Fruits are useful in wounds ulcers, inflammations, gastropathy, anorexia, helminthiasis, flatulence, haemorrhoids, jaundice, hepatopathy, splenopathy, pharyngodynia, hiccough, cough, neuropathy, vesical and renal calculi, cephalalgia, epilepsy, ophtha-lmopathy, skin diseases, leprosy, intermittent fever, cardiac disorders, stomatitis, neuropathy and general debility.

Compositae

Anacyclus pyrethrum (L.) Lag.

Vernacular Name: Kukka-ya-Kwet-Ka-Ya (Burmese); Pellitory (English); Akrakara (Hindi); Agragrahi (Sanskrit).

Description: A perennial, procumbent plant with numerous stems and somewhat downy branches; leaves redical, stalked, smooth, pinnatisect into linear, subulate segments, cauline leaves sessile; branches one-headed, involucral scales lanceolate, acuminate, brown at the edge, receptacle convex; ray flowers white and purplish beneath.

Distribution: It is found in North Africa, elsewhere in the Mediterrancan region, in the Himalayas, in North India, and in Arabian countries.

Part Used: Roots.

Phytochemistry: Roots contain anacyclin, pellitorine, enetriyne alcohol, hydrocaroline, inulin, seasamin, deca-, dodeca-, and tetradeca-trans-2, trans-4-dienamide, tyramine amides and polyacetyenic compounds.

Uses: Roots are good tonic for the nervous system, it has anti-inflammatory activity and useed as a gargle in dental caries, odontalgia, pharyngitis and tonsillitis; it is also used in the treatment of paralysis, hemiplegia, cephalalgia, chronic ophthalmia, epilepsy, cholera, rheumatism and typhus fever.

***Blumea balsamifera* (L.) DC.**

Vernacular Name: Phon-ma-thein (Burmese); Nigail camphor (English); Kakaronda, Kalahad (Hindi).

Description: Subshrub 1-2 m high; steams striate-furrowed, much-branched at the top; leaves alternate, bearing some linear segments at the base; margins toothed; inflorescence in terminal corymb of many heads; flowers yellow; achene hairy; all parts of the plants are clothed with a velvety tomentum and smell of camphor.

Distribution: Found in tropical and sub-tropical zones of Asia, especially the Indian Subcontinent and Southeast Asia.

Part Used: Leaves.

Phytochemistry: Essential oil from the leaves contains L-borneol, d-camphor and cineol as major contents.

Uses: Leaves have antibacterial, antefebrile and anodyne properties; used in coryza, fever, influenza, cough and dyspepsia; effective remedy for the common cold, infected wounds, respiratory infections, and stomach pains.

***Cyanthillium cinereum* (L.) H.Rob.**

Vernacular Name: Kadu-byan (Burmese); Ash-coloured fleabane (English); Sahadeyi, Sadodi (Hindi); Sahadevi (Sanskrit).

Description: An erect or somewhat decumbent annual herb, 12-75 cm in height with cylindric, striate, more or less pubescent branched stem; leaves simple, alternate, variable in shape, ovate, or lanceolate, shortly mucronate, irregularly toothed or shallowly crenate-serrate; flowers many, pinkish-violet, in small heads, involucral bracts oblong, lanceolate, acuminate, awned, silky on the back; fruits oblong, lanceolate, acuminate, awned, silky on the back; fruits oblong achenes, slightly narrowed at the base; clothed with appressed white hairs.

Distribution: Native to tropical Africa and to tropical Asia (India, Indochina, Indonesia, Myanmar, etc.) and has become naturalized in Australia, Mesoamerica, tropical South America, the West Indies, and the US State of Florida.

Parts Used: Whole plant.

Phytochemistry: Plant contains β-Amyrin acetate; β-amyrin benzoate; lupeol and its acetate; β- sitosterol, stigmasterol, a-spinasterol and carbohydrates.

Uses: Roots are useful in diarrhoea, stomachalgia, cough, inflammation, skin diseases, leprosy, strangury, renal and vesical calculi, chronic and intermittent fevers; leaves are useful in humid herpes, eczema, ringworm, guineaworms and elephantiasis; flowers are useful in conjunctivitis, and fever; seeds are useful in roundworms, threadworms, cough, flatulence, leucoderma, psoriasis, chronic, skin diseases, colic and dysuria.

Sphaeranthus indicus L.

Vernacular Name: Mon-di (Burmese); East Indian globe-thistle (English); Mundi (Hindi); Mudi, Sravani (Sanskrit).

Description: A spreading aromatic herb with glandular hairy stem and branches; with purple or pink flowers; stems with toothed wings; leaves alternate, toothed; flowers in close terminal, globose or ovoid heads.

Distribution: Distributed throughout South Asia.

Parts Used: Whole plant.

Phytochemistry: Essential oil contains methyl chavicol, α-ionone, d-cadinene, p-methoxycinnamaldehyde, a-terpinene, citral, geraniol, geranyl acetate, β-ionone, sphaerene, indicusene and sphaeranthol; leaves contain alkaloids.

Uses: Roots are useful in diabetes, leprosy, fever, pectoralgia, cough, gastropathy, hernia, haemorrhoids, helminthiasis and dyspepsia; oil is useful in scrofula; powdered leaf is good for skin diseases and is considered as a nervine tonic; flowers are highly esteemed as an alternant, depurative, refrigerant and tonic.

Taraxacum campylodes G. E. Haglund

Vernacular Name: Bur-dudli, dandelion (Burmese); Common dandelion, Blow ball (English); Dudhal, Kanphul (Hindi); Dugdhapheni (Sanskrit).

Description: A perennial herb with thick tap root and abundant milky juice in all parts; leaves radical, sessile, variable in shape, narrowly oblong, irregularly pinnatifid, lobes linear or triangular, toothed; flowers yellow in ligulate heads; fruits glabrous achenes, flattened, ribbed minutely, spiny on the upper half, crowned with white pappus hairs.

Distribution: Found in temperate regions of the world.

Parts Used: Whole plant.

Phytochemistry: Plant contains taraxacerine, taraxasterol, homotara-xasterol, saponin, cerylalcohol, lactuce-roltaraxacin, and choline; root contains inulin, tannin and ethereal oil; leaves contain vitamin C and flowers contain xanthophylls and vitamin A.

Uses: The plant is useful in chronic ulcers, tuberculosis, dyspepsia, colic, flatulence, verminosis, constipation strangury, nephropathy, fever, skin diseases, leprosy, inflammations, gout stiff joints, insomnia, hypochondria, jaundice, biliary calculi and other hepatic diseases; root is diuretic, laxative, cholagogue, anti-rheumatic and aperient; inulin content is used for loss of appetite and stomach upsets; also used for gallstones, cholecystits, jaundice, atonic dyspeosia with constipation, muscular rheumatism and oligurea.

Cucurbitaceae

Cucumis sativus L.

Vernacular Name: Thakhwa (Burmese); Cucumber, Common cucumber, (English); Khira, (Hindi); Trapusah (Sanskrit).

Description: A hispid hairy trailing or climbing annual, leaves simple, alternate deeply cordate, 3-5 lobed, both surfaces hairy, margin denticulate; flowers yellow, males clustered, bearing anthers cohering, connective crusted or elevated above the cell, females solitary, thickly covered with very bulbous-based hairs; fruits cylindrical pepo of varying sizes and form; seeds cream or white, hard and smooth.

Distribution: Originally from South Asia, but now grows on most continents.

Parts Used: Fruits, seeds.

Phytochemistry: Aerial parts contain a 14α-methyl-phytosterol.α-and β-amyrin, multiflorenol, isomultiflorenol, 24, methylenecy-cloartenol and tirucallol; cytokinin-binding protein and isopentenyl adenosine trialcolhol.

Uses: Fruits are useful in hyperdipsia, fever, insomnia, cephalalgia, bronchitis, jaundice, haemorrhages, strangury and general debility; seeds are useful in constipation, intermittent fevers, strangury, renal calculus and urodynia.

Luffa acutangula (L.) Roxb.

Vernacular Name: Bur-tha-we (Burmese); Ridged gourd, Ribbed gourd, (English); Tori, Kali tori, (Hindi); Dhamargavah, Svadukosataki, (Sanskrit).

Description: A large monoecious climber with 5-angled glabrous stems and 3-fid tendrils; leaves orbicular-cordate palmately 5-7 lobed, scabrid on both sides, veins and veinlets prominent beneath; flowers yellow, males in 12-20 flowered axillary racemes, female flowers solitary, in the axils as the males, ovary strongly ribbed; fruits oblong-clavate with 10-sharp oblong, much compressed not winged.

Distribution: It ranges from central and eastern Asia to southeastern Asia; grown as a houseplant in places with colder climates.

Parts Used: Leaves, fruits, seeds.

Phytochemistry: Arginine, glycine, threonine, glutamic acid, leucines, serine, alanine, γ-amino butyric acid and pipecolic acid, oleic acid and linoleic acid.

Uses: Leaves are used in splenitis, haemorrhoids, leprosy, granular-conjunctivitis and ringworm; fruits are demulcent, diuretic, tonic and nutritive, and used as an excellent vegetable; seeds are useful in dermopathy.

Momordica cochinchinensis (Lour.) Spreng.

Vernacular Name: Samon-new (Burmese); Gulkakra, gangerua, (Hindi); Karkataka, (Sanskrit).

Description: Twining perennial with tendrils. Leaves alternate, deeply 3-5 lobed; margins toothed; leaf-stalk glandular. Flowers unisexual, nonoecious, pale-yellow, solitary in the axil of the leaves. Fruit big, oval-shaped, densely aculeate, red when ripe. Seeds round, compressed, the margins sculptured.

Distribution: Found throughout the Southeast Asian region from South China to Northeastern Australia, including Thailand, Laos, Myanmar, Cambodia and Vietnam.

Part Used: Seed membranes, kernels, roots.

Phytochemistry: Seed membrane countains β-carotene and lycopene; seed kernels yield fatty oil; roots contain triterperiod saponins.

Uses: Seed membrane is used in treating infantile rickets, xerophthalmia and nyctalopia; juice of kernels triturated in alcohol or vinegar is applied locally for furunculosis and phlegmon.

Trichosanthes dioica Roxb.

Vernacular Name: Palwal (Hindi); Patola (Sanskrit); Pointed gourd (English).

Description: The plant has harsh broad, sinate or cordate five-lobed and serrated leaves upper surface downy, tendrils bifid; flowers are white and fringed; fruit is oblong, narrow and pointed at both the ends, 4-8 cms long, glaucous, of a greenish-white colour and marked with few irrregulr, longitudinal strips; stem is rather quadrangular, twining and striated, sometimes woody and scabrous; seeds are minute, compressed and corrugate on the margin.

Distribution: Distributed in tropical Asia and Australia.

Parts Used: Fruits and Roots.

Phytochemistry: Vitamin A, vitamin C, tannins, saponins, alkaloids, mixture of novel peptides, proteins tetra and pentacyclic triterpenes, etc.

Uses: Used in constipation, fever, skin infections, wounds; unripe fruit and the tender shoots and capsules are laxative; the vegetable is given to improve appetite and digestion; the root is a strong purgative.

Trichosanthes tricuspidata Lour.

Vernacular Name: Kyi-ah (Burmese); Indrayan, nanakal, (Hindi); Kakansa, Dhvamksanasa, (Sanskrit).

Description: A large, branched woody tendril climber, older part of stem light grey and younger smooth green, tendrils 2, commonly cleft into 3; leaves palmately 3-5 lobed, dark green above, pale beneath with dark-coloured circular gland scattered along the lower side; flowers unisexual, males in axillary racemes, bracts broadly ovate, many-nerved, fringed and gland-dotted, female flowers solitary; and axillary; fruits globous, red when ripe with 10 orange streaks; seeds numerous, smooth.

Distribution: It ranges from the Eastern Himalayas in India and southern China through southern Japan, Myanmar, Malaysia, and tropical Australia.

Parts Used: Roots and fruits.

Phytochemistry: Leaves contain methylsterol, cyclotrichosantol and cycloeucalenol; root contains triterpene- trichotetrol; punicic acid; cycloeucalenol, hexacosanoic acid, trichonin, santholin, β-sitosterol, α-spinosterol, cyclotrichosantol, bryonolic acid, cucurbitacins B and D and trichotetrol.

Uses: Roots are good for hemicranias, carbuncles and gonorrhoea; fruits are very valuable in curing otitis and rhinitis and also recommended for hemicranias, inflammations, weakness of lymph, asthma, ophthalmia, epilepsy and leprosy.

Cupressaceae

Juniperus communis L.

Vernacular Name: Bur-pha-yaungban (Burmese); Common Juniper (English).

Description: A dense shrub, or less procumbent; leaves 5-13 mm. long, in whorls of 3, linear, sharply pointed, spreading nearly at right angles from the branchlets, convex on the back, concave and glaucous, bluish-whish on upper surface with a pale line running the whole surface, jointed at the base and continued down the stem with a large gland on the decurrent portion; cones unisexual, axillary; fruit 7.5-10 mm long, subglobose, fleshy, blue-black, glaucous; seeds 1-3.

Distribution: Throughout the cool temperate Northern Hemisphere from the Arctic south in mountains in North America, Europe and Asia; found in the Atlas Mountains of Africa.

Parts Used: Berries.

Phytochemistry: Plant containsgeijerone C12 terpenoid, junionone, desoxypodophyllotoxin, α-pinene, myrcene, cineole; p-cymene, α-and γ-cadinene, limonen, β-pinene, y-terpinene; terpinen-4-ol, α-thujene, borneol, caryophyllene, epoxydihydrocaryophyllene, β-elemem-7-α-ol, quercetin, amentoflavone, and isoquercitrin.

Uses: Berries used as a female contraceptive; catarrhal inflammation, suppression of urine, chronic pyelitis, strangury, bloody scanty urine with prostatic discharge and renal hyperaemia; urogenital irritation.

Dioscoreaceae

Tacca leontopetaloides (L.) Kuntze

Vernacular Name: Bur-Pin-baca (Burmese); Polynesian arrowroot (English); Surna (Sanskrit).

Description: Coarse, stemless, perennial, tuberous herb upto 1 m high; leaves large, petiolate, the petioles transversely striated, arising from a tuber up to 8 cm in diameter; flowers small, bell-shaped, greenish, 6-parted, borne in a many-flowered umbel surrounded by leafy bracts and numerous long filaments, supported by an erect, hollow, transversely striated 1 m long stem arising from the tuber; fruit a yellow, ribbed, globose berry up to 2.5 cm in diameter containing numerous longitudinally ridged seeds.

Distribution: Native to tropical Africa, South Asia, Southeast Asia, northern Australia, New Guinea, Samoa, Micronesia, and Fiji.

Parts Used: Leaves and timbers.

Phytochemistry: Alkaloids, β-sitosterol, ceryl alcohol and taccalin.

Uses: Inside of the root squeezed in water to rinse to injured eyes; the starch from the tubers used as a remedy for diarrhoea and dysentery; crushed leaf stalks rubbed onto bee and wasp stings.

Euphorbiaceae

Baliospermum solanifolium (Burm.) Suresh

Vernacular Name: Nat-Cho (Burmese); Danti (Hindi); Danti (Sanskrit); Wild croton, Wild sultan seed (English).

Description: A stout undershrub 0.9-1.8 m in height with herbaceous branches from the roots; leaves simple, sinuate-toothed, upper ones small, lower ones large, sometimes palmately 3-5 lobed; flowers numerous, in axillary racemes with male flowers above and a few females below; fruits capsules, 8-13 mm long, obovoid, seeds ellipsoid, smooth, mottled.

Distribution: Found in East-Asia - southern China, Indian subcontinent, Myanmar, Thailand, Laos, Cambodia, Vietnam, Malaysia, and Indonesia.

Part Used: Roots, leaves, seeds.

Phytochemistry: Roots contain phorbol esters, diterpene hydrocarbon, tigliane skeleton, balliospermin; leaves contain β-sitosterol β-D-glucoside and hexacosanol.

Uses: Roots, seeds, leaves and seed oil are used to treat jaundice, constipation, piles, anaemia, conjunctivitis; roots are purgative, anthelmintic, carminative, rubefacient and anodyne; used in the treatment of abdominal pain, constipation, calculus, general anasarca, piles, helminthic infestation, scabies and skin disorders; root paste is applied externally to painful swellings and piles; decoction of the leaves is used as a purgative; leaves are used to relieve asthma.

***Croton tiglium* L.**

Vernacular Name: Kanako (Burmese); Purging croton, Purgative croton, True croton, Croton-oil plant (English); Jamalgota (Hindi); Jayapala (Sanskrit).

Description: Medium-sized tree, 3-6 m in height leaves alternate, toothed, glabrous, pinkish-violet when young, inflorescence in terminal raceme; flowers unisexual, monoecious; female flowers without petals; capsule ovoid, obtusely 3-lobed, yellow; seeds small with a hard brownish-yellow shell.

Distribution: Grows wild in hills and damp forests.

Parts Used: Seeds.

Phytochemistry: Seeds contains fatty oil (stearinic, palmitic acids, glycerides of crotonic and tiglic acids, crotonoleic acid, glyceryl crotonate, crotonic resin, and various carcinogenic phorbol derivatives.

Uses: Seeds are used in treating flatulence, dyspepsia, constipation, colic, oedema, dyspnoea and persistent cough; should be used cautiously because of its high toxicity; should not be administered to pregnant women.

***Mallotus philippensis* (Lam.) Müll. Arg.**

Vernacular Name: Bur-pothidian, tawthidin (Burmese); Kamala tree (English); Sindur, Kamala, Rohini (Hindi); Recanakah, Kampillakah (Sanskrit).

Description: A small tree about 10 m in height with grey or pale brown bark with irregular fissures; leaves alternate, longer than board, three ribbed at the base ovate or ovate-lanceolate, entire or shortly serrate, red glandular beneath, the transverse nerves prominent; flowers dioecious, small in spikes, the males clustered sessile or very shortly pedicellate in erect long terminal spikes usually several together, the females nearly sessile in short spikes; fruits globose, three-lobed capsules covered over with dense reddish-brown glandular pubescence, seeds black, smooth, sub-globose.

Distribution: Occurs in Southeast Asia, Afghanistan and Australia.

Parts Used: Fruits.

Phytochemistry: Plant contains rottlerin, kamlolenic, conjugated dienoic, linoleic, oleic, lauric, myristic, palmitic and stearic acids.

Uses: Fruits are useful in constipation, flatulence, wounds, ulcer, cough, renal and vesical calculi, haemorrhages, poisonous affections, scabies, ringworm herpes and other parasitic skin affections.

Gentianaceae

Swertia chirata Buch.-Ham. ex Wall.

Vernacular Name: Pwe-danug-hin-ga (Burmese); Clearing nut tree (English); Nirmali, Nelmal (Hindi); Katakah, Ambuprasadah, Katakatphalam (Sanskrit).

Description: A medium-sized deciduous, glabrous tree about 12 m in height with cracked and scaly black bark and irregularly fluted trunk; leaves simple, opposite, elliptic, acute, chartaceous, upto 15 cm long and 6.25 cm broad, transverse nerves about 4 pair joining the second pair of ribs to the mid rib, glabrous, shining; flowers white, fragrant, axillary cymes; fruits ovoid or globose, glabrous berries, black when ripe; seeds one or two, yellow, circular, not much compressed, 8 mm in diameter, shining with short appressed silky hairs.

Distribution: Native to India; found in Nepal, Bhutan, Pakistan, Bangladesh, Myanmar and Tibet.

Parts Used: Seeds.

Phytochemistry: Plant contains sawertiamarine, mangeferin, amarogenitine, 1,5,8-trihydroxy-3-methoxyxanthone, 1-hydroxy-2,3,5,7-tetramethoxyxanthone, 1-hydroxy-3,5,8-trimethoxyxanthone, 1-hydroxyl-2,3,4,6-tetramethoxyxanthone, 1-hydroxy-2,3,4,7-tetra-methoxyxanthone, 1,8-dihydroxy-3,5-dimethoxyxanthone, 1,7-dihydroxy-3,8-dimethoxyxanthone, 1,3,5,8-tetrahydroxyxanthone, balanophonin, oleanolic acid, maslinic acid, and sumaresinolic acid.

Uses: Seeds are useful in cancer, malaria, stomach problem, appetite, skin diseases, regular itching, internal worm infestations, constipation, liver problems, blood circulation, diarrhoea, dyspepsia, dropsy, bronchitis, internal bleeding, respiratory problems, vomiting, diabetes, joint pain, vaginal infection and kidney problems.

Hydrangeaceae

Dichroa febrifuga Lour.

Vernacular Name: Bur-yin-bya (Burmese); Dichroa (English); Basak, Bansuk, Aseru (Hindi).

Description: Small shrub, 1-2 m high, stems glabrous, greenish or pale-violet; leaves opposite, oblong, acuminate; margins toothed petiole and main nerve usually violet; inflorescence in axillary or terminal divaricate cyme: flowers and berries bluish or lilac, seeds small pear-shaped.

Distribution: Grows in China, Nepal, India, Myanmar, Vietnam and Pakistan.

Parts Used: Roots, leaves.

Phytochemistry: Roots contain alkaloids, febrifugine, isofebrifugine, α-dischronic β-dichroine, γ-dischoroine and 4-ketodihydroquinazoline.

Uses: Roots are indicated in the treatment of pyrexia, malaria, intermittent fever, cough and oliguria; roots and leaves are impregnated with alcohol and torrefied.

Hypoxidaceae

Curculigo orchioides Gaertn.

Vernacular Name: Blackmushli, Mutha-li (Burmese); Golden eye-grass (English); Kalimusli, Musli (Hindi); Musali (Sanskrit).

Description: A herbaceous tuberous geophilous perennial with a short or elongate root-stock bearing several fleshy lateral roots; leaves simple, crowded on the short stem, sessile or short petioled with sheathing leaf bases, liner or linear-lanceolate often producing adventitious buds at the tips when in contact with soil; flowers bright yellow, the scape usually very short and hidden among the bases of the leaves underground, the lowest flowers bisexual, the upper male fruits capsules, derived from inferior tricarpellary syncarpous ovary, 1-4 seeded; seeds black, oblong, deeply grooved in wavy lines.

Distribution: Native to China, Japan, Indian Subcontinent, Myanmar, Papuasia.

Parts Used: Roots.

Phytochemistry: Plant contains curculigoside A-D, and curculigine A-D; saponins, curculigo-saponin, sapogenins, curculigenin, phenolic glycosides, corchioside, curculigoside, chlorophenyl glycosides, curculigol, 31-methyl-3-oxo-20-ursen-28-oic acid, 25-hydroxy-33-methylpentaricontan-6-one, N-acetyl-N-hydroxy-2-carbamic acid methyl ester, 3-acetyl-5-carbomethyoxy-2H-3,4,5,6-tetrahydro-1,2,3,4,6-oxatetrazine, N,N,N',N'-stigmasterol, cyclorartenol, and sucrose.

Uses: Roots are useful in leucorrhoea, haemorrhoids, pruritus, skin diseases, asthma, bronchitis, jaundice, diarrhoea, cuts and wounds, dyspepsia, colic, vomiting, ophthalmia, lumbago, and gonorrhoea.

Juglandaceae

Juglans regia L.

Vernacular Name: Bur-thit-kya (Burmese); Common walnut, Persian walnut, European walnut (English); Akhor, Akhrot (Hindi); Aksotah (Sanskrit).

Description: A large deciduous monoecious tree with tomentose shoots; bark grey, longitudinally fissured; leaves alternate, imparipinnate; leaflets entire, aromatic; flowers small, yellowish-green, male in pendulous slender catkins, female in 1-3 flowered terminal catkins; fruits drupes, 5 cm long with leathery exocarp, woody wrinkled, hard endocarp enclosing 4-lobed, corrugated, oily, edible seed.

Distribution: Native to the mountain ranges of Central Asia, extending to China, Nepal, Bhutan, Tibet, northern India, Pakistan and Sri Lanka, Myanmar, Afghanistan, Turkmenistan, Iran, Azerbaijan, Armenia, Georgia and eastern Turkey.

Parts Used: Leaves, bark, fruits.

Phytochemistry: Various part of the plant contain palmitic, stearic, oleic, linoleic and linolenic acids; juglansin, cystine and tryptophane.

Uses: Leaves are useful in herpes, eczema, scrofula and syphilis; fruits are useful in rheumatism; fruit oil is considered useful against tapeworm; used to strengthen and lubricate the muscles; kernels possesses aphrodisiac properties and are recommended in colic and dysentery.

Lamiaceae

Callicarpa macrophylla Vahl

Vernacular Name: Sigye (Burmese); Priyamgu, Daiya, (Hindi); Beautyberry (English); Priyangu, Phallini (Sanskrit).

Description: An erect shrub, 1.2-2.4 m in height with branches having tomentose tips; leaves simple, opposite, ovate or ovate-lanceolate, acuminate, base cuneate or rounded, white tomentose beneath, main lateral nerves 12-16 pairs; flowers rose-coloured, crowded in axillary peduncled globose cymes; fruits white drupes.

Distribution: Native to the Indian subcontinent.

Parts Used: Flowers, fruits.

Phytochemistry: Leaves contain fatty acids and their ethyl esters, 7,3,3'-trimethylquercetin, 7-O-glucuronides of luteolin and apigenin, ursolic, 2α-hydroxyursolic, L-α-amino-β-propionic and crategolkic acids, β-sitosterol and its β-D-glucoside.

Uses: Flowers and fruits are useful in rheumatoid arthritis, burning sensation, cephalalgia, diaphoresis, foul ulcer, dyspepsia, flatulence, colic, diarrhoea, dysentery, haemorrhages, haemoptysis, poisonous bites, skin diseases, diabetes, vomiting, fever and general weakness; leaves are useful in gout and arthralgia.

Clerodendrum japonicum (Thunb.) Sweet

Vernacular Name: Bugyini (Burmese); Japanese Glorybower (English); Dhago Phool (Hindi).

Description: Small shrub about 1 m high; stems and branches glabrous, quadrangular; leaves opposite, long-petioled, round cordate, entire or slightly dentate; inflorescene in terminal cymose panicle flowers tinged red, stamens and style very long, exserted; berry globose.

Distribution: East Asia to Southern China, India, Nepal, Bangladesh, Myanmar, Thailand, Laos, Vietnam, Malaysia, Indonesia, and Philippines.

Part Used: Leaves, roots.

Phytochemistry: Plant contains flavonoids, alkaloids and terpenoids.

Uses: Leaves are applied to boils; decoction of the inflorescence is used in the treatment of gonorrhoea, haematochezia and nosebleed; leaves are diuretic; decoction is used in the treatment of gonorrhoea; juice from the leaves is an ingredient of a herbal bath for children with furuncles; the root is used in jaundice.

Mentha piperita L.

Vernacular Name: Bala-shin, peper-ment (Burmese); Peppermingl, Gamathi phudina, (English); Paparaminta, gamathi, phudina, (Hindi).

Description: A strongly scented, perennial herb; leaves ovate or oblong-lanceolate, serrate; flowers light purple, thick terminal spikes.

Distribution: Indigenous to Europe and the Middle East, the plant is now widespread in cultivation in many regions of the world.

Part Used: Leaves.

Phytochemistry: Peppermint oil (menthol, menthone, menthyl acetate, isomenthone, neomenthol, cineole, limonene, pipertione, and pulegone).

Uses: Used for irritable bowel syndrome; peppermint oil used as a carminative, cholagogue, antibacterial, and secretolytic; externally, peppermint oil has been used for muscle pain, nerve pain and relief from itching; flowering tops are to prepare beverages like peppermint tea and in the formulation of liqueurs and bitters; peppermint oil is used to flavour non-alcoholic beverages, ice-creams, candies, baked goods, gelatins and puddings, chewing gum and some mint prepa-rations.

Origanum majorana L.

Vernacular Name: Taw-Yon-Yon-bwe (Burmese); Sweet marjoram (English).

Description: Woody perennial herb growing to 50 cm; leaves are smooth, simple, petiolated, ovate to oblong-ovate, 0.5–1.5 cm long, 0.2–0.8 cm wide, with obtuse apex, entire margin, symmetrical but tapering base, and reticulate venation; texture is extremely smooth due to the presence of numerous hairs.

Distribution: Native to countries bordering the Mediterranean; cultivated as a culinary herb, and for its essential oil.

Part Used: Aerial parts.

Phytochemistry: Essential oil (borneol, camphor, pinene, etc.).

Uses: Plant has stimulant and antispasmodic properties; treats flatulence, colic and respiratory problems; good general tonic, helping to relieve anxiety, headaches and insomnia.

Plectranthus amboinicus (Lour.) Spreng. (Syn. Coleus amboinicus Lour.)

Vernacular Name: Yawa-ni-ywet (Burmese); Countryborage, Indian borage, (English); Patta ajavayin, Patharcur (Hindi); Karpuravalli, Sugandhavalakam (Sanskrit);

Description: A large succulent aromatic perennial herb with hispid villous or tomentose fleshy stem; leaves simple, opposite, broadly ovate, crenate, fleshy, very aromatic; flowers pale purplish intense whorls at distant intervals in a long slender raceme; fruits orbicular or ovoid nutlets.

Distribution: Native to Southern and Eastern Africa, South Africa, Swaziland, Angola, Mozambique, Kenya and Tanzania; widely cultivated and naturalised elsewhere in the Old and New World tropics.

Parts Used: Leaves.

Phytochemistry: Thymol, carvacrol and camphor.

Uses: Leaves are useful in cephalalgia, otalgia, anorexia, dyspepsia, flatulence, colic, diarrhoea and cholera especially in children, halitosis, convulsions, epilepsy, cough, chronic asthma, hiccough, bronchitis, renal and vesical calculi, strangury, hepatopathy and malarial fever.

Premna serratifolia L.

Vernacular Name: Bur-taungtan-gyi (Burmese); Headache tree (English); Agetha, arani, ustabunda (Hindi); Agnimanthah (Sanskrit).

Description: A large shrub or a small tree up to 90 m in height with yellowish lenticellate bark, spinous large branches and yellowish-brown woody aromatic root; leaves simple, opposite, sometimes, whorled, elliptic-ovate, membranous when young, coriaceous when mature, entire or irregularly toothed, primary lateral nerves 4-6 pairs; flowers small, greenish-yellow or greenish-white with a strong disagreeable odour in corymbiform cymose panicles; fruits globose drupes, black when ripe with the persistent saucer-shaped calyx surrounding its base.

Distribution: Found in Asia and tropical east Africa.

Parst Used: Roots, leaves.

Phytochemistry: Root contains acteoside, 11,12,16-trihydroxy-2-oxo-5-methyl-10-demethyl-abieta-1[10],6, 8,11,13-pentene; premnine, ganiarine and ganikarine; p-methoxy cinnamic acid, linalool, linoleic acid, β-sitosterol, luteolin, premnazole, aphelandrine, pentacyclic terpene betulin, caryophellen, premnenol, premna spirodiene, and clerodendrin-A.

Uses: Roots are useful in inflammations, cardiac disorders, hepatopathy, cough asthma, bronchitis leprosy, skin diseases, dyspepsia, flatulence, colic, anorexia, constipation, haemorrhoids, fever, and diabetes; leaves are useful in dyspepsia, flatulence, colic, agalactia, cough and catarrh, fever, rheumatalgia, neuralgia, haemorrhoids and tumours.

Rotheca serrata (L.) Steane & Mabb. (Syn. Clerodendrum serratum (L.) Moon)

Vernacular Name: Bebya (Burmese); Bharangi (Hindi); Bharngi, Kharasakah (Sanskrit).

Description: A slightly woody shrub with bluntly quadrangular stems and branches; leaves usually three at a node, sometimes opposite, oblong or elliptic, coarsely and sharply serrate; flowers blue, many, in long cylindrical thyrsus with a pair of acute bracts at each branching and a flower in the fork; fruit a 4-lobed purple, dupe, somewhat succulent with one pyrene in each lobe.

Distribution: Found in warm temperate regions and native of South Africa, Southern Asia and Malaysia and distributed in various regions of tropical America, Northern Australia, India, Myanmar and Sri Lanka.

Part Used: Roots, leaves.

Phytochemistry: Herb contains serratin, lupeol, saponin, catchin, olionalic acid, carbohydrates, flavonoids, phenolics, steroids, and terpenes.

Uses: Herb is used to curing a common cold, cough, tuberculosis, rhinitis, asthma, chronic respiratory disorders and their underlying symptoms.

Thymus serpyllum L.

Vernacular Name: Bur-samon-phyu-yaing (Burmese); Breckland thyme, Wild thyme (English).

Description: The plant is a variable, aromatic, hairy, prostrate and evergreen perennial herb; 10-25 cm high and is one of the most effective ground covers, forming dense mats; rootstock is woody and the leaves are small, elliptic or oblong-ovate or lanceolate, 6-15 mm long, subsessile and glandular; flowers are polygamous pink in globose, terminal whorls; male flowers are larger in size.

Distribution: Native to most of Europe and North Africa.

Part Used: Whole plant.

Phytochemistry: Leaves and floral tops yield volatile oil; scutellarin heteroside, luteolin 7-glucoside, and apigenin; leaves yield ursolic acid and ascorbic acid; oil contains thymol.

Uses: Herb possesses anti-spasmodic, anti-septic, expectorant, carminative, anthelmintic and stimulating properties; leaves and floral shoots used in whooping cough, epilepsy, suppression of urine and menstrual catarrh; infusion of the leaves used in itching and skin eruptions; leaves used in biliousness and their juice is considered emetic; leaf juice applied over bald patches of alopecia; seeds are considered antifebrile and anthelmintic; remedy for stomach complaints, cough and colds.

Lauraceae

Cinnamomum tamala (Buch.-Ham.) T.Nees & Eberm.

Vernacular Name: Thitkyabo, sing-num (Burmese); Indian cassia lignea, (English); Tejpatt (Hindi); Tamalapatram (Sanskrit).

Description: A moderate-sized evergreen tree 7.5 m in height dark brown or blackish rough bark and pinkish or reddish-brown blaze; leaves simple, opposite, sub-opposite or alternate, ovate-lanceolate or ovate-oblong, acuminate, coriaceous, glabrous, 3-nerved from base to apex; flowers pale yellowish in axillary and terminal, lax puberulous panicles; fruits ovoid, fleshy, black drupe, supported by enlarged perianth tube.

Distribution: Native to India, Bangladesh, Nepal, Bhutan, China and nearby areas.

Parts Used: Leaves.

Phytochemistry: Plant contains cinnamic aldehyde, linalool, eugenol, eugenol acetate, β-caryophyllene, benzaldehyde, camphor, candinene, α-terpineol, α and β-pinene, p-cymene, limonene, geraniol, ocimene, γ-terpinene, β-phellandrene, benzyl cinnamate, and benzyl acetate.

Uses: Dried bark is used to treat stomachache; leaves are used in the treatment of colic and diarrhoea; showed antidiabetic and fungicidal effects.

Lecythidaceae

Barringtonia acutangula (L.) Gaertn.

Vernacular Name: Kyeni, Kyi (Burmese); Indian oak (English); Samudraphal, Hijjal (Hindi); Samudraphalah (Sanskrit).

Description: A medium-sized glabrous tree 10-15 m in height with pale grey slender young branches and rough dark brown bark; leaves simple, alternate, obovate-oblong or elliptic-cuneate, the margins minutely denticulate or crenulate, main nerves 10-13 pairs; flowers fragrant, dark scarlet, in pendulous many-flowered racemes; fruits bluntly quadrangular, narrowed towards the ends, crowned by a small persistent calyx.

Distribution: Native to coastal wetlands in southern Asia and northern Australasia, from Afghanistan east to the Philippines and Queensland.

Part Used: Roots, leaves, fruits.

Phytochemistry: Seed contains barringtogenin and berringtogentin; bark and leaves contain sapogenin; 3,3'-dimethoxy ellagic acid, dihydromyticetin, gallic acid, bartogenic acid and stigmasterol, olean-18-en-3beta-O-E-coumaroyl ester and olean-18-en-3β-O-Z-coumaroyl ester, 12,20(29)-lupadien-3-o[12], racemosol A and isoracemosol, barringtoside A, 3-O-β-D-xylopyranosyl(1-->3)-[β-D-galactopyranosyl (1-->2)]-β-D-glucuronopyranosyl barringtogenol C; barringtoside B, 3-O-β-D-xylopyranosyl(1-->3)-]β-D-galactopyranosyl(1-->2)]-β-D-glu-curonopyranosyl-21-O-tigloyl-28-O-isobutyryl barringtogenol C; barringtoside C, 3-O-α-L-arabinopyranosyl(1-->3)-[β-D-galactopy-ranosyl(1-->2)]-β-D-glucuronopyranosyl barringtogenol C.

Uses: Plant used in cold, dyspnea, pain, headache, eye disorders, diarrhoea, urinary disorders, insect bites and skin disorders.

Leguminosae

Acacia catechu (L.f.) Willd.

Vernacular Name: Tun-Sa-se, chat (Burmese); Cutch tree (English); Katha, Khair, Khaira (Hindi); Khadirah (Sanskrit).

Description: A moderate sized deciduous tree, 9-12 m in height with dark grayish or brown rough bark and hooked short spines, leaves bipinnately compound, leaflets 30-50 pairs, main rachis pubescent with a large conspicuous near the middle of the rachis; flowers pale yellow, sessile in peduncled spikes; fruits flat brown pods, shiny and with a triangulate beak at the apex and narrowed at the base; seeds 3-10 per pod.

Distribution: Found in Asia, China, India and the Indian Ocean area.

Part Used: Bark and heartwood.

Phytochemistry: Heartwood contains catechin and catechutannia acid; wood contains epicatechin, afzelchin, catechin tetramer, dicatechin, gallochin, gossypetin, isorhamnetin, and kaempferol.

Uses: Bark is useful in leprosy; heartwood is beneficial in cough and diarrhoea, applied externally to ulcers, boils and eruptions of the skin.

Acacia leucophloea (Roxb.) Willd.

Vernacular Name: Tanaung (Burmese); White babool, Distiller's acacia (English); Safed babul, Sarb ki kikar (Hindi); Arimedah (Sanskrit).

Description: A moderate-sized tree upto 3 m in height with spreading branches, crooked stems, white spines and pale yellowish-grey to nearly white bark with pale red inside; leaves bipinnate, 2.5-5 cm long, main rachis pubescent with a cup-shaped gland between each pair of pinnae, 5-15 pairs of pinnae of 12-30 pairs linear-oblong, obtuse; flowers in large terminal tomentose panicles, heads numerous, globose; fruits sessile, thin flat, slightly curved pods, covered with pale brown tomentum; seeds 10-20 per pod.

Distribution: Found in East Asia - Indian subcontinent, Myanmar, Thailand, Vietnam, Malaysia, and Indonesia.

Part Used: Bark.

Phytochemistry: Plant contains polyphenols, carbohydrates, isookanin, and cyanin.

Uses: Bark is used in infections, fiver, snake bites, bronchitis, cough, vomiting, wounds, ulcers, diarrhoea, dysentery, internal and external haemorrhages, dental caries, stomatitis, and intermittent fevers and skin diseases.

Adenanthera pavonina L.

Vernacular Name: Yway-gyi (Burmese); Coralwood tree, Red sandalwood tree (English); Baragunci (Hindi); kucandanah, Tamrakah (Sanskrit).

Description: A medium-sized, unarmed deciduous tree about 20 m in height with a clear bole of 6.0 m having greyish brown bark with longitudinal fissures; leaves bipinnate, the pinnae 3-6 pairs, opposite, leaflets many, alternat, ovate-oblong, obtuse, glabrous, unequal sided, dark green above; flowers pale yellow, in short peduncled racemes; fruits falcately curved pointed pods, the valves spirally twisted after dehiscence; seeds shining brilliant scarlet, lenticular-globose.

Distribution: Found in East Asia - China, India, Sri Lanka, Bangladesh, Myanmar, Thailand, Cambodia, Laos, Vietnam, Malaysia, Indonesia, Philippines, Australia, and Solomon Isles.

Part Used: Bark, leaves seeds, heartwood.

Phytochemistry: Plant contains O-acetylethanolamine, 1-octacosanol, galactitol, 2,4-dihydroxybenzoic acid, ampelopsin, butein, dihydrorobinetin, robinetin, echinocystic acid, oleanolic acid, dauco-sterol, β-sitosterol, stigmasterol, 2-amino-4-ethylidenepentanedioic acid, γ-methyleneglutamine, O-acetylethanolamine and 1H-imidazole.

Uses: Plant is antibacterial and haemaglutinin; decoction of the leaves is used in the treatment of rheumatism and gout; bark is used to treat leprosy; decoction of the bark and leaves is used to treat dysentery, diarrhoea and tonsillitis; wood is used as a tonic; pulverized wood, mixed with water, is taken orally for treating migraines and headaches.

Albizia odoratissima (L.f.) Benth.

Vernacular Name: Thit-mogyi (Burmese); Black siris, Kala Siris (English); Kala siris (Hindi); Bhusirisah (Sanskrit).

Description: A medium-sized unarmed tree about 20 m in height with dark coloured young shoots and grey, rough, irregularly cracked bark with dark patches; leaves abruptly pinnate, alternate, main rachis with a gland on the upper side near its basal part and often with similar glands at the bases of the first two pairs of pinnae, leaflets unequal sided, rounded, or semicordate at base, obtuse or rounded at the apex dark green, slightly pubescent above; flowers white, fragrant, sessile, numerous, in small globose 5-10 or more flowered heads, in corymbiform spreading panicles; fruits shortly stalked pods, brown, slightly reticulately veined; seeds flat yellow.

Distribution: Native to India, Bangladesh, Bhutan, Nepal, Myanmar, Laos, Thailand, Vietnam, Sri Lanka and China.

Part Used: Bark.

Phytochemistry: Bark contains D-catechin, leucocyanidin, melacacidin, leucoanthocyanidin, lebbecacidin, friedelin and β-sitosterol; seeds contain protein, calcium, phosphorus, iron, niacin, and ascorbic acid, amino acid (arginine, histidine, leucine, isoleucine, methionine, phenylalanine, threonine, tyrosine, and valine); flowers contain lupeol, α-and β-amyrin.

Uses: Bark is useful in ulcer, leprosy, skin diseases, erysipelas, cough, bronchitis, diabetes and burning sensation.

Cullen corylifolium (L.) Medik. (Syn. Psoralea corylifolia L.)

Vernacular Name: Ne-hle (Burmese); Malay tea scur pea; bawchee seed plant (English); Babchi, bavanchi, bukchi (Hindi); Bakuchi, Kushthanashini, sugandhakantak (Sanskrit).

Description: An erect annual herb, up to 90 cm high, wide blackish glands, leaves distinctly petioled, roundish-oval, 6-9 cm long by 5-7 cm wide; apex slightly acute; base truncate or slightly cordate, dentate at margin, about-faces conspicuously dotted with black dots, nearly glabrescent; flowers 10-30 in abundant dense long-peduncled heads; calyx cupuliform, teeth 5, lanceolate, the lowest longest; corolla yellow, little exserted; stamens 10, ovary sessile; fruit an oval pod, short, dry, indehiscent, black, glabrous, surrounded by the persistent calyx.

Distribution: Found in China, India, Myanmar, and other Asian countries.

Parst Used: Seeds and fruits.

Phytochemistry: Seeds and fruits contain flavonoids (neobavaisoflavone, isobavachalcone, bavachalcone, bavachinin, bavachin, corylin, corylifol, corylifolin and 6-prenylnaringenin), coumarins (psoralidin, psoralen, isopsoralen and angelicin) and meroterpenes (bakuchiol and 3-hydroxybakuchiol); leaves contain genistein.

Uses: Fruits used in diarrhoea, joint pain, cough, psoriasis, leucoderma, leprosy, and inflammatory diseases.

Desmodium oojeinense (Roxb.) H.Ohashi (Syn. Ougeinia oojeinensis (Roxb.) Hochr.

Vernacular Name: Bur-sit-gyi (Burmese); Sandan (English); Sandan, Tinnas, Panjan, Panan (Hindi); Tinisah, Syandanah (Sanskrit).

Description: A medium-sized deciduous pretty tree, 6-12 m in height with a short crooked trunk and dark brown deeply cracked bark; leaves pinnately 3-foliate, leaflets large, rigidly coriaceous, the terminal broadly elliptic or roundish, sometimes trapezoidal, main nerves 4-8 pair, prominent; flowers numerous, white or pink in short fascicled racemes from the nodes of old branches; fruits linear, elongate, flat pods, light brown in colour, seeds 2-5 per pod.

Distribution: Throughout Southeast Asia; India, Myanmar, Bangladesh and Vietnam.

Parts Used: Bark.

Phytochemistry: Heartwood contains homoferreirin, and ougenin.

Uses: Bark is useful in inflammations, diarrhoea, dysentery, urorrhagia, diabetes, verminosis, leprosy, leucoderma, haemorrhages, fevers, ulcers, gonorrhoea and anaemia.

Erythrina variegata L. (Syn. Erythrina indica Lam.)

Vernacular Name: In-Kathit (Burmese); India bean, India coral-bean, Mocha wood tree, East Indies coral tree (English); Dadap, Mandara (Hindi); Mandar, parijata (Sanskrit).

Description: Tree reaching about 10 m in height; stems branched, smooth, sparsely covered with short prickles; leaves alternate, trifoliate, the terminal largest; leaflet-stalk glandular, inflorescence in dense axillary and terminal raceme; flowers red, appearing before the leaves; pod long, black, constricted between the seeds; seeds reniform, red or brown.

Distribution: Native to the tropical and subtropical regions of eastern Africa, the Indian Subcontinent, northern Australia, and the islands of the Indian Ocean and the western Pacific Ocean east to Fiji.

Parts Used: Leaves.

Phytochemistry: Leaves and stems contain erythrinaline; seeds contain hypaphorine and migarrhin.

Uses: Leaves are sedative; used for the relief of insomnia and anxiety; fresh leaves are used externally as poultice in haemorrhoids and metroptosis; powdered leaves are topically applied for wounds and ulcers; stem bark is used against rheumatism in the form of a decoction.

Glycyrrhiza glabra L.

Vernacular Name: Thinbaw-nywe-cho (Burmese); Liquorice (English); Mulhatti, Jethi-madh, (Hindi); Madhuka, Yashti-madhu (Sanskrit).

Description: A hardy herb or undershrub; rootstock thick with may branched stems or roots; leaves pinnate; flowers lavender or violet in axillary spikes; pods compressed.

Distribution: Cultivated in India, Iran, Italy, Afghanistan, China, Pakistan, Myanmar, Iraq, Azerbaijan, Uzbekistan, Turkmenistan, Turkey, and England.

Part Used: Roots and underground stems.

Phytochemistry: Plant contains a volatile oil (estragole, anethole, eugenol, indole, y-nonalactone and cumic alcohol); glycyrrhizin, glabrene, glabridin, and phytoestrogens.

Uses: Licorice used in acidity, heartburn, indigestion, ulcer, liver problems, high cholesterol, respiratory problems, irritation, inflammation and poor immunity; topically used to fight dermatitis, eczema and psoriasis.

Mucuna pruriens (L.) DC.

Vernacular Name: Bur-Khway-lay-pe (Burmese); Common cowitch, Cowhage (English); Gonca, Kaunc, Kivacc (Hindi); Atmagupta, Kapikacchuh (Sanskrit).

Description: A slender climbing annual with hairy branches; leaves trifoliate, leaflets broadly ovate, elliptic or rhomboid ovate, membranous, unequal at base, pubescent above and densely clothed with silvery grey hairs beneath; flowers purple, in axillary pendulous 6-30 flowered racemes; fruits turgid pods, longitudinally ribbed, curved, densely clothed with persistent pale brown or grey irritant bristles, seeds black, 4-6 per pod, ovoid.

Distribution: Native to Africa and tropical Asia; widely naturalized and cultivated.

Part Used: All parts.

Phytochemistry: Seeds contain dihydroxyphlalanine or dopa glutathione, lecithin, gallic acid, nicotine, prurieninine, and pruienidine.

Uses: Roots are useful in constipation, nephropathy, strangury, dysmenorrhoea, amenorrhoea, elephantiasis, dropsy, neuropathy, consumption, ulcers, helminthiasis, fever and delirium; leaves are useful in ulcers, inflammation, helminthiasis, cephalalgia; seeds are useful in gonorrhoea, and consumption.

***Pterocarpus santalinus* L.f.**

Vernacular Name: Natha-ni, Santagu-ni (Burmese); Red sandalwood (English); Lal chandan, Rakta-chandan (Hindi); Rakta chandana (Sanskrit).

Description: Trees about 12 m high; bark blackish brown, 1-1.5 cm thick, deeply cleft into rectangular plates by deep vertical and horizontal cracks; blaze pale yellow with numerous pink streaks exuding copious red sticky thick gum; branchlets drooping, hairless; leaves 3-foliolate, 10-18 cm long; rachis swollen at base; leaflets usually 3, rarely 4 or 5, broadly egg-shaped or orbicular, 7-12 x 4-8 cm, base round or slightly heart-shaped apex rounded or deeply notched, margin entire, leathery, shiny, hairless, distinctly stalked; flowers bisexual, stalked in axillary simple or sparingly branched racemes, yellow, about 2 cm long, fragrant; pods unequally orbicular, flat, about 5 x 4.5 cm including the wing, gradually narrowed into a short tip which is about 1 cm long; seeds 1 rarely 2, more or less kidney-shaped, 1-1.5 cm long, smooth, reddish-brown.

Distribution: Distributed in Indian subcontinent.

Parts Used: Heartwood.

Phytochemistry: Heartwood contains santalin, desoxysantalin, santal, pterostilbene, ptterocarpin and homopterocarpin.

Uses: Plant is used as an antipyretic, anti-inflammatory, anthelmintic, tonic, haemorrhage, dysentery, aphrodisiac, anti-hyperglycaemic and diaphoretic medicine.

***Pueraria tuberosa* (Willd.) DC.**

Vernacular Name: U-hmoon-u, kyway (Burmese); Indian kudzu(English); Viddarikand, Tirra, Sural, bilaikand (Hindi); Vidari, Vidarika (Sanskrit).

Description: A large woody deciduous climbing shrub with huge tuberous root; tubers globose or pot-like, about 25 cm across, white inside, starchy and mildly sweet; branchlets with dense appressed velvet hairs; leaves alternate, 3-foliolate; leaflets egg-shaped, about 18 x 16 cm, base rounded and unequal sided, apex acuminate, margin with fringed hairs, hairless above, densely pubescent below; leaf stalks up to 18 cm long; lateral nerves 7-8 pairs; flowers bisexual, about 1.5 cm across, bluish-purple to pale violet, shortly stalked, densely arranged in 15-25 cm long axillary panicles; pods linear, 2-5 cm long distinctly constricted between the seeds covered with dense bristly reddish-brown hairs; seed 3-6.

Distribution: Native to India, Pakistan, and Nepal; distributed to nearby areas.

Parts Used: Tubers.

Phytochemistry: Tubers contain tuberosin, hydroxytuberosone, anhydrotuberosin, 3-O-methylanhydrotuberosin, coumestan tuberostan, puerarone, puerarostan, puerarin and daidzein.

Uses: Tubers used for urinary and reproductive disorders; useful for lactating mothers; it helps in the diseases of the respiratory system to remove excess congestion and provide relief; used as a general tonic, for headaches, and as an aphrodisiac.

Saraca asoca (Roxb.) Willd.

Vernacular Name: Thawka-india (Burmese); Ashoka (English); Asok, Asoka (Hindi); Asokah, Gatasokah (Sanskrit).

Description: A medium sized handsome evergreen tree upto 9 m in height with numerous spreading and drooping glabrous branches; leaves pinnate, 30-60 cm long having 2-3 pairs of lancelalate leaflets; flowers orange or orange-yellow in dense corymbs, very fragrant; fruits flat black pods, leathery, compressed; seeds 4-8 per pod, ellespsoid, oblong and compressed; bark is dark brown to grey or black with a warty surface, fresh-cut ends are pale yellowish red.

Distribution: Throughout Indian subcontinent and adjacent areas.

Parts Used: Bark, leaves, flowers, seeds.

Phytochemistry: Bark contains tannin, catechol, sterol, haematoxylene, tannin, glucose, galactose and mannose.

Uses: Bark is useful in dyspepsia, fever, dipsia, colic, ulcers, menorrhagia, metropathy, leucorrhoea and pimples; leaves are used for treating stomachalgia; flowers are considered to be a uterine tonic; dried flowers are used in diabetes and haemorrhagic dysentery; seeds are used in bone fractures, strangury and vesical calculi.

Senna auriculata (L.) Roxb. (Syn. Cassia auriculata L.)

Vernacular Name: Peikthingat (Burmese); Tanner's cassia, Tanner's Senna (English); Tarval, Anwal (Hindi); Avarttaki, Hemapushpam (Sanskrit).

Description: A much-branched shrub with reddish-brown branches; leaves with subulate glands between all the 8-12 pair of leaflets and a pair of large obliquely cordate stipules at their bases; flowers bright yellow in sub-terminal axillary corymbs; fruits pods, flat thin, papery, pale brown, deeply depressed between the seeds, transversely veined; seeds 10-20 per pod, obovate, dark brown with hard shiny seed coat.

Distribution: Throughout central and south India, Sri Lanka, Myanmar; cultivated in the tropics.

Parts Used: All parts.

Phytochemistry: Plant contains cardiac glucoside, sennapicrin; leaves and bark contains anthraquinones, tannins, emodin, chrysophanol, and rubiadin.

Uses: Root is used in decoctions against fevers, diabetes, diseases of urinary system and constipation; leaves have laxative properties; dried flowers and flower buds are used as a substitute for tea in case of diabetes patients; powdered seed is also applied to the eye, in case of chronic purulent conjunctivitis; seeds are said to give relief in rheumatism, eye diseases, gonorrhoea, diabetes and gout.

Senna tora (L.) Roxb. (Syn. Cassia tora L.)

Vernacular Name: Dangywe (Burmese); Sickle senna, sickle pod, coffee weed, tavara (English); Chakunda, panevar (Hindi); Dadamari (Sanskrit);

Description: Annual shrubby weed, 30-90 cm in height; leaves alternate, pinnate, with 3 pairs of obovate leaflets; inflorescence in an axillary raceme, shorter than the leaf; flowers yellow: 1-3 pod slender, very long curved seeds numerous, dark-brown, shining.

Distribution: Found in South-East Asia and the South West Pacific.

Part Used: Seeds.

Phytochemistry: Plant contains anthraglucosides, emodin, chrysophanol and rhein; seeds contain fatty acids (oleic, lionlic, palmitic and lignoceric acids) and β-sitosterol.

Uses: Seeds are utilized as a laxative; effective for insomnia, headache, constipation, oliguria, cough, ophthalmia, dacryoliths, amblyopia, ocular congestion and hypertension; alcoholic maceration of leaves is used externally to treat eczema and dermatomycosis.

Uraria picta (Jacq.) DC.

Vernacular Name: Pathani-Pa-hyni (Burmese); Dabra, Prisniparni, Prishtaparni Chitraparni, (Hindi); Prasniparni, (Sanskrit).

Description: It is an erect, little-branched, perennial herb, 90-180 cm tall. Leaves are very variable, 1-3 foliolate, upto 30 cm long and blotched white. Leaflets are 4-6, rarely 9, linear, oblong or lanceolate, obtuse and mucronate with subulate stipules The flowers are purple in colour and are in dense, long, cylindrical, terminal racemes. Pods are 3-6 jointed, polished and whitish in colour.

Distribution: Widespread in the Tropics through tropical Africa, India, China, through Southeast Asia to Indonesia, the Philippines, Papua New Guinea and Australia.

Parts Used: Root.

Phytochemistry: Alkaloids, steroids, flavonoids, coumarins, carbohydrates and amino acids and triterpenoids.

Uses: Roots are used as a sedative and to invigorate the liver and spleen; pounded leaves are used in the treatment of gonorrhea; plant is said to be an antidote to the bites of phursa snakes in India.

Vigna unguiculata (L.) Walp. (Syn. Dolichos biflorus L.)

Vernacular Name: Pe-bizat (Burmese); Kulthi (Hindi); Kulitha (Sanskrit); Cowpea (English).

Description: A herbaceous, prostrate, climbing, or sub-erect to erect annual, growing 15-80 cm high; leaves are alternate, trifoliolate with petioles 5-25 cm long; lateral leaflets are opposite and asymmetrical, while the central leaflet is symmetrical and ovate; inflorescence racemose, flowers white, cream, yellow, mauve or purple; there are accessions with determinate and with indeterminate growth habit; pods 10-23 cm long with 10-15 seeds/pod; seeds are variable in size and shape, square to oblong and variously coloured, including white, brown, maroon, cream and green, commonly in the range of 5,000-12,000 seeds/kg.

Distribution: Native to West Africa; cultivated in the semiarid tropics covering Asia, Africa, southern Europe, and Central and South America.

Parts Used: Seeds.

Phytochemistry: Seeds contains amino acids, urease, lectins, proteins, N-acetyl glucosamine and N-β-glycosides.

Uses: Seeds are beneficial in treating kidney, bladder and gall stones; due to its anthelmintic properties, it is useful in treating amoebic diarrhoea, bowel haemorrhage and colic pains.

Lythraceae

Trapa natans L.

Vernacular Name: Bur-kywe-gaung (Burmese); Water chestnut, Caltrops, Singhara nut, (English); Simghara, Simghada, (Hindi); Smgtakah (Sanskrit).

Description: An aquatic floating herb with flexuose ascending stems and gren photosynthetic submerged roots; leaves simple, alternate, crowded at the upper part of stems in rosettes; rhomboidal, apex triangular, irregularly inciso-serrate in the upper part, dark green above reddish-purple beneath, petioles dilated near the apex into a large spongy float; flowers white, opening above the surface of water in the afternoon, axillary and solitary; fruits obovoid, bony, with ashort conical beak in the centre at the apex and a spreading, flattened, very sharp spinous horn on either side, indehiscent, one-seeded; seeds white, starchy.

Distribution: Native to warm temperate parts of Eurasia and Africa; cultivated in China and the Indian subcontinent.

Parts Used: Fruits.

Phytochemistry: Bicornin, thiamine, riboflavin, nicotinic acid, vitamin C, vitamin A, oxalates and β-amylase.

Uses: Fruits are useful in dipsia, dyspepsia, haemorrhages, haemoptysis, diarrhoea, dysentery, strangury, intermittent fevers, loeprosy, fatigue, inflammation, urethrorrhea, fractures, erysipelas, lumbago, pharryngitis, bronchitis and general debility.

Woodfordia fruticosa (L.) Kurz (Syn. W. floribunda Salisb.)

Vernacular Name: Pan-le-pat-ta-gye (Burmese); Fire-flame bush, Shiranjitea, Woodfordia, (English); Davi, Tavi, Dhay, (Hindi); Dhataki, Madaniyahetu, (Sanskrit).

Description: A much-branched, deciduous shrub, attaining a height of about 3-7 m with many long arching branches and reddish-brown bark; leaves simple, opposite, entire, ovate-lanceolate, acute, subcoriaceous, with black glandular dots on the undersurface; flowers numerous, axillary paniculate-cymose clusters; fruits ellipsoid, irregularly dehiscent capsules about 1 cm long; seeds obovoid, smooth, brown.

Distribution: Throughout South Asia, India, Myanmar, Pakistant and Afganistan.

Parts Used: Flowers.

Phytochemistry: Flowers contain ellagic acid, β-sitosterol polystachoside, octacosanol, myricetin-3-galactoside, cyaniding-3, 5-diglucoside, pelargonidin-3, 5-diglucoside, and chrysophanol-8-O-B-D-glucopyranoside.

Uses: Flowers are useful in leprosy, skin diseases, haemorrhages, menorrhagia, leucorrhoea, haemoptysis, erysipelas, diarrhoea, dysentery, foul ulcers, diabetes, bilious fever, hepatopathy and verminosis. Flowers are used in bowel complaints and haemorrhages and is also administered in menorrhagia and seminal weakness.

Malvaceae

Abroma augusta (L.) L.f.

Vernacular Name: Naga-ble (Burmese); Devil's Cotton (English); Ulatkambal (Hindi); Pivari, Pisacakarpasa, Yosini (Sanskrit).

Description: A large, spreading, quick-growing hariy shrub or a small tree, roots found up to 1.5m depth; leaves lower, cordate, 3-5 lobed; upper, ovate-lanceolate, non-lobed; flowers yellow, purple or dark red; capsules membranous, pentagonal, pyramidal, winged, 4 cm long, seeds many, small, blackish, covered with silky hair.

Distribution: Widely distributed in Asia.

Part Used: Root, root-bark, stem, leaves.

Phytochemistry: Stem-bark contains β-sitosterol and friedelin; seeds contain fixed oil (linoleic, oleic, hexadececenoic, palmitic and stearic acids); root contains a fixed oil, resins, and alkaloids.

Uses: Root and bark are uterine tonic and emmenagogue, contract the uterus and are used for treatment of congestive and nervous dysmenorrhoea, amenorrhoea, sterility and other menstrual disorders; powdered root act as an abortifacient and anti-fertility agent; leaves useful in treating uterine disorders, diabetes, rheumatic pain of joints, and headache with sinusitis; leaves and stem are demulcent and an infusion of fresh leaves and stems in cold water is very efficacious in gonorrhoea.

Helicteres isora L.

Vernacular Name: Bur-thunge-che (Burmese); East-Indian screw tree (English); Marodphali (Hindi); Avarttani, Mrgasrngi (Sanskrit).

Description: A large shrub or small tree up to 5 m in height with grey bark and young shoots clothed with stellate hairs; leaves simple, alternate, bifarious, obovate, obliquely cordate, serrate, scabrous above, pubescent beneath; flowers red, fading to lead colour, distinctly bilabiate, in axillary clusters of 2-6 together, staminal column fused with the gynophore, much exserted, suddenly deflexed; fruits greenish-brown beaked, corkscrew like 5-follicles, seeds numerous, angular testa wrinkled.

Distribution: Found in Asia including Indian Subcontinent, South China, Malay Peninsula, Java and Saudi Arabia; Australia.

Part Used: Roots, bark, fruits.

Phytochemistry: Gallic acid, caffeic acid, vanillin, p-coumaric acid, cucurbitacin b, isocucurbitacin b, rosmarinic acid and their derivatives; isoscutellarein and their derivatives; D-glucopyranosyl isorinic acid; helisterculins A and B; helisorin.

Uses: Roots and bark are useful in colic, scabies, empyema, gastropathy, diabetes, diarrhoea and dysentery; fruits are useful in ophthalmitis, colic flatulence diarrhoea, dysentery, verminosis, wounds ulcers, haemorrhages, epistaxis and diabetes.

Hibiscus tilliaceus L.

Vernacular Name: Bur-thinban (Burmese); Beach hibiscus (English); Bola, Chelwa (Hindi).

Description: Spreading medium-sized tree up to 15 m tall; leaves alternate, petiolate, broadly ovate to cordate and palmately veined; flowers large and showy, yellow with a deep maroon centre and conspicuous stamina column arising from base of ovary; fruit a subglobose capsule upto 25 cm long with numerous small seeds.

Distribution: Common plant in Eastern and Northern Australia, Oceania, Maldives, South Asia, and Southeast Asia.

Part Used: Leaves, fruits and bark.

Phytochemistry: Gemlofuran, hibiscolactone, hibiscones, hibiscoquinones, lapachol, gossypol, gossypetin, mansonnes, hypero-side, kaempferol, quercetin and 3-0-galactosides of quercetin and kaempferol, gossypetin glucosides, gossypitrin and gossytrin, cyaniding-2-sophoroside, β-sitosterol, p-coumaric and fumaric acids.

Uses: Leaves are wrapped around fractured bones and sprained muscles; juice from the leaves is used in treating gonorrhoea; fluid from the bark is used to promote menstruation; flowers are used in making a salve; bark and the young leaves are used to treat skin diseases.

Sida cordata (Burm.f.) Borss. Waalk.

Vernacular Name: Kat-si-nye "galay" (Burmese); Bananiyar, Bhyunli (Hindi); Nagabala, Bhumibala (Sanskrit).

Description: A procumbent, diffuse, much-branched hairy herb with a very short main stem and long slender trailing branches that occasionally root at places of contact with the soil; leaves long-petioled, cordate to roundish with stellate hairs; flowers yellow, solitary or in pairs in the axils; fruits schizocarp located within the persistent calyx; seeds brownish, glabrous.

Distribution: Found in most Asian countries like China, India, Nepal and Myanmar.

Parts Used: Whole plant.

Phytochemistry: It contains β-phenethylamines, quinazoline, carboxylated tryptamine, linoleic acid, malvalic acid, sterculic acid and gossypol.

Uses: Plant juice used for boils and pimples; the root is tonic; juice of the root is used to treat indigestion; paste of the root is applied as a poultice to remove pus from boils and wounds; gonorrhoea and other venereal diseases; juice of the leaves is used to treat cuts and wounds.

Sida rhombifolia L.

Vernacular Name: Kat-si-ne-Nagbala (Burmese); Jamglimedhi, (Hindi); Bala (Sanskrit); Jelly leaf (English).

Description: An erect, minutely hairy, branched undershrub with a firm woody stem and intricate branches; leaves short-petioled, obovate, truncate of more often retuse and serrate; flowers yellow, solitary and axillary; cocci; seeds black, smooth.

Distribution: Native to the New World tropics and subtropics.

Parts Used: Roots, leaves.

Phytochemistry: Ephedrine, cellulose, lignin, quinazolines, β-phenethylamines and tryptamine derivative, sterculic, malvalic, linoleic acids.

Uses: Roots and leaves are good for rheumatism, flatulence, colic, haemothermia, emaciation, seminal weakness, arthritis and diarrhoea.

***Sida spinosa* L.**

Vernacular Name: Kat-si-ne-Maybala (Burmese); Prickly side (English); Bariyara, Gulsakari (Hindi); Mahabala (Sanskrit).

Description: A small, erect, grey, pubescent, branched undershrub with a slender erect stem, the young shoots being covered over with soft, grey stellate down; leaves with 2 or 3 small, stiff, minute spiny projections at the nodes adjacent to the place of insertion; flowers pale yellow to creamy white, axillary and solitary on slender peduncles; fruits 5-6 or 3-chambered with one seed in each chamber; seeds brownish-black, smooth.

Distribution: Native to certain tropical areas such as the Neotropics and parts of tropical Asia, that has become a widespread alien species in many temperate zones.

Parts Used: Whole plant.

Phytochemistry: Quinazoline, β-phenethylamines and tryptamine derivatives.

Uses: Roots are diaphoretic, antiperiodic, anaphrodisiac and tonic; used in malarial fever, haemorrhoids, swellings and in irritability of the bladder; leaves are useful in gonorrhoea, gleet and scalding urine.

***Sterculia foetida* L.**

Vernacular Name: Bur-Ietkok, Le-tpan-shaw (Burmese); Pinari (English); Jamglibadam (Hindi); Putidaru, Vatkhadirah (Sanskrit).

Description: A medium-sized to large tree about 18 m in height with reddish-brown heartwood, whitish bark and whorled horizontal branches; leaves digitate, crowded at the ends of branches, leaflets 7-9, subsessile, oblong-lanceolate, acute or acuminate, pubescent when young, glabrous when mature, stipules caducous; flowers dull orange, fetid smelling in erect racemose panicles, unisexual, male flowers with curved long stamina column, female flowers with stout gynophores and deflexed style; fruits 1-5 oblong, boat-shaped, bright red follicles; seeds 10-15 ovoid-oblong black.

Distribution: Found in India, Myanmar, Taiwan, Thailand, the Philippines, United States, Indonesia, Ghana, Australia, Mozambique, and Togo.

Parts Used: Heartwood.

Phytochemistry: Thiamine, riboflavin, nicotinic acid and vitamin C.

Uses: Heartwood is useful in convulsion, neuropathy, helminthiasis, leprosy, erysipelas, skin diseases, fever, ulcers and cephalalgia.

Theobroma cacao L.

Vernacular Name: Bur-Kokoe (Burmese); Cacao, Cocoa, (English).

Description: Evergreen tree growing to 8 m; has pale brown bark, glossy oval leaves, clusters of small yellow flowers, and large, pear-shaped, red-yellow seedpods; fruit is ovoid, 15–30 cm long and 8–10 cm wide, ripening yellow to orange, and weighs about 500 g when ripe.

Distribution: Native to Mexico and Central America, cacao is now a major crop throughout the tropics.

Parts Used: Seeds.

Phytochemistry: Seeds pulp contains xanthenes, fixed oil and endorphins.

Uses: Used as a nervous system stimulant; seeds are taken as a heart and kidney tonic; used to treat angina, and as a diuretic.

Marsileaceae

Marsilea polycarpa Hook.& Grev.

Vernacular Name: Hmo-nado (Burmese); Caupatiya, Sunsuniya (Hindi); four-leaf clover; European waterclover (English); Sunisanah, Catuspatri (Sanskrit).

Description: A creeping herbaceous perennial with slender long dichotomously branching rhizome rooting at the nodes; leaves quadrifoliolate, circinate when young, petioles long, slender, flexible, lamina divided into four leaflets, sporocarps ben-like borne on short or long stalks inserted a short distance above the base of the petiole.

Distribution: Found in central and southern Europe, Caucasia, western Siberia, Afghanistan, south-west India, China, Japan, Myanmar, Vietnam and some parts of the United States.

Parts Used: Whole plant.

Phytochemistry: Plant contains steroids, saponins, tannins, amino acids, reducing sugar, alkaloids, phenolics, vitamin C, hexadecanoic acid ethyl ester and phytol.

Uses: Plant is anti-inflammatory, diuretic, depurative, febrifuge and refrigerant; used to treat snakebite and applied to abscesses.

Meliaceae

Azadirachta indica A.Juss.

Vernacular Name: Tamakha, Tama Nimbah, Prabhadrah (Sanskrit); Neem tree, Margoes tree, India lilac (English); Nim (Hindi).

Description: A medium to large sized tree, 15-20 m in height with a clear bole of 7.0 m having grayish to dark grey tubercled bark; leaves compund imparipinnate, leaflets, opposite, serrate, very oblique at base; flowers cream or yellowish-white in axillary panicles, stamina tubes conspicuous, cylindric, widening above, 9-10 lobed at the apex; fruits one-seeded drupes with woody endocarp greenish-yellow when ripe, seeds ellipsoid, cotyledons thick, fleshy and oily.

Distribution: Native to India and the Indian subcontinent including Nepal, Pakistan, Bangladesh, Mauritius and Sri Lanka; typically grown in tropical and semi-tropical regions.

Part Used: Bark, leaves, flowers, seeds.

Phytochemistry: Stem bark contains tannins; leaves contain nimbin, nimbinene, 6-desacetylnimbinene, nimbandiol, nimbolide, quercetin, β-sitosterol, n-hexacosanol and nonacosane; thiamine, niacin, carotene and amino acids (glutamic acid, tyrosine, aspartic acid alanine, proline, and glutamine); fruits contain gedunin, 7-deacetoxy-7α-hydroxy gedunin, azadiradioe, azadirone, 17β-hydroxy-azadiradione, 17-epiazadiradione and nimbiol.

Used: Bark is useful in malarial fever, wounds, ulcer, burning sensation, tumour, tubercular glands, anorexia, vomiting dyspepsia, intestinal worms, hepatopathy, cough, bronchitis, urorrhea, diabetes, inflammation, amenorrhoea, lumbago, haemorrhoids, otalgia, syphilis and fatigue; leaves are useful in leprosy, skin diseases, leucoderma, pruritus, ophthalmopathy, intestinal worms, dyspepsia, ulcers, tuberculosis, boils, eczema and malarial and intermittent fevers.

Melia azedarach L.

Vernacular Name: Bur-than-baw-tama, Kala-tama (Burmese); Mahanimba (Sanskrit); Margosa tree, China berry, Barbados lialc, Pride of India, Persian lialc, Beds tree, Hoop tree (English); Drek, Bakain (Hindi).

Description: A tree 10 cm high or more; bark pale-brown, lenticellate, longitudinally furrowed; leaves alternate, odd-pinnate; leaflets opposite, serrate, glabrous on both sides; inflorescence in an axillary or terminal biparous cyme; flowers white-lilac; drupe ovate or ellipsoid-globose with 4 seeds, yellow when ripe.

Distribution: Native to Indomalaya and Australasia.

Parts Used: Stem, root bark.

Phytochemistry: Stem bark and root bark contain alkaloid (margosine), sterols and tannins; leaves contain paraisine and rutin; seeds contain fatty acids (stearic, palmitic, oleic and linoleic acids).

Uses: Root bark is used in treating ascariasis, oxyuriasis; vaginal injection of the decoction is effective for trichomonas infections.

Menispermaceae

Cocculus hirsutus (L.) W.Theob.

Vernacular Name: Kywet-nabaung (Burmese); Broom creeper, Ink berry, (English); Patalagarudi, Jaljamni (Hindi); Patalagarudha, Chilihindah (Sanskrit).

Description: A straggling scandent shrub with softly villous young parts; leaves simple; alternate, ovate-oblong, obtuse, speculate, subcordate or truncate at the base; softly villous on both surface, petioles densely villous; male flowers small, in axillary cymose panicles; female flowers 2-3 together in axillary clusters, petals thick and fleshy, divided at the apex into two triangular lobes, claw hairy; fruits transversely rugose, purplish-black drupes.

Distribution: Native to India, Pakistan, Myanmar and tropical Africa.

Part used: Roots, leaves.

Phytochemistry: Plant contains cohirsine, haiderine, jamtine N-oxide, trilobine, isotrilobine, syringaresinol, protoquercitol, hirsudiol, nonacosan-10-ol, cohirsitinine, hirsutine, shaheenine, cohirsinine, jamtinine, coclaurine, magnoflorine, sitosterol, ginnol and monomethyl ether of inositol.

Uses: Roots are useful in poisonous bites, leprosy, skin diseases, pruritus, dyspepsia, flatulence, colic, strangury, cough, bronchitis, gout, cephalalgia, intermittent fevers, spermatorrhoea, urethrorrhea, burning sensation, tubercular glands, fractures, hypertension and general debility; leaves are useful in eczema, gonorrhoea, prurigo, impetigo cough, ophthalmia, cephalalgia and neuralgia.

Moraceae

Artocarpus lacucha Buch.Ham.

Vernacular Name: Myauklaung (Burmese); Monkey jack (English); Barhal (Hindi); Lakucha (Sanskrit).

Description: A large deciduous tree, up to 20 m in height; bark dark-brown, exfoliating in small round woody peels, reddish inside, softy fibrous with faint streaks of white lates; leaves elliptic, oblong or ovate, broadly cuneate, pinnatifid, serrate, thick like a skin, rough, having venis in 8-12 pairs; inflorescence axillary, solitary; male heads ellipsoid; perianth of 2-3 free segments; bracts stoutly stalked; female heads with styles exerted to 1.0-1.5 cm via low papilla emerging between peltate bracts; fruit a syncarp, sub-globose, shallowly lobed, 6-12 cm long, yellow and drying to brown; surface pubescent, irregularly papillate, with numerous persistent bracts.

Distribution: Distributed throughout the Indian Subcontinent and Southeast Asia.

Part Used: Seeds and bark.

Phytochemistry: Wood contains galangin, artocarpin, cycloartocarpin, norcycloartocarpin, norartocarpin, oxyresveratol and resorcinol; bark contains β-sitosterol, cycloartenol, cycloartenone, α-amyrin and inpeol acetate.

Uses: Externally powdered bark is applied to sores to draw out purulent matter; topical application of the paste of the bark is useful in boils and small pimples; seeds are purgative.

Streblus asper Lour.

Vernacular Name: Ok-hne (Burmese); Siamese rough-bush(English); Sihora, Daheya (Hindi); Sakhotah, patrollekhataruh (Sanskrit).

Description: A small, rigid evergreen tree, upto 15 m in height with hairy scabrid twigs, milky white latex and light grey or greenish rough bark; leaves simple, alternate, rhomboid-elliptic, acute or acuminate, more or less crenate, scabrid on both surfaces, lateral nerves 4-6 pairs, prominent beneath; connected by intramarginal loop; flowers unisexual, dioecious, male heads globose, minute, yellowish-green, female flowers solitary or 2-4 together, very small, fruits one-seeded succulent berries, yellow when ripe; seeds smooth, round, greenish-white.

Distribution: Native to dry regions in Cambodia, Thailand, India, Sri Lanka, Myanmar, Malaysia, and Vietnam.

Parts Used: Roots, bark, leaves, and latex.

Phytochemistry: Plant contens cardenolides, cardenolide glycosides, kamaloside, asperoside, pyranoside, streboloside, indroside, lucknoside, cannodimethoside, strophalloside, strophanolloside, glucokamaloside sarmethoside and α, β-punsaturated lactone.

Uses: Various parts of the plant are used for the treatment of filariasis, leprosy, toothache, diarrhoea, and cancer; roots are useful in ulcers, sinusitis, inflammations, elephantiasis, boils, haemorrhages, haemoptysis, cough, bronchitis, convulsions, diarrhoea, dysentery, fever, syphilis and haemorrhoids; bark used in papermaking.

Oleaceae

Jasminum arborescens Roxb.

Vernacular Name: Bur-sabui (Burmese); Tree Jasmine (English); Bela, Chameli (Hindi); Nagamalli (Sanskrit).

Description: A large suberect shrub with hairy climbing branchlets; leaves simple opposite, ovate or ovate-oblong, acuminate, grey-tomentose on both surfaces; flowers white, fragrant, in lax trichotomous pubescent panicles; fruit single, ellipsoid, black, often curved berry.

Distribution: Throughout North-West Himalaya, Peninsular India, Myanmar and Sri Lanka.

Parts Used: Leaves.

Phytochemistry: Alkaloids, glycoside, flavonoid and triterpenes.

Uses: Leavesastringent and stomachic; juice of the leaves, with pepper, garlic and other stimulants, is used as an emetic in obstruction of the bronchial tubes due to viscid phlegm.

Jasminum multiflorum (Burm.f.) Andrews

Vernacular Name: Bur-taw-sabai (Burmese); Downy jasmine (English); Kundphul (Hindi); Kundah (Sanskrit).

Description: A large scandent tomentose shrub with young branches clothed with velvety pubescence; leaves simple, opposite, ovate, more or less pubescent beneath, base rounded or often cordate main nerves 4-6 pairs, petioles densely villous; flowers white, slightly fragrant, sessile in dense terminal capitates cymes at the extremities of short axillary branches; fruits black, globose berries surrounded by suberect calyx teeth.

Distribution: Native to India, Nepal, Bhutan, Laos, Myanmar, Thailand and Vietnam.

Parts Used: Leaves, flowers.

Phytochemistry: Nerolidol, benzyl benzoate, jasmolactone, nerolidol, jasmine, hexyl benzoate, hexenyl benzoate, β-farnesene and α-cadinol.

Uses: Dried leaves are good for indolent ulcer; flowers are useful in inflammations, rheumatism and cephalalgia.

Jasminum sambac (L.) Aiton

Vernacular Name: Bur-sabui (Burmese); Arabian jasmine, Tuscan jasmine, Sambac jasmine, (English); Bel, Moghrs, (Hindi); Mallika, (Sanskrit).

Description: A suberect or sometimes climbing shrub with pubescent branchlets; leaves simple, opposite, variable in shape, usually broadly ovate or elliptic, acute, entire, nearly glabrous, main nerves 4-6 pairs; flowers white, very fragrant, solitary or in 3-flowered terminal cymes, calyx lobes curried especially in fruits; fruits berries, black when ripe.

Distribution: Native to a small region in the eastern Himalayas in Bhutan and neighbouring India and Pakistan; cultivated in many places, especially across much of South and Southeast Asia.

Parts Used: Roots, leaves, flowers.

Phytochemistry: β-Farnesene, nerolidol, benzyl alcohol, linalool, benzaldehyde, benzyl alcohol and α-cadinol.

Uses: Roots and leaves are useful in ophthalmopathy; leaves are useful in skin diseases, ulcer and fever; flowers are useful in ophthalmopathy, pruritus, cephalalgia, otopathy, stomatopathy, leprosy, ulcers, fever, vomiting, hiccough, insanity and galactorrhoea.

Nyctanthes arbor-tristis L.

Vernacular Name: Bur-seikphalu (Burmese); Night jasmine, Coral jasmine (English); Harsimgar, Parja (Hindi); Parijatah, Sephalika (Sanskrit).

Description: A hardy large shrub or small tree upto 10 m in height with grey or greenish-white rough bark and sharply quadrangular strigose young branches; leaves simple opposite, ovate, acute or acuminate, scabrous above with short bulbous hairs, densely pubescent beneath with appressed hairs, main nerves few, conspicuous beneath; flowers small, white with bright orange corolla tubes, 3-7 in each head, in trichotomous cymes; fruits capsules, compressed, separating into one-seeded segments.

Distribution: Native to South Asia and Southeast Asia.

Part Used: Leaves, flowers, seeds.

Phytochemistry: Flowers contain d-mannitol, tannin and glucose; oil consists of the glycerides of linoleic, oleic, lignoceric, stearic, palmitic and myristic acids; β-sitosterol, tannic acid, methyl salicylate, and mannitol.

Uses: Leaves are useful in inflammation, dyspepsia, helminthiasis, pruritus, dermatopathy, chronic fever, bronchitis, asthma, cough strangury, constipation, hepatopathy, haemorrhoids, greyness of hair and baldness; flowers are useful in inflammations, ophthalmopathy, flatulence, colic, dyspepsia, splenomegaly, greyness of hair and baldness; seeds are useful in baldness, scurvy and affections of the scalp.

Schrebera swietenioides Roxb.

Vernacular Name: Bur-thitswe (Burmese); Weaver's Beam tree (English); Moka, Banpalas (Hindi); Muskakah, Ksarasresthah (Sanskrit).

Description: A moderate-sized deciduous tree upto 20 m in height with thick grey bark; leaves imparipinnate, leaflets opposite, 3-4 pairs, ovate, entire, unequal-sided, petioles thickened at the insertion of leaflets; flowers yellowish-brown in terminal trichotomous, corymbose, compound cymes; fruits pendulous, pear-shaped, 2-valved capsules; seeds 8, ending in long wings.

Distribution: Throughout India and Myanmar.

Parts Used: Roots, bark, leaves, fruits.

Phytochemistry: Plant contains D-mannitol, β-sitosterol, flavanol glycosides, astragalin, nicotiflorin, oleanolic acid, nyctanthic acid, tannic acid, ascorbic acid, methyl salicylate, an amorphous glycoside, an amorphous resin, trace of volatile oil, carotene, friedeline, lupeol, mannitol, glucose, fructose, iridoid glycosides, and benzoic acid.

Uses: Roots, bark and leaves are useful in dyspepsia, colic, flatulence, skin diseases, leprosy, diarrhoea, urethrorrhea, splenomegaly, anaemia, helminthiasis, boils, burns and rectal disorders; fruits are useful in curing hydrocele.

Orchidaceae

Orchis mascula (L.) L.

Vernacular Name: Salabmissi (Burmese); Purple Orchid, Salep, (English).

Description: Perennial herb growing to 60 cm; narrow leaves often have purple-black blotches; flowering stem bears purple or purple-pink flowers; two tuberous roots, one larger than the other; inflorescence is 7.5–12.5 centimetres (3–5 in) long and it is composed of 6 to 20 flowers gathered in dense cylindrical spikes.

Distribution: Native to Europe, the Middle East and North Africa; found in clearing in woodland, scrubland and grassland.

Parts Used: Tuber.

Phytochemistry: Alkaloids and volatile oil.

Uses: Used in diarrhoea and irritated gastrointestinal tracts in children.

Vanda tessellata (Roxb.) Hook. ex G.Don (Syn. Vanda roxburghii R.Br.)

Vernacular Name: Kyi-baung-myo (Burmese); Nai-Rasna, Banda, Rasna, (Hindi); Rasna, (Sanskrit).

Description: A small epiphytic herb; roots white, shiny, silvery grey coloured; develops two kinds of roots, the clinging roots and aerial roots; leaves coriaceous in two rows oblique at apex; flowers axillary racemose or solitary.

Distribution: This is a common orchid epiphytic on mango and other trees; grow all over Asia including India and Sri Lanka.

Parts Used: Whole plant.

Phytochemistry: Plant contains alkaloids, glycoside, tannins, saponins, sterols, fatty oils, resins and colouring matter.

Uses: Plant used in rheumatic disorders, local swellings, indigestion; roots are useful in inflammations, rheumatic disorders, osteoarthritis and allied disorders; indigestion; used as a laxative and stimulant to the liver; external application in rheumatic pains.

Oxalidaceae

Biophytum sensitivum (L.) DC. (Syn. Oxalis sensitiva L.).

Vernacular Name: Hit-gayoon (Burmese); Lajjalu, Laksana (Hindi); Viparitalajjalu (Sanskrit);

Description: A slender erect annual with a rosette of leaves on top of the stem; leaves abruptly pinnate, sensitive, leaflets opposite, 6-12 pairs, the terminal pair being the largest, oblong, apiculate at the apex, glabrous, pale beneath; flowers yellow, dimorphic, peduncles many, slender upto 10 cm long, fruits ellipsoid capsules; seeds prominently ridge, transversely striate.

Distribution: Found in Nepal, India, Myanmar and other Southeast Asian countries; tropical Asia, Africa and America.

Parts Used: Whole plant.

Phytochemistry: Plant contains amentoflavone, cupressuflavone, and isoorientin.

Uses: Plant is antibacterial, anti-inflammatory, antioxidant, antitumor, radioprotective, chemoprotective, antimetastatic, antiangiogenetic, wound-healing, immunomodulatory, anti-diabetic, and cardioprotective in nature; useful in strangury, urinary calculi, hyperdipsia in bilious fever, wounds, abscesses, gonorrhoea, asthma, phthisis, stomachalgia and snake bite.

Oxalis corniculata L.

Vernacular Name: Hmo-na-do (Burmese); Wood sorrel or yellow sorrel (English); Amrul ask, chukka tripati (Hindi); Amilika (Sanskrit).

Description: Small creeping perennial herb which form roots at nodes; Leaves alternate, long-petiolate, trifoliate, the leaflets obcordate with a conspicuous notched apex, each leaflet upto 2 cm long; flowers yellow, 5-merous, borne in axillary few-flowered inflorescences; fruit a sub-cylindrical capsule up to 20 cm long containing numerous tiny black seeds.

Distribution: Widely naturalized in Asia, South America, Africa and Australia.

Part Used: Whole plant.

Phytochemistry: Plant contains glyoxylic acid, oxalic acid, vitexin, isovitexin, vitexin-2-O-β-D-glucopyranoside, glycolipids; vitamin C, phospholipids; fatty acids, α and β tocopherols; 2-heptenal, 2-pentylfurn and transphyto.

Uses: Leaves are edible, with a tangy taste of lemons; used in a fever; being very high in Vitamin C and potassium, it helps to relieve scurvy, skin disorders, stomach complaints, eye disorders, jaundice and diabetes.

Papaveraceae

Papaver rhoeas L.

Vernacular Name: Bein-ni (Burmese); Red poppy (English); Lalpost, Post, Posteka bija (Hindi).

Description: An erect annual, forming a long-lived soil seed bank that can germinate when the soil is disturbed; grows up to about 70 cm in height; flowers are large and showy, 50 to 100 mm across, with four petals that are vivid red, most commonly with a black spot at their base; flower stem is usually covered with coarse hairs that are held at right angles to the surface; capsules are hairless, obovoid, less than twice as tall as they are wide, with a stigma at least as wide as the capsule.

Distribution: Native to Europe, North Africa and temperate regions of Asia; naturalized in North and South America.

Parts Used: Flowers.

Phytochemistry: Flowers contain rhoeadine meconic acid, mekocyanin, mucilage and tannins.

Uses: Plant is analgesic and sedative; used as a mild pain reliever; effective for irritable coughs, nervous overactivity, insomnia, general irritability, and asthma.

Papaver somniferum L.

Vernacular Name: Bein-(Poppy) (Burmese); Opium poppy, white poppy (English); Afim, afyun, kashkash, post (Hindi); Ahifen, chosa, khasa (Sanskrit).

Description: Erect annual herb, about 1 m high; stems slender, glabrous; leaves alternate, sessile, amplexicaual, irregularly lobed and toothed; flowers large, handsome of various shades: white, scarlet or lilac, solitary ath the top, with long stalk; capsule globular, longitudinally grooved, seeds numerous, small and black.

Distribution: Native range is probably the eastern Mediterranean; cultivated many parts of the globe.

Parts Used: Seeds.

Phytochemistry: Seeds contain morphine, codeine, thebaine, nacrocotine, narceine and papaverine; organic acids; meconic, lactic, malic, tartaric, citric, acetic and succinic; proteins, dextrose and pectin.

Uses: Seeds have hypnotic, analgesic, antitussive and antidiarrhoeal properties; effective in dyspnoea consequent upto heart failure; used in treating chronic cough pertussis, and diarrhoea.

Passifloraceae

Passiflora foetida L.

Vernacular Name: Bur-taw-suka(Burmese); Stinking passion flower, tagua passion floer, hispid granadilla (English).

Description: Perennial climber with tendrils; stems hollow, tubular; leaves alternate, trilobed, flowers white, solitary in the axil of leaves; coronal violet; fruit globose, enveloped by remains of pinnatifid bracts, yellow when ripe, edible; seeds numerous; whole plant is densely hirsute.

Distribution: Native to the southwestern United States, Mexico, the Caribbean, Central America, and much of South America; distributed to tropical regions around the world, such as Southeast Asia and Hawaii.

Parts Used: Whole plant.

Phytochemistry: Leaves contain pachypodol, 4,7-dimethylnaringenin, ermanin, 4,7-O-dimethyl naringenin, 3, 5-dihydroxy-4, 7-dimethoxy flavanone; stem contains apigenin, ermanin, indole alkaloid, harman, hydrogen cyanide and cyanogenic glycosides.

Uses: Plant has sedative properties; used in neurasthenia, insomnia, nightmares and anxiety.

Phyllanthaceae

Phyllanthus reticulatus Poir.

Vernacular Name: Bur-badibyu (Burmese); Black-Honey Shrub, black-berried featherfoil, potato-bush (English); Pancoli, Makhi (Hindi); Pulika, Krsnakamboji (Sanskrit).

Description: A large glabrous or pubescent subscandent shrub, 1.5-4.5 m in height with smooth or lenticellate branches; leaves elliptic to oblong or obovate, obtuse or acute, thin, main nerves few; flowers unisexual, axillary, male in fascicles of 2-6, female solitary; fruit a purplish blackberry, globose, smooth and shining, seeds irregularly 3-gonous, finely granulate.

Distribution: Tropical and subtropical regions of Africa, through Asia to Australia.

Parts Used: Whole plant.

Phytochemistry: Stem and leaves contain sitosterol, friedelin and betulinic acid; stem bark contains pentacosane, 21-α-hydroxyfriedelan-3-one, taraxerol and lupene-24-diol.

Uses: Plant used in diabetes, cancer, viral infection, malaria, burns and skin allergies; used in bleeding gums, smallpox, asthma and syphilis.

Phyllanthus urinaria L.

Vernacular Name: Mye-zee-byua, Mye-shiw-sha (Burmese); Lal-bhuni anvalah, hazar main, (Hindi).

Description: Diffuse annual or perennial herb, 20-30 cm height; stems glabrous, usually purplish; leaves alternate, very short-petioled, prominently arranged in two rows, resembling a pinnate leaf; flowers minute, yellowish, unisexual, solitary in the axils; both sexes on the same plant; capsule without stalk, globose, slightly compressed, echinate.

Distribution: Asian origin (India, Indonesia, China, Myanmar); widely found in all tropical regions of the world; United States.

Parts Used: Whole plant.

Phytochemistry: Plant contains corilagin, rutin, brevifolincarboxylic acid, isostrictiniin, geraniin, gallic acid, ellagic acid, methyl ester dehydrochebulic acid and methyl brevifolin carboxylate.

Uses: Plant has antibacterial and demulcent properties; useful in sore, boils, impetigo, infantile cheek eczema, tongue thrush, arthralgia, snake and centipede bites, post-partum haematometra, fever, ophthalmia and hepato-biliary diseases.

Pinaceae

Cedrus deodara (Roxb. ex D.Don) G.Don

Vernacular Name: Nat-htinyu (Burmese); Deodar, Himalayan cedar, True cedar (English); Debdar, Deyodar (Hindi); Devadaru (Sanskrit)

Description: A large handsome evergreen conifer tree reaching up to 85 m in height, with almost rough, black, furrowed bark and spreading branches; shoots dimorphic; leaves needle-like, triquetrous, sharp, pointed; male cones solitary, cylindric at the ends of branchlets, female cones solitary at the ends of branchlets, composed of imbricating thin woody placental scales; seeds pale brown, wings longer than the seeds.

Distribution: Native to the western Himalayas in eastern Afghanistan, northern Pakistan, India and Myanmar.

Parts Used: Leaves, heartwood, oil.

Phytochemistry: Bark contains taxifolin; wood contains cedeodarin, dihydromyricetin, cedrin, cedrinoside and deodarin; needle contains essential oil (α-terpineol, linalool, limonene, anethole, caryophyllene and eugenol); lignans and phenolic sesquiterpene himasecolone, isopimaric acid, (−)-matairesinol, (−)-nortrachelogenin and a dibenzylbutyrolactollignan.

Uses: Leaves are useful in inflammations and tuberculosis; heartwood is useful in inflammations, dyspepsia, cephalagia, haemorrhoids, insomnia, epilepsy, hiccough, bronchitis, tubercular glands, diabetes, urethrorrhea, renal and vesical calculi, elephantiasis, fever, cardiac disorders, leucoderma, skin diseases, proptosis; oil is useful in leprosy, syphilis, skin diseases, wounds and ulcers, fever and strangury.

Piperaceae

Piper cubeba L.f.

Vernacular Name: Myeet-tet, nga-yok (Burmese); Cubebs, Tailed pepper, Tailed cubebs, Jaca pepper (English); Kababcini (Hindi); Kankolam (Sanskrit).

Description: A perennial woody climber with ash grey climbing stems and branches, rooted at the joints; leaves simple, ovate-oblong, smooth, base cordate or rounded, tip pointed, lower surface densely provided with minute sunken glands; flowers small, unisexual, in smooth scaly spikes, female spikes often curved; fruits subglobose berries, stalked, somewhat apiculate.

Distribution: Native of Indonesia; found in India, Malaysia and Myanmar.

Parts Used: Berries.

Phytochemistry: Berries contain monoterpenes (sabinene 50%, α-thujene, and carene) and sesquiterpenes (caryophyllene, copaene, α- and β-cubebene, δ-cadinene, germacrene), the oxides 1,4- and 1,8-cineole, cubebol and cubebic acid.

Uses: Cubeb paste used as a mouthwash; dried cubebs internally used for oral and dental diseases, loss of voice, halitosis, fevers, and cough; paste of the cubeb berries externally used on male and female genitals to intensify sexual pleasure during coitus.

Plantaginaceae

Plantago ovata Forssk.

Vernacular Name: Eiktha-pha-gula (Burmese); Ispaghula, Blond psyllium, Spogel seeds (English); Isabgol, Isabgul (Hindi); Snigdhabijah, Snigdhajirakah (Sanskrit).

Description: A very short-stemmed softy hairy annual herb; leaves simple, narrowly linear or filiform, finely acuminate, entire or distantly toothed usually 3-nerved; flowers small, in cylindric or ovoid spikes; fruits ellipsoid obtuse capsules, the upper half coming off as a blunt conical lid, seeds smooth, yellowish-brown, ovoid-oblong, boat-shaped.

Distribution: Native to Western Asia and Southern Asia; growing wild in the southwestern United States.

Parts Used: Seeds.

Phytochemistry: Linolenic, linoleic, oleic, palmitic, stearic, lignoceric acids; arabinoxylan, hemicellulose.

Uses: Seeds mainly used in constipation, high blood cholesterol, irritable bowel syndrome, gastric inflammations and urinary tract infection.

Plumbaginaceae

Plumbago indica L.

Vernacular Name: Kant-gyok-ni (Burmese); Rosy-flowered leadwort, Fire plant (English); Lalcitra, Raktacitra (Hindi); Citrakah, Dahanah (Sanskrit).

Description: A pretty subscandent perennial shrub with semi-woody striate stems and flexible branches; leaves simple, alternate, oblong, short-cuneate at the base passing into a very short amplexicaul, exauriculate, reddish petiole; flowers bright red, in long terminal spikes, the calyx ribs covered with stipitate, bifarious and subsessile glands, corolla tube slander, 4 time as long as the calyx; stout cylindrical, irregularly bent, light yellowish-brown with smooth surface having short transverse shallow fissures at the regions of the bents.

Distribution: Native to India; distributed to most Asian countries.

Parts Used: Roots,

Phytochemistry: P. indica contains an orange yellow pigment named plumbagin, sitosterol, fatty alcohol probably arachidic alcohol, tannin and an amorphous brown pigment. The flowers contain 3-rhamnosides of pelargonidin, cyaniding delphinidin and kaempferol.

Uses: Root is used externally or internally in rheumatism and paralytic afflictions; it is powerful sialogogue and a remedy for secondary syphilis, leprosy and leucoderma; milky juice of the plant is used in ophthalmia and in scabies.

Eleusine indica (L.) Gaertn.

Vernacular Name: Sinnigomyet, Sin-nga-myet (Burmese); Crowfoot grass, Dog's tail grass, Crabgrass, Wire grass (English); Mandla (Hindi).

Description: Annual herb; stems erect or occasionally prostrate in tufts, 30-60 cm high leaves loricate, narrow, arranged in two rows; the blade glabrous; sheath hariy; inflorescence in the whorl of 5-7 digitate spikes, with single spikelet or two spikelets separated below; fruit oblong, nearly trigonal.

Distribution: Distributed throughout the warmer areas of the world; tropical and subtropical regions.

Parts Used: Whole plant.

Phytochemistry: Plant contains prussic acid (cyanogenetic glucoside).

Uses: Decoction of the fresh plant used as a diuretic and for dysentery, Also used for the treatment of fever, Dandruff, Post-partum, Sprains and lumbago, Hemoptysis, hypertension, hemoptysis, asthma, jaundice etc.

Hordeum vulgare L. (Syn. H. sativum Jessen.)

Vernacular Name: Bur-Byit-gyonbarley (Burmese); Barley, (English); Jav, Jau, (Hindi); Yacah, (Sanskrit).

Description: An erect, stout, tufted annual herb upto 100 cm in height; leaves few, flaccid, linear upper ones close to the spike, sheath smooth, ligule short, membranous, spike flattened, 2-ranked, 20-30 cm long, terminal spikelets sessile, arranged in threes on two sides of the rachis, all fertile or lateral ones barren; fruit an elliptic short pointed grooved caryopsis.

Distribution: Grown in temperate climates globally.

Parts Used: Grains.

Phytochemistry: Arginine, histidine, lysine, tyrosine, tryptophan, phenylalanine, cystine, methionine, theronine, leucine, isoleucine, valine, glycine.

Uses: The grains are useful in cough, asthma, strangury, amentia, fever, bronchiotitis, urocystitis urethritis, dyspepsia, gastropathy, ulcers, burns cephalalgia, anaemia and in the diet of invalids.

Imperata cylindrica (L.) Raeusch. (Syn. I. arundinacea Cyrill).

Vernacular Name: Bur-thetkegale (Burmese); Lalang-grass, alang-alang, bedding grass, (English); Dabh, siru, ulu, (Hindi); Darbha, (Sanskrit).

Description: Perennial herb, 0.60-1.50 m. high; rhizome hard coriaceous, descending deeply into the soil; clumps erect, with wiry hairs; leaves narrow and long, prominently nerved; scabrous on the upper surface margins sharp-edged; inflorescence of many spikes silver-white, densely silky.

Distribution: Native to east and southeast Asia, India, Myanmar, Micronesia, Melanesia, Australia, and eastern and southern Africa.

Parts Used: Rhizomes.

Phytochemistry: The rhizomes contain glucose, fructose and organic acids.

Uses: Rhizomes used in the treatment of urodynia, pollakiuria, haematuria and fever; combined with corn silk enhances its diuretic action.

***Oryza sativa* L. (Syn. *Oryza plena* (Prain) N.P.Chowdhury)**

Vernacular Name: Bur-sabah (Burmese); Paddy, Rice, (English); Dhan, Caval, (Hindi); Dhanyan, Vrihi, Sali, (Sanskrit).

Description: An annual or perennial grass with tuft of fibrous roots and swollen nodes; leaves simple with sheathing bases, long and narrow, slightly pubescent with spiny hairs on the margins; flowers spikelets in terminal compound panicles, lamma punctuate or granulate without wing on the back, lemma and palea surrounding the kernel, golden yellow, reddish-purple, brown or smoky black becoming straw-coloured on ripening, grains narrowly oblong, free within the lemma and palea.

Distribution: Cultivated throughout the world.

Parts Used: Roots, grains.

Phytochemistry: Protein, fat, carbohydrates, fibre, ash, thiamine, riboflavin, nicotinic acid, pantothenic acid, pyridoxine, choline chloride.

Uses: The roots are useful in burning sensation, dipsia, bilious fever, strangury and diabetes; grains are useful in pneumonosis especially pulmonary consumption, diarrhoea and colonopathy.

***Panicum sumatrense* Roth (Syn. *P. miliare* Lam.)**

Vernacular Name: Myek-lu (Burmese); Little millet, (English); Savan, Kumku, Kutki, (Hindi); Syamakah, (Sanskrit).

Description: A slender erect branched annual, 30-90 cm in height, usually leafy upto the panicle; leaves linear, tapering from a broad base, upto 60 cm long, glabrous or finely hairy; panicles very compound, contracted, often nodding; spikelets glabrous, not gaping lower floral gum 9-nerved, upper floral gum 3-5 streaked; fruits caryopsis, glabrous, striated, brown.

Distribution: In the temperate zones of Asia: the Caucasus, China, East Asia and also in the tropics of the continent: India, Indochina and Malaysia.

Parts Used: Grains.

Phytochemistry: Arginine, lysine, methionine, phenylalanine, threonine, tryptophan and valine.

Uses: Grains are useful in burning sensation, diarrhoea, dysentery, dyspepsia, wounds, ulcer and gonorrhoea.

***Chrysopogon zizanioides* (L.) Roberty (Syn. *Vetiveria zizanioides* (L.) Nash)**

Vernacular Name: Myet-myit-hmway (Burmese); Vetiver, Khas-khas, khus-khus, (English); Khas, Khas-Khas, (Hindi); Usirah, Sevyah, Sugandhimulah, (Sanskrit).

Description: A densely tufted perennial grass with aromatic roots and rhizomes; leaves narrow, linear, erect, sheaths compressed, ligules reduced to a scrious rim, midrib slender, lateral nerves close, spikelets grey, green or purplish in a panicle of numerous slender racemes, sesile spikelets linear-lanceolate, glabrous, pedicelled spikelets 2-flowered; fruits oblong, grains, slightly oblique at the apex.

Distribution: Native to India; widely cultivated in the tropical regions of the world; major vetiver producers include Haiti, India, Indonesia, and Réunion.

Parts Used: Roots.

Phytochemistry: Essential oil containing β-vetivone, isobisabolene, khusol and azulene.

Uses: Plant has antiseptic properties to treat acne and sores; roots are useful in hyperdipsia, burning sensation, ulcers, skin diseases, nausea, obstinate vomiting, dyspepsia, flatulence, colic, anaemia, haemorrhages, haemoptysis, cough, asthma, hiccough, strangury, bilious fever, gout, lumbago, sprains, halitosis, cephalalgia, spermatorrhea, hysteria, insomnia, diarrhoea, hyperhidrosis, amentia, cardiac debility, amenorrhoea, dysmenorrhoea, helminthiasis, spasmodic affections, erysipelas and emaciation.

Polygonaceae

Fagopyrum esculentum Moench

Vernacular Name: Shari-Man-buck Kout (Burmese); Buckwheat (English); Kotu, Phaphra (Hindi)

Description: An annual herb, 3-5 ft. high with alternate, hastate, acute leaves, flowers in axillary or terminal cymes, pinkish-white, fragrant, fruit a 3-cornered achene.

Distribution: Native to central and northern Asia; cultivated extensively in temperate regions of India.

Part Used: Leaves, flowers.

Phytochemistry: Plant contains bioflavonoids, rutin, catechin-7-O-glucoside, salicylaldehyde, 2,5-dimethyl-4-hydroxy-3(2H)-furanone, (E,E)-2,4-decadienal, phenylacetaldehyde, 2-methoxy-4-vinylphenol, (E)-2-nonenal, decanal and hexanal.

Uses: Plant used for blood circulatory problems; helps in strengthen varicose veins and heal chilblains; commonly taken in combination with other herbs for high blood pressure.

Persicaria chinensis (L.) H. Gross (Syn. Polygonum chinense L.)

Vernacular Name: Bur-boktaung(Burmese); Chinese knotweed(English).

Description: Undershrub about 1 m. high; stems cylindrical, glabrous, much-branched; leaves alternate, ovoid, entire with a tubular ocrea and sometimes a black marking like a V on the upper face; inflorescence in a terminal panicled cyme; flowers small, white or pink, nutlet trigonous, pulpy, dull-black when ripe.

Distribution: Widespread across China, Japan, the Indian Subcontinent, Indonesia, Myanmar, Malaysia, and Vietnam.

Parst Used: Leaves.

Phytochemistry: Plant contains rubin, rheum-emodin, oxy-methyl-anthraquinone, anthraquinone, glucosides, and myristyl alcohol.

Uses: Leaves have antibacterial properties; effective in the treatment of furunculosis, impetigo, ulceration of the helix, scalp scabs, cold sore, eczema, and streptococcal skin infections; decoction of dried leaves used for boils and dysentery; leaves residue applied to the bite as a plaster on snake-bite.

Rheum emodi Wall.

Vernacular Name: Pito-mon-la (Burmese); Hindirevaandchini, doler, (Hindi); Sandhint, pitamulika, (Sanskrit).

Description: A robust perennial herb with thick, long, roots; 1-15 m tall and grows on rocky lands. Roots are stout and the stem is hollow. Lower leaves are 20-25 cm in diameter, orbicular and are broadly ovate. Flowers are small and dark purple in colour. Fruits are purple and angled.

Distribution: Originally belongs to southeastern Russia; found in China, Myanmar, India and Central Africa.

Part Used: Rhizomes.

Phytochemistry: Plant contains hydroxyanthracene derivatives, emodin, chrysophanol, rhein, aloe-emodin, physcion and their glycosides.

Uses: Rhizomes are used for cleaning teeth and are sprinkled over ulcers for quick healing; infant's stomach problems; cuts, wound and muscular swellings, tonsillitis and mumps.

Rumex crispus L.

Vernacular Name: Kala-chinbanng (Burmese); Curled dock, (English).

Description: A smooth, perennial herb, with deep, spindle-shaped, yellow root. Stem, 90-120 cm. high, angular, furrowed, somewhat zigzag. Leaves lanceolate, petiolate, whorled, acute, wavy-curled, crisp margined, smooth and light green; the radical leaves long-petioled, truncate, or the radical leaves long-petioled, truncate, or scarcely heart-shaped at the base; the cauline, acute at both ends and nearly sessile. Flowers with 6 perianth sepaloid leaves, petals none, bisexual, green small, numerous and on short rather stout pedicels, pedicels 5-10 mm. long visibly jointed at the base, forming crowded and dense whorled racemes. Inflorescence stalk jointed at the base. Inner fruiting perianth segments, green, cordate of truncate, entire each with 3 tubercles, unequal, the largest one ovoid, very turgid, tubercular, round at ends. The nut, 2.5 mm. by 3.0 mm. ovoid, pointed and slimy brown.

Distribution: Native to Europe and Western Asia; US, Mexico, New Zealand, and India.

Parts Used: Root.

Phytochemistry: Roots contain chrysophanol, emodin, nepodin, physcion, catechol, brassidinic acid, oxalic acid, chrysophanic acid, and rumicin.

Uses: Roots used in respiratory problems, tubercular diathesis, cold, cough, and itching of skin; a poultice of the powdered root to sores, rashes and skin infections, and use an infusion of the root for athlete's foot.

Primulaceae

Embelia ribes Burm. f.

Vernacular Name: Swa-bwasa, min-gaung-gharung (Burmese); Embelia (English); Vayvidamg, Bhabhiramg (Hindi); Vidangah, Vellah (Sanskrit).

Description: A large scandent shrub with long, slender, flexible branches and bark studded with lenticels; leaves simple, alternate, elliptic-lanceolate, short and obtusely acuminate, entire, coriaceous, glabrous on both sides, shiny above, silvery beneath gland-dotted, glandular pits near the midrib beneath; flowers white or greenish-white in terminal and axillary, lax panicled racemes; fruits globular dull red to nearly blackberries, longitudinally striated with short slender persistent pedicel and style base; seed single globose, hollowed at the base, white-spotted, testa membranous, albumen pitted.

Distribution: Found in India, Sri Lanka, Singapore, Southern China, Myanmar, Indonesia and some parts of Africa.

Parts Used: Roots, leaves, fruits.

Phytochemistry: Plant contains embelin, christembine, homoembelin, vilangine and quercitol.

Uses: Roots are useful in colic, flatulence and dyspepsia; leaves are useful in pruritus, skin diseases and leprosy; fruits are useful in skin diseases, leprosy, pruritus, nervous debility, amentia, dyspepsia, flatulence, colic, constipation, strangury, tumours, asthma, bronchitis, dental caries, odontalgia, hemicranias, dyspnoea, cardiopathy psychopathy, ringworm infestation and fever.

Ranunculaceae

Aconitum ferox Wall.ex Ser.

Vernacular Name: Aseik-pin, myo (Burmese); Indian aconite (English); Bish, Bacchanag (Hindi); Vatsanava (Sanskrit).

Description: A perennial herb; roots tuberous, paired; daughter tuber ellipsoid to ovoid-oblong. 2.5-4.0 cm in length, with filiform root fibres; leaves scattered, orbicular-cordate to reniform, palmately 5-lobed, resembling those of melon leaves; blade and petiole pubescent; peduncle straight, bearing flowers on both sides. Flowers pale dirty blue, borne in a dense terminal raceme, 10-25 cm long, helmet-vaulted with a short sharp beak, resembling a pea flower; follicles oblong; seeds long, pyramidal to obovoid, winged along the raphe.

Distribution: Native to India; found many parts of Asia.

Parts Used: Underground stem and root.

Phytochemistry: Roots contain pseudaconitine, chasmaconitine, indaconitine, bikhaconitine, veratroyl pseudacounitie and diacetyl pseudaconitine.

Uses: Herb is used as antipyretic, analgesic, appetizer, digestive and anti-rheumatic; used in fever, respiratory disorders, asthma, chronic obstructive pulmonary diseases, hypertension, cardiac disorders, palpitation, urinary disorders, seminal ejaculations, neuralgia, inflammation, joint pains and an enlarged spleen.

Aconitum heterophyllum Wall.ex Royle

Vernacular Name: Di-Wok (Burmese); Atis roots, Indian Atis (English); Atis (Hindi); Ativisa (Sanskrit).

Description: A Perennial, erect, showy herb; stem long, simple or branched from base, glabrous and puberulous above; leaves broad, ovate or orbicular or somewhat 5-lobed and toothed, upper 3-fid or entire; flowers 2.5 cm long, helmet-shaped, bright, blue or greenish-blue with purple veins.

Distribution: Found in India, Pakistan and Nepal and Myanmar.

Part Used: Underground stem and root.

Phytochemistry: Root and stem contain atisine, atidine, histisine, hetisine, hetidine, heterophllisine, heterophylline, heterlophylline, isoatisine, dihydroatisine, hetisinone and benzoyl heteratisine.

Uses: Plant used as an anthelmintic, anti-inflammatory, antipyretic, analgesic, astringent and febrifuge; useful in treating coughs, diarrhoea and indigestion; drug for infants in dentition, diarrhoea, fever and vomiting; used as an antidote to poisoning.

Aconitum napellus L.

Vernacular Name: Aseik-Pin-myo (Burmese); Indian aconite, Monk's hood (English); Bachnag, Mithavis (Hindi); Vatsanabhah (Sanskrit).

Description: A Perennial herb; stem often decumbent in small states, glabrous or slightly pubescent; leaves variable in size, palmatipartite, segments linear; racemes simple or sparingly compound; bracts entire or 3-fid; flowers borone on the upper portion of the peduncle and on both sides, bright, or dull greenish-blue; helmet thrice as long as high; follicles 3-5, having many seeds inside.

Distribution: Native to Asia and North America.

Part Used: Root.

Phytochemistry: Roots contain aconitine, mesaconititne, neoline, songoramine, deoxyaconitine, hypaconitine, napelline, 12-epidehydro-nepelline, 12-epiacetyldehydronapelline, 1,14-diacetylneoline and vitamin C.

Uses: Roots are useful in cephalagia, neuralgia, painful inflammations, strangury, cough, asthma, bronchitis, hypotension, dyspepsia, flatulence, colic, cardiac debility, dysmenorrhoea, amenorrhoea, spermatorrhoea, leprosy, skin diseases, sciatica, gout, paralysis, hepatopathy, splenopathy, pharyngodynia, diabetes, nasal discharges, and fever.

Coptis teeta Wall.

Vernacular Name: Shwe-gyo-myit (Burmese); Gold thread, Coptis, Mouth Root, Vegetable Gold (English); Mamira, Mamiran (Hindi); Supita, Pita (Sanskrit).

Description: A rhizomatous, herbaceous perennial with horizontally growing golden yellow woody underground rhizome bearing many fibrous roots; leaves ternatisect, leaflets 5-7.5 cm, ovate-lanceolate, pinnatifid, the terminal largest, the rachis 15-30 cm long; flowers small, white, on slender leafless scapes, fruits many-seeded follicles; seeds black, tests crustaceous.

Distribution: Found in China, India, Myanmar, Tibet and Nepal.

Parts Used: Rhizomes.

Phytochemistry: Plant contains berberine, coptine, coptisine, jatrorrhizine, berherastine, and palmatine.

Uses: Rhizomes are are useful in ophthalmopathy, cough, bronchitis, atonic dyspepsia, flatulent colic, helminthiasis, diabetes, intermittent fevers, stranger, leprosy, leucoderma, dropsy and general debility.

Naravelia zeylanica (L.) DC.

Vernacular Name: Bur-Khwa-nyo, sat-gyo-Yit (Burmese); Dhanavalli (Sanskrit).

Description: A scandent or climbing shrub with tuberous roots, wiry stem and strong tendrils; leaves 3-foliate, opposite, terminal leaflet modified into a 3-branched tendril, leaflets ovate-lanceolate, serrate or crenate, prominently nerved; flowers yellow, fragrant in axillary and terminal panicles, sepals downy, petals linear-clavate, elongate; fruits aggregate of achenes, ending in twisted feather tails.

Distribution: Throughout South East Asia.

Part Used: Whole plant.

Phytochemistry: Plant contains alkaloids, phytosterols, flavonoids, terpenoids, tannins and carbohydrates.

Uses: Vine crushed and inhaled to cure headache; fresh stems chewed in toothache; plant paste consumed with *Borassus flabellifer* for chest pain; young leaves paste applied on skin diseases and ulcers, and on forehead for cold and headache; leaves pounded with roots of *Eranthemum palatiferum* applied in bone fracture; crushed roots inhaled against cold and fever.

Ranunculus sceleratus L.

Vernacular Name: Tanga-nagayok (Burmese); Blister butter cup, Celery-leaved butter cup (English).

Description: An erect glabrous, much-branched annual, up to 70 cm. in height; leaves alternate, petiolate, upper ones simple, crenate, or of 3 linear-oblong segments; lower ones usually 3-partite, segments cuneate and again variously lobed or notched; cauline sessile; stem hollow, furrowed and glabrous; flowers bracteates, bisexual, numerous, in cymose inflorescence with sepals 5, petaloid; petals 5, obovate; nectary present at the base of each petal; stamens many polyandrous; receptacle hairy.

Distribution: Native to temperate and boreal North America and Eurasia.

Parts Used: Whole plant.

Phytochemistry: Plant contains anemonin and protoanemonin.

Uses: Plant used in boring and gnawing pain, skin vesicular eruptions, blister, sore, burning, corns, and bruised pain in the chest.

Rhamnaceae

Ventilago maderaspatana Gaertn.

Vernacular Name: Thwe-det, Ta-yaw-nyo (Burmese); Red creeper, (English); Pitti, (Hindi); Dinesavalli, Raktavalli, (Sanskrit).

Description: A large, much-branched, woody climber with branches hanging down and dark grey bark having vertical cracks exposing the inner vermilion surface; leaves simple, alternate, oblong-lanceolate or elliptic-ovate, obtuse or acute, entire or crenate, base rounded or acute, main nerves-4-8 pairs; flowers greenish with an offensive odour in drooping pubescent terminal panicles, petals much smaller than the sepals; fruits yellowish drooping pubescent terminal panicles, petals much smaller than the sepals; fruits yellowish globular nuts, supported by the persistent calyx, wing linear-oblong; 1-nerved, rounded at the apex.

Distribution: Throughout Indo-Malesia, Myanmar and nearby areas.

Parts Used: Bark.

Phytochemistry: Root bark contains trihydroxy-methyl-anthranol-monomethyl ether; emodinmonomethyl ether; lupeol, β-sitsterol and its glucosides in fruits, leaves and stem; friedelin in stem bark.

Uses: Bark is useful in dyspepsia, colic, flatulence, erysipelas, leprosy, scabies, pruritis, and other skin diseases, fever and general debility.

Rosaceae

Fragaria vesca L.

Vernacular Name: Shan-zee (Burmese); Wild strawberry, European strawberry (English).

Description: Perennial herb, with runners; height: 5–20 cm; runners up to 2 m long; stem hairs ascending oblique–spreading; leaves in basal rosette, quite long-stalked; blade palmate, with 3 leaflets; leaflets elliptic, wedge-shaped base, with toothed margins, both sides smoothly and sparsely haired; all margin teeth roughly the same size; corolla regular, white, 12–18 mm broad; petals 5, quite round, touching each other or covering each other's edges, 4–6 mm long; calyx 5-lobed; with epicalyx; stamens 20; gynoecium separate, pistils several; receptacle glabrous; inflorescence a lax cyme; fruit red, juicy, shiny, easily breaking off at its base, 0.5–1 cm long berry, small achenes all over surface of berry; sepals in fruiting stage descending–spreading.

Distribution: Native to Europe and temperate regions of Asia.

Part Used: Leaves, fruit.

Phytochemistry: Leaves contain ellagitannin agrimoniin, flavonoids, tannins and a volatile oil; fruit contains methyl salicylate and borneol.

Uses: Leaves are used as a gargle for sore throats, and in a lotion for minor burns and grazes; fruit is considered to have cooling and diuretic properties, and used in tuberculosis, gout, arthritis and rheumatism.

***Prunus cerasoides* Buch.-Ham.ex D.Don.**

Vernacular Name: Bur-Pannu (Burmese); Himalayan wild cherry, Bird cherry (English); Patmakath, Patmakh, Paddam (Hindi); Padmakah, Padmagandhi (Sanskrit).

Description: A medium-sized or large tree with pale redwood and smooth brown bark peeling off in horizontal strips exposing a shining copper-coloured surface; leaves simple, alternate, glossy, ovate, long, acuminate, doubly-serrate, one or more conspicuous glands on the petiole; flowers white, pink or crimson in fascicles or umbels; fruits ovoid or globose drupes, yellow or red, stone ovoid, rugose and furrowed.

Distribution: Native to India, Myanmar, China, Thailand, Laos, Vietnam; found in East Asia, South Asia and Southeast Asia.

Parts Used: Heartwood.

Phytochemistry: Plant contains hydrogen cyanide, tannins, amygdalin, prunasin, and prussic acid.

Uses: Juice of the bark is applied externally to treat backaches; bark paste used in skin diseases, vomiting, nausea and gastritis; also used in respiratory and digestion problems.

Rubiaceae

***Catunaregam spinosa* (Thunb.) Tirveng. (Syn. *Randia dumetorum* (Retz.) Lam.)**

Vernacular Name: Bur-say-than-baya (Burmese); Emetic nut tree, Common emetic nut (English); Mainphal, Madan (Hindi); Madanad (Sanskrit).

Description: A large deciduous armed shrub upto 9 m in height with strong straight spines, horizontal rigid branches and dark brown or grey rough scaly bark; leaves simple, usually fascicled on the suppressed branches, obovate, obtuse, wrinkled, shining or more or less pubescent above and on nerves beneath, main nerves 6-10 pairs; flowers at first white, later turning yellow, fragrant, solitary or 2-3 together at the ends of short leaf-bearing branchless; fruits globose or broadly ovoid, smooth or obscurely longitudinally ribbed berries, yellow when ripe, crowned with the large calyx limb; seeds many, angular, embedded in the pulp.

Distribution: Throughout Indian subcontinent.

Part Used: Bark, fruits.

Phytochemistry: Saponin, valerianic acids, vax, resin, colouring matter, d-mannitol and essential oil. A new glycoside and lupeol have been isolated from the ripe fruit. Bark gave randialic acid-A and randialic acid-B; whereas toot contains scopolatin and d-mannitol. The pulp contains citric and tartaric acid tannins, pectins, pectin, mucilage, saponins, dumetaronins and free sugars.

Uses: Bark used in cough, asthma, skin diseases, ulcers, flatulence, and colic; fruit has anti-inflammatory, emetic and abortifacient; bark is applied externally to relieve the pain of bruises and rheumatism.

Paederia foetida L.

Vernacular Name: Pe-bok-new (Burmese); Chinese moon-creeper, king's tonic, strinking opalberry (English); Gandhali, Somaraji (Hindi); Prasarani (Sanskrit).

Description: Perennial climbing plant; leaves opposite, margins entire, purple beneath; inflorescence in axillary cyme; flowers small white with pale-violet spots; fruit sub-ovoid, slightly compressed glabrous; the whole plant is covered with pubescence; it has a fetid smell when crushed.

Distribution: Native to temperate, and tropical Asia; naturalized in the Mascarenes, Melanesia, Polynesia, and the Hawaiian Islands; found in North America.

Parts Uscd: Lcaves.

Phytochemistry: The leaves contain an essential oil (methyl mercaptan and linalool); leaf and stem contain hentriacontance, ceryl alcohol and hentriacontanol; herb also contains saponin, quinones, fatty acids, alkaloids (α-paederine and β-paederine), vitamin C and triterpenoids.

Uses: Leaves are used in rheumatic pain, infertility, paralysis, chronic abdominal pain, arthritis, inflammation, colic and CNS related problems.

Rubia cordifolia L.

Vernacular Name: Pain-Jai-ni (Burmese); Indian madder (English); Mamjith, Majith (HIndi); Manjistha, Yojanavalli (Sanskrit).

Description: A very variable prickly or scabrous climbing perennial herb with long cylindric, flexuose, smooth reddish, roots and 4-angled rough stems and branches; leaves very variable 2-8 in a whorl often only 4, cordate-ovate-lanceolate, 7-5, ribbed, scabrous above, base rounded or slightly cordate, petioles very long, provided with sharp recurved prickles on the edges, stipules absent; flowers white, greenish yellowish or reddish, sweet-fleshy, smooth, purplish-black when ripe; seeds two, small.

Distribution: Many regions of Asia, Europe and Africa.

Parts Used: Roots.

Phytochemistry: Roots contain purpurin, munjistin, xnthopurpurin, pseudopurpurin, alizarin and its glucoside.

Uses: Roots effective as blood detoxifier; treat skin disorders like hyperpigmentation, scabies, acne and allergies; maintain overall health and is beneficial in the treatment of liver diseases, gall stones and amenorrhea.

Smilacaceae

Smilax china L.

Vernacular Name: Sein-na-maw cineroot (Burmese); China root, (English); Cobcini, (Hindi); Dvipantaravaca, Madhusnuhi, (Sanskrit).

Description: A hard tendril climber with sparsely prickled or unarmed stems and thick tuberous rhizomes; leaves simple, alternate, elliptic, rounded at the base, prominently nerved; flowers many, small, white in umbels; fruits red berries.

Distribution: Throughout Asia (China, Japan, India and Myanmar).

Parts Used: Rhizomes.

Phytochemistry: Tannin, resin and cinchonin, smilacin, steroidal saponin, flavonoid glycosides, diosgenin, and rutin.

Uses: Rhizomes are useful in syphilis, leprosy, skin diseases, epilepsy, insanity, scrofula, flatulence, dyspepsia, colic, neuralgia, constipation, helminthiasis, psoriasis, fever, strangury, seminal weakness and general debility.

Solanaceae

Hyoscyamus niger L.

Vernacular Name: Khorasani, samon-byu, samon-nant (Burmese); Black henbane; common henbane (English); Khurasaniajavayan (Hindi); Dipya, parasikaya (Sanskrit).

Description: An erect, biennial or annual herb 30-80 cm high, villous, viscid and with a disagreeable, heavy odour; stem robust, 0.3-0.9 m; basal leaves spreading, petiolate, oblong-ovate, 15-20 cm coarsely sinuate-toothed; cauline leaves smaller, sessile, clasping, oval-oblong, sinuate, pinnatifid; lobes irregular, triangular, lanceolate; flowers in terminal, scorpioid cymes, or axillary, nearly entirely sessile; june; calyx tubular-campanulate, segments 5, mucronate, persistent, accrete; corolla infundibular, 5-lobed, irregular, dirty-yellow veined with violet, the throat purplish-black; stamens 5, barely exserted; ovary 2-celled; fruit a capsule, enclosed in the persistent and enlarged calyx; seeds small, compressed nearly ovoid, slightly reniform, 1 mm in diameter, brownish-grey, surface reticulate.

Distribution: Cultivated in Asia, Europe, North Africa, England, Germany and Russia.

Part Used: Seeds.

Phytochemistry: Hyoscyamine, scopolamine, and other tropane alkaloids.

Uses: Gastric or intestinal cramps, diarrhoea; prolapse of the rectum, neuralgia, cough, hysteria; skin inflammation and boils; pain due to dental caries. Hyoscyamus has anodyne, narcotic, and mydriatic properties; employed as a sedative in nervous affections and irritable conditions, such as asthma and whooping cough.

Solanum aculeatissimum Jacq. (Syn. Solanum khasianum C.B. Clarke)

Vernacular Name: Ban-bhindi, Kandiari (Hindi); Kantkari, Bari kandiari (Sanskrit); Dutch eggplant, love-apple (English).

Description: Plant is a stout, much-branched, under shrub, upto 1.5 m tall, with almost straight prickles; leaves are ovate, lobed and the lobes are lanceolate or triangular, hirsute and prickly on both the surfaces; flowers are white in colour and are in lateral, 1-4 flowered racemes; berries are yellowish or greenish globose and 2.5 cm in diameter; seeds are smooth, brown and compressed.

Distribution: Found in South Asia, Africa and South America.

Parts Used: Fruits.

Phytochemistry: Berries contain solasodine, solakhasianin and diosgenin.

Uses: Used as antimicrobial.

Solanum erianthum D. Don

Vernacular Name: Bur-daung-sat-pya (Burmese); Vidari (Sanskrit).

Description: Small plant, 1-3 m. high stems and twig covered with a dense yellowish tomentum. Leaves alternate, entire, velvety-pubescent on both sides. Inflorescence in an axillary or terminal cyme; flowers white.Berry globose, many-seeded, yellow when ripe.

Distribution: Native to southern North America and northern South America; introduced to other parts of the world and has a nearly pantropical distribution.

Parts Used: Leaves.

Phytochemistry: Leaves contain essential oil, saponins, alkaloids, solanine and solasodins.

Uses: A locally applied poultice of pounded and heated fresh leaves is indicated for the relief of haemorrhoids and scrofula; plaster made of concentrated fresh leaf juice cures dermatomycosis and impetigo; internally applied decoction of the leaves is used as a taeniafuge in buffaloes and oxen.

Solanum melongena L.

Vernacular Name: Bur-Khayan (Burmese); Brinjal, Egg plant (English); Bengan, Badanjan (Hindi); Varttaki (Sanskrit).

Description: An erect herbaceous, armed or unarmed perennial; leaves simple, large entire lobed; flowers blue, in clusters of 2-5; fruits large, white, yellow or dark purple berries of different shapes capped with the thick persistent calyx; seeds many, yellow or cream, discoid.

Distribution: Cultivated throughout the world.

Parts Used: Roots, leaves, unripe fruits.

Phytochemistry: Solasodine, nasunin, arginine, aspartic acid and chlorogenic acid.

Uses: Roots are useful in inflammations, neuralgia, cardiac debility and ulcers in the nose; leaves are useful in cholera, bronchitis, asthma, odontalgia and fever; unripe fruits are aphrodisiac, cardiotonic and haematinic.

Symplocaceae

Symplocos paniculata Miq.

Vernacular Name: Bur-makauk-galay (Burmese); Lodh, Lodar, Lodhra, Lojh (Hindi).

Description: A deciduous shrub or a tree, upto 12 m in height; bark is soft, friable, of light brown colour; external surface is corky of much fissured transversely; while the internal part is of a lighter colour and of fibrous texture; taste is faintly balsamic and astringent; leaves are obovate, serrate, 6.25 cm x 2.5-terminal panicles; fruit is drupaceous, blue on maturity, 1-seeded.

Distribution: Throughout India, Pakistan, southwest China, Japan, Korea and Myanmar.

Parts Used: Bark.

Phytochemistry: Bark contains loturine and colloturine; seeds contain fatty acids.

Uses: Bark is astringent, cooling and tonic; useful in the treatment of menorrhagia, bowel complaints, eye diseases and ulcers; also used as a gargle for giving firmness to spongy and bleeding gums; juice of the bark is applied externally to sprains and muscular swellings.

Symplocos racemosa Roxb.

Vernacular Name: Bur-daukyat (Burmese); Lodh (Hindi); Rodhra (Sanskrit).

Description: Medium-sized trees, 10-15 m tall, about 1 m girth; branchlets hairless; leaves alternate, elliptic-oblong, 8-15 x 3-6 cm, base wedge-shaped to rounded, apex acute to acuminate, margin toothed with rounded to saw-like teeth, obscurely recurved, papery, hairless, distinctly stalked; lateral nerves 8-11 pairs; flowers bisexual, shortly stalked, in 6-17 cm long axillary racemes, about 1 cm across, white with pink tinge, fragrant; drupes ellipsoid to ovoid, about 1.5 x 0.6 cm, hairless, ripening dark blue, turning yellow when dry; seeds 1-2, oblong.

Distribution: Found throughout the tropical and sub-tropical countries.

Parts Used: Bark.

Phytochemistry: Oturine, loturidine, colloturine, sugars, oxalic acid, phytosterol, 3-monoglucofuranoside of 7-O-methyl leucopelargonidin.

Uses: Stem bark is used to treat haemorrhage, acne and pimples, leucorrhea, wounds, skin disorders, ascites, hoarseness of voice, fever, menstrual disorders, liver diseases, menorrhagia, bowel complaints, ulcers and to give firmness to spongy and bleeding gums.

Thymelaeaceae

Aquilaria malaccensis Lam.

Vernacular Name: Akya (Burmese); Aloe wood, eaglewood (English); Agar (Hindi); Aguruh, Krsnaguruh (Sanskrit).

Description: A large evergreen tree about 21 m in height, leaves linear-lanceolate to avate-oblong, 5-9 cm long, silky glossy faintly parallel nerved; flowers small, greenish on very slender pilose petioles in shortly peduncled umbles, on younder branchlets, perianth about 5mm long, slightly hairy outside, stamens alternate the perianth, filaments red at the apex, ovary tawany-tomentose; fruits slightly compressed, yellowish tomentose capsules.

Distribution: Widely distributed in Asia; India, Bhutan, Myanmar. Sumatra and Malaysia

Part Used: Stem, oil.

Phytochemistry: Stem contains selinene, dihydroselinene, sequiterpene-hydrocarbons, agarol, sesquiterpene alcohol, hydroxy ketone, decene and rhombic as 5-isopropyl-7 methyl-5,5,5a, 6,7, 8-hexahydro-3H-naphtho; agarofurans, cadinanes, eudesmanes, valencanes and eremophilanes, guaianes, prezizanes, vetispiranes.

Uses: Wood is useful in skin disease, leprosy, foul ulcers, hypothermia, inflammation, rheumatoid arthritis, cough, and asthma; oil in rheumatoid arthritis, cough, asthma, bronchitis, leprosy, skin diseases and foul ulcers.

Ulmaceae

Holoptelea integrifolia Planch. (Syn. Ulmus integrifolia Roxb.)

Vernacular Name: Bur-my aukseil (Burmese); Indian elm, Kanju (English); Cilbil, Kanju (Hindi); Cirabilvah, Putikaranhah (Sanskrit).

Description: A medium-sized to large glabrous deciduous tree, 15-25 m in height with whitish or yellowish-grey bark exfoliating in irregular flakes and with an offensive smell when freshly cut; leaves simple, alternate, distichous, elliptic, entire, glabrous with rounded or cordate base and 5-7 pairs of main nerves; flowers greenish-yellow, male and hermaphrodite mixed in short racemes of fascicles near the leaf scars; fruits suborbicular samara with membranous reticulately veined wings.

Distribution: Found in India, Sri Lanka, Myanmar, and China.

Part Used: Bark, leaves.

Phytochemistry: The seed cake contains N, 10.3; lysine, 3.3; glutamic acid, 13.0; and histidine, 1.3%.

Uses: Decoction of the bark is applied to cure rheumatism; bark and leaves are useful in inflammations, dyspepsia, colic, helminthiasis, vomiting, diseases, leprosy, diabetes, haemorrhoids and rheumatism.

Valerianaceae

Nardostachys jatamansi (D.Don) DC.

Vernacular Name: Zawagyi-mok-seik (Musk-root) (Burmese); Jatamansi, (Hindi); Jatamasni, (Sanskrit).

Description: Erect herb, rhizome thick covered with a bundle of fine, reddish-brown fibers, fragrant; leaves radical, 15-20 cm narrowed into the petiole, entire flowers capitates, and pinkish white.

Distribution: Grows in the eastern Himalayas, Nepal, Sikkim and Bhutan and other parts of South Asia.

Part Used: Rhizome, oil from rhizome.

Phytochemistry: The plant contains acaciin, ursolie acid, octacosanol, kanshone A, nardosinonediol, nardosinone, aristolen-9beta-ol, oleanolic acid and beta-sitosterol.

Uses: Oil is used as a perfume, an incense, a sedative, and a herbal medicine said to fight insomnia, birth difficulties, and other minor ailments.

Valeriana officinalis L.

Vernacular Name: Saung-daw-gu (Burmese); Heal-all valerian, Great wild Valerian, (English).

Description: A deciduous herb, with a tuberous, short, rhizome, bearing numerous slender fleshy tapering, pale-brown rootlet and one or more stolons; leaves both radical and cauline, the radical being on long petioles whereas the cauline are opposite or alternate, exstipulate and pinnatisect with 11-12 segments with clasping petioles; basal leaves also divided into numerous segments; leaflets sessile, lanceolate, dentate; flowers numerous, small, white or pink, in recemes of cymes; calyx many-toothed; corolla 5-lobed, pink; fruit oblong-ovate 4-ridged, 1-seeded achenes with a feathery persistent calyx; fruits ciliate.

Distribution: Native to Europe and parts of Asia; North America.

Parts Used: Rhizome.

Phytochemistry: Actinidine, skyanthine, chatinine, valerianine, valerine, caffeic acid, chlorogenic acids. β-sitosterol, choline; α-& β-pinene, camphene, borneol, eugenol, isoeugenol, β-bisabolene, caryophyllene, valeranone, ledol, pacifigorgiol, valerianol, valerenal acid, valerenic acid; valtrate, valtrate, isovaleroxyhydrin, acevaltrate, valechlorine; didrovaltrate, homodidro valtrate, deoxydidro-valtrate, homodeoxydidro-valtrate, isovaleroxy-hydroxydidro-valtrate; isoval-trates, bornyl valerate and bornyl formate.

Uses: Plant used for treating insomnia; hysteria, over sensitiveness, nervous affections; cold draught exposure.

Violaceae

Viola odorata L.

Vernacular Name: Nilapan, Viola (Burmese); Sweet violet, (English); Banaphasha, (Hindi); Banaphasha, (Sanskrit).

Description: The plant is a creeping herb, colour brownish, without any stem, consists of leaves and flowers-yellow, white, blue purplish or pink with long filiform stalks; roots dry, thread, fibrous pale yellow knotty, slender as thick as a quill; leaves cordate at the base, heart-shaped carpels three nerved and brown.

Distribution: Distributed to Northern and West Asia; North Africa and Europe.

Parts Used: Flowers.

Phytochemistry: Flowers and root contain voiline; colouring matters ionone and ionine; glucoside, methol salicylic ester and saponins.

Uses: Root is laxative, good febrifuge, tonic, expectorant, diuretic; it alleviates thirst, relieves inflammation; oil used for abdominal pain; relieves cough and acts as a sedative; root is an emetic; syrup, made from the flower petals, is good medicine for cough, hoarseness of voice; syrup, when mixed with almond oil, is a gentle laxative for children and also helps to soothe irritative cough and to relieve sore throat.

Viscum album L.

Vernacular Name: Kyi-baung (Burmese); Bhulu-ka-banda, Ujjral, Banda, Kahbung, Mistletoe.

Description: A large, epiphytic, profusely branching shrubby perennial, forming rounded clumps over the host trees; branches dichotomous or whorled; leaves about 5 dy 10 cm, sessile, cuneate-oblong, or oblanceolate, with 3-5 longitudinal, basal nerves; flowers are dioecious, sessile; in clusters of 3-5, supported by cocave bracts; perianth-segments 3-4, triangular, deciduous; fruit small, ellipsoid white, smooth, almost transparent, mucilaginous.

Distribution: Native to Europe and western and southern Asia..

Parts Used: Whole plant.

Phytochemistry: Fatty acids, nicotine alkaloids, amines, flavonoids, chalcone derivatives and flavonone derivatives; terpenoids-β-amyrin, betalinic acid, lupecol and ester combinations; oleanoilic acid resin acid, ursolic acid, betasitosterol, sterol A; phytosterol glucoside, viscotoxin; mucilage, polyoles; sugars, starch, syringing and tannin, viscin, lutin, α and β carotene and vitamin C.

Uses: Leaves, stems and berries constitute a drug used for hypertension, arteriosclerosis, tumours and srthrosis; hysteria, epilepsy and ear diseases; cardiotonic, antiseptic, antispasmodic, diuretic, purgative and emetic with narocotic action; decoction or tincture in intermitted fevers; useful in splenic and hepatic enlargements; in lumbago and piles.

Zingiberaceae

Amomum subulatum Roxb.

Vernacular Name: Hpa-la-Gyi (Burmese); Greater cardamom, Napal cardamom (English); Badi llayaci, Bari llayaca (Hindi); Sthulaila, Bhadraila (Sanskrit).

Description: A herb with stem upto 90-100 cm in height, leaves oblong-lanceolate, bright green, glabrous on both surfaces; flowers white in globose shortly peduncled spikes, bracts reddish brown, lip ovate-cuneate, emarginated, yellowish-white, filaments very short, anther crest small, truncate, entire, fruits redish brown, densely echinate globose capsules; seeds many, held together by a viscid sugary pulp.

Distribution: Throughout Asia including China, India, Myanmar, Pakistan and Srilanka.

Part Used: Seeds

Phytochemistry: Seeds contain glycosides, petunidin, 3.5-diglucoside, leucocyanidin, sublin, chalcone, and alpinetin.

Uses: Seeds are userful in dysentery, skin diseases, pruritus, wounds, ulcers, cephalalgia, odontalgic, neuralgia, cardiac debility, liver congestion, splenomegaly, cough, bronchitis, strangury, fever proptosis, hyperdipsia and gonorrhoea.

Curcuma amada Roxb.

Vernacular Name: Thayat-Kin (Burmese); Mango-ginger, (English); Ama-haldi, (Hindi); Amrardrakam, Karpuraharidra, (Sanskrit).

Description: A rhizomatous aromatic herb with a leafy tuft, 60-90 cm in height; leaves long, petiolate, oblong-lanceolate, tapering at both ends, glabrous, green on both sides; flowers white or pale yellow in spikes in the centre of the tuft of leaves, lip semi-elliptic yellow 3-lobed, the mid lobe emarginated.

Distribution: Native to India cultivated throughout Asia.

Parts Used: Rhizomes.

Phytochemistry: Cis-and trans-dihydroocimenes, myrcene, ocimene, linalool, nonan-2-one, citronellal, α-and β-terpineol, β-and δ-elemenes, α-and β-curcumenes, β-terpineol, β-co-paene, bisabolene, α-zingiberene, and ar-turmerone; curzerenone, epicurzerenone, α-pinene and car-3-ene.

Uses: Rhizomes are useful in dyspepsia, flatulence, colic, bruises, wounds, chronic ulcers, skin diseases, pruritus, fever, constipation, strangury, hiccough, cough, bronchitis, sprains, gout, halitosis, otalgia and inflammations.

Hedychium spicatum Sm.

Vernacular Name: Bur-Ngwe-ban (Burmese); Kapurkachali, (Hindi); Shathi, (Sanskrit).

Description: It is an annual herb, with a horizontal root-stock, and tuberous root fibres, leaves are 30 cms or more in length; inflorescence spiked flowers ascending and dense yellow coloured; the capsule is globose.

Distribution: Native to China, the Himalayas, Myanmar, Thailand and Ethiopia.

Parts Used: Rhizomes.

Phytochemistry: Rhizome contains essential oil, starch, resins, organic acids, albumen and saccharine; essential oil has ethyl ester of p-methoxy cinnamic acid, d-sabirene cineole, sesquiterpenes and pentadecan methyl paracumarine acetate; β-sitosterol and its β-D-glycoside.

Uses: Plant used in nausea, vomiting, bronchial asthma, halitosis, diminished appetite, hiccups, local inflammation; root stalk is useful in local inflammatory, nausea, asthma, bronchitis, hiccups, and pain; rhizome is carminative, stimulant, tonic and an antiasthmatic agent.

***Kaempferia galanga* L.**

Vernacular Name: Bur-Kun-sa-gamon (Burmese); East Indies galingale, (English); Chandramula, (Hindi); Chandramulika, sugandhavacha, (Sanskrit).

Description: Perennial herbaceous plant; rhizome includes little ovate tubers; leaves2-3 with a broad blade, spreading flat on the ground, appearing annually in the rainy season, hairy beneath; flowers white tinged with violet, sessile, arising from the axil.

Distribution: Found primarily in open areas in Indonesia, southern China, Taiwan, Cambodia, and India; widely cultivated throughout Southeast Asia.

Parts Used: Rhizomes.

Phytochemistry: Rhizomes contain an essential oil consisting of p-methoxy transcinnamate ethyl, p-methoxy transcinnamic acid, transcinnamic acid, p-methoxystrene, p-coumaric acid, n-pentadecne, 3, careen, borneol and camphene.

Uses: Rhizome is recommended for dyspepsia and useful for pectoral and abdominal pains, headache, toothache and cold; externally in massage for the treatment of rheumatism.

***Kaempferia rotunda* L.**

Vernacular Name: Bur-mye-Padauk (Burmese); Indian crocus, (English); Bhuyicampa, Bhumichampa, (Hindi); Bhumicampaka, Bhucampaka, Hallakah, (Sanskrit).

Description: A handsome aromatic herb with tuberous root-stalk and very short stem; leaves simple, few, erect, oblong or ovate-lanceolate, acuminate, 30 cm long, 10 cm wide, variegated green above, tinged with purple below; flowers fragrant, white, lip purple or lilac in crowded spikes, opening successively.

Distribution: Native to China, India, Nepal, Indochina; widely cultivated in Asia.

Parts Used: Tubers.

Phytochemistry: Oil contains cineol and methyl chavicol; flower contains the toxin benzyl benzoate.

Uses: Tubers are useful in gastropathy, dropsy inflammations, wound, ulcers, blood clots, tumours and cancerous swellings.

***Zingiber zerumbet* (L.) Roscoe ex Sm.**

Vernacular Name: Mahabari-linne (Burmese); Wild ginger, (English); Narkachur, (Hindi); Karpooraharidra, (Sanskrit).

Description: Erect herb to 2 m tall arising from thick yellowish aromatic underground rhizome; leaves parallel-veined, ligulate, lanceolate, up to 30 cm long arising in 2 ranks from unbranched, fleshy aerial stems; flowers white to yellowish, borne on a fleshy spike with each flower arising from under a green to reddish bract, the inflorescence many-flowered but usually only 1-3 flowers open at once; fruit small capsule with tiny seeds; flowers December-April.

Distribution: Found throughout Asia.

Parts Used: Rhizome.

Phytochemistry: Camphene, camphor and other monoterpenoids; gingerol, zingeberol, zingerone, sesqui-terpenoids including zerumbone and zerumbone epoxid, oxalic acid, kaempferol derivatives, and flabonoids such as afzelin, flavonoid glycosides, essential oils, chlorgenic acid, ferulic acid.

Uses: It is used as a cough remedy and to treat the bacterial diseases, thrush, and diabetes; rhizome is used as stimulant, antihypertensive, carminative and flavouring agent; to treat dyspepsia wounds, haemorrhoids and flatulent colic for the cure of stomach troubles and fever; peptic ulcer and related stomach problems as well as mouth infections.

Chapter 4

TRADITIONAL MEDICINAL FLORA OF MAURITIUS

INTRODUCTION

The herbal drugs are usually single-ingredient drugs with one target action in the body. There are, of course, other effects, which are usually unplanned, called side effects. Steroid drugs, for example, are normally used to suppress inflammation in the body but their unwanted side effects include diabetes, unwanted hair growth, and depression. Herbal medicines are totally different; they have many ingredients and each herb usually has several different actions in the body. Herbs may also have side effects, but these are normally of minor and hence they are unnoticed.

Herbalists treat people, not symptoms. This means that herbs are chosen for their action on the body rather than to treat a particular symptom. Of course, herbs can be used to relieve symptoms, but usually in the context of treating the whole body. Herbs are chosen for a prescription on the basis of the total character of each herb. The mixture and the balance of the action in a herb are important consideration. However, in order to understand how herbs work, we can consider some of the important constituents that contributed to the action. For the purpose of these guidelines Herbal medicine should be regarded as:

> Herbal medicines can be defined as medicinal products which contain aerial or underground parts of plants, other plant material, their combination (crude state) or plant preparation as active ingredients. Plant material includes juices, gums, fatty oils, and any other substance of this nature. Herbal medicines may contain excipients in addition to the active ingredients. Medicines containing plant material combined with chemically defined active substances, including chemically defined, isolated constituents of plants, are not considered to be herbal medicines. Exceptionally, in some countries, herbal medicines may also contain, by tradition, natural organic or inorganic active ingredients which are of plant origin.

In the past few decades, the use of herbal medicine has been significantly increased. As a result of WHO's promotion of herbal medicine, countries have been seeking the assistance of WHO in identifying safe and effective herbal medicines for use in national health care systems. In 1989, one of the many resolutions adopted by the World Health Assembly in support of national traditional medicinal programmers drew attention to herbal medicines as being of great earlier resolution on pharmaceutical production in developing countries, this called on the Director-General to provide assistance to the health authorities of Member States to ensure that the drugs used are those most appropriate to local circumstances, that they are rationally used, and that the requirements for their use are assessed and accurately as possible. Moreover, the declaration of Alma-Ata in 1978 provide for inter alia. In developed countries, the resurgence of interest in herbal medicine has been due to the preference of many consumers for products of natural origin. In addition, manufactured herbal medicines from their countries of origin often follow in the wake of migrants from countries where traditional medicines play an important role.

In both developed and developing countries, consumer and health care providers need to be supplied with up-to-date and authoritative information on the beneficial properties, and possible harmful effect, of all herbal medicines.

The Fourth International Conference of drug regulatory Authorities, held in Tokyo in 1986, organized a workshop on the regulation of herbal medicines moving in international commerce. Another workshop on the same subject was held as part of the fifth international conference of Drug Regulatory Authorities, held in Paris in 1989, both workshops confined their consideration to the commercial exploitation to traditional medicines through over-the –counter labelled products. The Paris meeting concluded that the World Health Organization should consider preparing model guidelines containing basic elements of legislation designed to assist those countries who might wish to develop appropriate legislation and registration.

The objective of these guidelines, therefore, is to define basic criteria for the evaluation of quality, safety, and efficacy of herbal medicines and thereby to assist national regulatory authorities, scientific organization and manufacturing to undertake an assessment of the documentation/ submission/dossiers in respect to such product. As a general rule in this assessment, traditional experience means that long-term use, as well as the medical, historical and ethnological background of those products, shall be taken into account. Depending on the history of the country the definition of long-term use may very but would be at least several decades. Therefore the assessment shall take into account a description in the medical/pharmaceutical literature or similar sources, or documentation of knowledge on the application of herbal medicine without a clearly defined limits. Marketing authorization for similar products should be taken into account.

These efforts concentrate on herbal medicines, but might at stage be the basis for the assessment of other traditional medicine not covered by these guidelines. In the meantime, it is up to the national authorities to adopt the guidelines for the assessment of traditional medicines and other herbal drugs.

Prolonged and apparently uneventful use of a substance usually offers testimony of its safety. In a few instances, investigation of the potential toxicity of naturally occurring substances widely used as ingredients in these preparations has revealed the previously unsuspected potential for systematic toxicity, carcinogenicity and teratogenicity. Regulatory authorities need to be quickly and reliably informed of these findings. They should also have the authority to respond promptly such alerts, either by withdrawing or varying the licenses of registered products containing the suspect substance or by rescheduling the substance in order to limit its use to medical prescription. The species described in this chapter are including both wild and cultivated in Mauritius. The detailed description of these plant species including Phytochemistry, taxonomy and medicinal uses is given below.

Acanthaceae

Asystasia gangetica (L.) T.Anderson

Vernacular Name: Chinese violet, coromandel, creeping foxglove (English)

Description: Spreading herb or groundcover, reaching 600 mm in height; stems root easily at the nodes; leaves simple and opposite; fruit explosive capsule which starts out green in colour, but dries to brown after opening.

Distribution: Asian and African countries; tropical Americas and Hawaii.

Part used: Whole plant

Phytochemistry: Flowers contain apigenin 7-O-glucosyl (3'→6") luteolin 7"-O-glucoside; aerial parts contain asysgangoside, salidroside, benzyl β-D-glucopyranoside, roseoside, ajugol, apigenin 7-O-neohesperidoside; stem and leaves contain cardiac glycosides and anthraquinones

Uses: Leaves used in asthma; effective as an antioxidant, antimicrobial and antidiabetic.

Acoraceae

Acorus calamus L.

Vernacular Name: Vacha, Ugragandha (Sanskrit); Bach, Garbacc (Hindi); Sweet flag, Calamus (English)

Description: Semi-aquatic rhizomatous perennial herb; rhizome creeping, much-branched, as thick as the middle finger, cylindrical of slightly compressed, light brown or pinkish-brown; leaves bright green, distichous, ensiform, based equitant, thickened in the middle, margins wavy; flowers light brown densely packed in sessile cylindric spadix; fruits oblong turbinate with a pyramidal top; seeds free, pendant from the apex of the cell.

Distribution: India, Mauritius elevated up to 1800 m hight from sea level.

Parts Used: Rhizomes.

Phytochemistry: Volatile oil (calamus oil); asarone and its β-isomer; calamine, calamenone, methyleugenol, eugenol and α-pinene, camphene, palmitic acid, butyric acids, asaronaldehyde, calamol, azulene, acorone, calarene, calacone, calacorene, acorenone, acolamone, isoacolamone, epishyobunone, shyobunone, isoshyobunone, acoragemacrone, isocalamendiol and preiso-calamendiol.

Uses: Rhizome used as intellect promoting, emetic, laxative, carminative, stomachache, emmenagogue, diuretic, alexeteric, expectorant, anodyne, antispasmodic, aphrodisiac, anticonvulsant, resuscitative, anti-inflammatory, sudorific, antipyretic, sialagogue, insecticidal, tranquillising, nervine tonic, sedative and tonic.

Amaranthaceae

Achyranthes aspera L.

Vernacular Name: Chirchria, Latjira (Hindi); Prickly Chaff Flower (English); Apamarga (Sanskrit).

Description: Herbaceous plant about 1-2 m high; stem erect, pubescent, swollen at the nodes; leaves opposite, short-petioled, margins undulate; flowers numerous, stiffly deflected against the pubescent rachis in elongate terminal spike, 20-30 cm long; utricle oblong-cylindrical, enclosed in the hardened perianth, brown; seeds obeying-ovoid.

Distribution: India, Mauritius.

Part Used: Whole plant.

Phytochemistry: Roots contain betaine, achyranthine, hentriacontane, ecdysterone and oleanolic acid glycosides.

Uses: Used as anti-inflammatory, uterine stimulant; root astringent and its paste applied to clear opacity of cornea and wound healing; roots effective in cancer, stomach troubles, stones in the gall bladder.

Alternanthera sessilis (L.) R.Br. ex DC.

Vernacular Name: Matsyaksi, Patturah (Sanskrit); Gudrisag (Hindi).

Description: Much branched prostrate herb, branches often purplish, frequently rooting at the lower nodes; leaves simple, opposite, somewhat fleshy, lanceolate, oblanceolate or linear-oblong, obtuse or subacute, sometimes obscurely denticulate, glabrous, shortly petiolate; flowers small, white, in axillary clusters; fruits compressed obcordate utricles; seeds suborbicular.

Distribution: India, Mauritius, in moist places up to 2100 m from sea level.

Part used: Whole plant.

Phytochemistry: β-Sitosterol, stigmasterol, campesterol, α-spinasterol, oleanolic acid rhamnoside, 24-methylene cycloartenol, cycloeucalenol, lupeol, 5- α stigmasta-7-enol and its palmitate.

Uses: Used in burning sensation, diarrhoea, leprosy, skin disease, dyspepsia, haemorrhoids agalactia, splenomegaly and fever.

Alternanthera ficoidea (L.) Sm. (Syn. Alternanthera tenella Colla)

Vernacular Name: Joseph's Coat, Joyweed (English).

Description: Low-growing perennial herb that typically grows on erect to procumbent stems 15-30 cm tall. These stems are often bent at the nodes, and the plant is deep-rooted. Species plants have opposite elliptic to broadly ovate leaves (to 2.5 cm long), but it is the brightly coloured cultivars that have become popular garden plants, featuring green leaves blotched with yellow, orange, red, brown, copper or purple, often with red veining. Insignificant white apetalous flowers appear either sessile or on short pedicels in small axillary clusters.

Distribution: Common and widespread throughout the tropics, India, Island of Palau, Philippines, Mauritius, Japan and many other Pacific islands, America, Mexico, Argentina.

Parts used: Leaves and shoots.

Phytochemistry: Leaves contain 7-O-β-D-glucopyranosyl-6-C-β-D-glucopyranosyl-apigenin; 6-C-β D glucopyranosylapigenin; 8-C-β-D-glucopyranosylapigenin; (E)-3-(4-hydroxyphenyl)-prop-2-enoic acid; (E)-3-(3,4-dihydroxyphenyl)-prop-2-enoic acid and (E)-3-(4-hydroxy-3-methoxyphenyl)-prop-2-enoic acid.

Uses: Leaves are edible as raw or cooked; used as antiviral, antioxidant, antidiabetic etc.

Amaranthus dubius Mart. ex Thell.

Vernacular Name: Red spinach, Chinese spinach (English); Chauli, Chavleri Sag (Hindi).

Description: An annual herb usually grows to a size of 80-120 cm. Stems erect, green, branched, 0.3-1 m; leaves: petiole of proximal leaves equaling or longer than a blade, becoming shorter distally; blade rhombic-ovate or ovate to elliptic, base broadly cuneate, margins entire, apex slightly acuminate to obtuse and faintly emarginate, mucronate; stipules absent. Inflorescences terminal panicles and axillary spikes; panicles erect or often drooping, green, dense, branched, leafless at least distally. Bracts lanceolate, shorter than 2 mm, shorter than tepals, apex spinescent, midrib prominent, glabrous, usually white-yellow. Pistillate flowers: tepals 5, oblong-spatulate to oblong, not clawed, apex acute, often very shortly mucronate, and yellowish; style branches strongly spreading, shorter than the body of fruit; stigmas 3. Staminate flowers usually clustered at tips of inflorescence branches, sometimes gathered in proximal glomerules; tepals 5, equal or subequal; stamens 5, about 2 mm long. Utricles ovoid or subglobose, 1.5-2 mm, slightly shorter than tepals, smooth to irregularly wrinkled, dehiscence regularly circumscissile; seeds dark reddish-brown to black, subglobose or lenticular, 0.8-1 mm diameter, shiny and smooth.

Distribution: Native to South America, Mexico and the West Indies; also distributed to Asia and Africa.

Part used: Leaves and shoots.

Phytochemistry: Leaves contain hydrocyanic and oxalic acids; proteins; fibres and minerals.

Uses: Leaves are eaten as a vegetable; leaves used as medicine for young children, lactating mothers and for patients with fever, haemorrhage, anaemia, constipation or kidney complaints; the whole plant used in stomachache.

Amaranthus spinosus L.

Vernacular Name: Tenduliyah (Sanskrit); Prickly amaranth (English); Cauleyi, Kateli (Hindi).

Description: An erect, glabrous, spinous herb, varying in colour from green to red or purple, 30-60 cm in height with grooved branches and sharp divaricate spines in the leaf axils; leaves simple, alternate, ovate, Ianceolate or oblong, entire, glabrous above, main nerves numerous, conspicuous below; flowers small, sessile, yellowish-white or pale green, numerous, in dense axillary clusters and in terminal or interrupted spikes; fruits ovoid capsules, membranous, circumscissile about the middle.

Distribution: Throughout India, Mauritius in wastelands.

Part Used: Whole plant.

Phytochemistry: β-Sitosterol, stigmasterol, campesterol, cholesterol, α-spinasterol, oleanolic acid, higher alkanes and their methyl derivatives, higher aliphatic alcohols, acid and esters, amino acids, etc.

Uses: The plant used as alexeteric, laxative, diuretic, stomachic, antipyretic, febrifuge, sudorific, galactagogue, haematinic, appetizer and tonic. It is also used in hyperdipsia, burning sensation, hallucination, leprosy, eczema, bronchitis, leucorrhoea, menorrhagia, haemorrhoids, abscesses, boils, burns, strangury, nausea, flatulence, colic anorexia, fever, intermittent fever, agalactia and anaemia.

Amaranthus tricolor L. (Syn. Amaranthus gangeticus L.)

Vernacular Name: Tanduliya (Sanskrit); Chinese spinach (English); Lalsag (Hindi).

Description: A diffuse or ascending herb, rooting at the nodes; branches slender, not spiny; leaves small, elliptic; flowers unisexual, white, borne in clusters; male produced more than the female.

Distribution: Distributed in throughout Asia.

Part Used: Whole plant.

Phytochemistry: Leaf and seed contain protein, fat, fibre, carbohydrates, minerals, calcium, phosphorus, iron, thiamine, riboflavin, vitamin, energy and carotene; amino acids (arginine, cystine, histidine, leucine, leucine, isoleucine, lysine, methionine, phenylalanine, threonine, tryptophan, tyrosine, and valine); leaves also contain amarantin, isoamarantin and betanin.

Uses: The plant is used as astringent, diuretic, appetizing; cooked leaves are edible; decoction of the whole plant used in dysentery and haemoptysis; leaves are effective antidote to poison; root-paste, mixed with warm water, when given internally, induces vomiting and thereby purges out toxic matter from the bowel; powdered root is effective in onychia whereas root paste in combination with honey and rice water, is effective in leucorrhoea.

Celosia argentea L. (Syn. Celosia cristata L.)

Vernacular Name: Feather cockscomb, quil grass (English)

Description: An annual erect tropical herb, 0.3-1.0 m high, glabrous, more or less branching; leaves variable, linear or lanceolate, 8-10cm long by 2-4cm wide; tip acuminate, entire glabrous; base tapering sharply into a snort petiole or leaves sessile; flowers at first pinkish, afterwards glistening white, crowded and imbricate, in close cylindrical blunt or acuminate terminal spikes, capsules 3-4 mm long, ellipsoid, tapering at the apex into the style circumscissile; seeds black subreniform, shining flat small taste bitter.

Distribution: On slopes and roadsides in dry, exposed areas in Africa and India.

Part Used: Seeds.

Phytochemistry: Vitamins, fats, carbohydrate, oxalic acid.

Uses: Acute conjunctivitis; keratitis, chronic uveitis, dizziness due to hypertension.

Dysphania ambrosioides (L.) Mosyakin & Clemants (Syn. Chenopodium ambrosioides L.)

Vernacular Name: Wormseed, Jesuit's tea, Mexican-tea (English)

Description: It is an annual or short-lived perennial herb, growing to 1.2 m tall, irregularly branched, with oblong-lanceolate leaves up to 12 cm long; flowers small and green, produced in a branched panicle at the apex of the stem.

Distribution: Native to Central America, South America, and southern Mexico; distributed to Asia and Europe.

Part Used: Whole plant.

Phytochemistry: Essential oil contains ascaridole, limonene, p-cymene, α-pinene, myrcene, terpinene, thymol, camphor and trans-isocarveol; ascaridole; monoterpene peroxide boldo.

Uses: It is a leaf vegetable; herbal tea; culinary use in traditional Mexican dishes; medicinally used as an anthelmintic to treat ascarids and hookworms in humans and livestock; it is believed to prevent flatulence; used for amenorrhea, dysmenorrhea, malaria and chorea.

Anacardiaceae

Spondias pinnata (L. f.) Kurz

Vernacular Name: Wild mango, Hogplum (English); Amar, Jungli-aam (Hindi).

Description: A medium-sized aromatic, deciduous tree, up to 27 m tall and 2.5 m in girth; leaves compound, crowded at the ends of branches, leaflets large having parallel nerves meeting in an intramarginal never; bark thick, surface light grey to greyish brown, shallowly furrowed or cracked longitudinally, brittle, their fracture splintery; flowers many, in terminal spreading panicles; fruits fleshy drupes with woody endocarp surrounded by longitudinal interwoven fibres, young fruits green in colour, turning light yellow or greenish-yellow on ripening.

Distribution: Throughout India, Pakistan, Mauritius, and Bangladesh wild as well as cultivated.

Parts Used: Root, bark, leaves, fruits.

Phytochemistry: Fruits containβ-amyrin, oleanolic acid, glycine, cystine, serine, alanine and leucine; aerial parts contain lignoceric acid, β-sitosterol and its glucoside.

Uses: Roots useful in regulating menstruation; bark used in dysentery, diarrhoea, vomiting and muscular rheumatism; leaves used in dysentery and otalgia; ripe fruit effective as a tonic, constipating and antiscorbutic, also useful in bilious dyspepsia and diarrhoea.

Annonaceae

Annona muricata L.

Vernacular Name: Soursop, custard apple (English).

Description: Small tree upto 7 m tall; leaves alternate, petiolate, the blades leathery and oblong-lanceolate; flowers 3-merous, sepal and petals fleshy, and greenish in colour; fruit a fleshy syncarp with a green exocarp covered with conspicuous long pseudo-spines, mesocarp a somewhat fibrous juicy sweet-sour flesh surrounding several large smooth black seed; flowers and fruit are usually available throughout the year.

Distribution: Cultivated at lower altitude in various Asian countries.

Part Used: Leaves and fruits.

Phytochemistry: Annomonicin, annomontacin, annomruicins, annonacins and derivaties, annonacinone, anomuricine, anomurine anonaine anoniin, atherospermine, atherosperminine, corepoxylone, corossolone, coclaurine, coreximine, epomuricenins, gigantetrocins, gigantetrocins, deacetyluvarcin, gomothalamicin, howiicins, montanacin, muricatacin, muricatetrocins, muricatocins, murihexocins, murisolin, rolliniastain, solamin, muricine,muricinine, reticuline, lipids, monotetrahydro-furan acetogenins, β-sitosterol, stigmasterol and tannins.

Uses: Infusion of the leaves used in treating stomach ailments; effective as antimalarial, smooth muscle relaxant, uterine stimulant, anti-crustacean, antiparasitic, cytotoxic, cardiac depressant, antiamoebic, antibacterial, antifungal, hypertensive, spasmogenic, vasodilator and insecticide.

Annona reticulata L.

Vernacular Name: Ramphala, Lavali, Mruduphalam (Sanskrit); Bullock's heart (English); Luvani, Anta, Nagnewa (Hindi).

Description: ASmall deciduous or semi-deciduous tree of 6.1-7.6 m in height; leaves acuminate, 12.7-30.3 cm long and 3.8-5.1 cm wide; flowers cordate or subglobose, brownish or yellowish, with flat areoles, 7.6-12.7 cm diam; flowers in the summer and fruits towards the end of rainy season; ripe fruits are sweet and edible.

Distribution: A native of West Indies; distributed in Asia.

Parts Used: Fruits, seeds, leaves and bark.

Phytochemistry: Anonaine, norushinsunine, tannic acid, oxoushinsunine and reticuline have been reported.

Uses: Unripe and dried fruits are used in diarrhoea and blood dysentery; leaves considered insecticide, anthelmintic and external useful as suppurant.

***Annona squamosa* L.**

Vernacular Name: Sitaphalam (Sanskrit); Custard apple, Sugar apple (English); Sitaphal (Hindi).

Description: A small tree about 6.0 m in height; leaves simple, alternate, bifarious, oblong-lanceolate or elliptic obtuse, glabrous above, lateral nerves 8-11 pairs ascending; flowers yellowish green, solitary, leaf-opposed or 2-4 on short extra axillary branchlets, sepals and petal three each; fruits yellowish-green, globose with well market areoles easily breaking into large pieces; seeds hard, brownish-black, smooth.

Distribution: Cultivated throughout India and other parts of the Asia including Mauritius.

Parts Used: Root, leaves, fruits, seeds.

Phytochemistry: Higenamine, anonaine, roemerine, norcorydiene, corydine, norisocorydine, isocorydine, glaucine, oxoushinsunine, norushinsunine, norlaureline, reticuline, amino acids, epicatechin, catechin, camphor, borneol, squamolone, five kaurane related diterpenes, α-and β- pinene, limonene, β-farnesene, β-sitosterol, rutin, hyperposide and squamocin, a new cytotoxic bis-tetrahydrofuran containing acetogenin.

Uses: The roots are purgative; used in mental depression and spinal disorders; leaves are suppurative and insecticidal and are useful in destroying lice, proptosis in children and as a poultice to produce suppuration; fruits are sedative, stimulant, expectorant, maturant and tonic; useful in anaemia, burning sensation, vomiting, cough, malignant tumours, and for strengthening muscles; seeds are abortifacient and insecticidal.

Apiaceae

***Coriandrum sativum* L.**

Vernacular Name: Dhanyakam, Dhanaka (Sanskrit); Coriander (English); Dhaniya (Hindi).

Description: A glabrous, aromatic, herbaceous annual 30-90 cm in height; leaves decompounds, lower ones long-petioled and upper ones short petioled or sub-sessile, imparipinnate sect into linear-setaceous lobes; flower small, white or pinkish purple in compound terminal umbels; fruits yellowish-brown, globular and ribbed separating into two halves (mericarps) each containing a seed.

Distribution: Cultivated throughout Asia.

Part Used: Leaves, fruits.

Phytochemistry: Essential oil (Linalool, linalyl acetate, thymol, β-caryophyllene, α-pinene, borneol, limonene, β-phellandrene, citronellol, 1,8-cineole and geranyl acetate).

Uses: Leaves are analgesic, anti-inflammatory and styptic; useful in halitosis, pharyngopathy, epistaxis, ulemorrhagia, chronic conjunctivitis, cough, suppuration, haemorrhoids, jaundice and odontalgia; seeds used as anti-inflammatory, anthelmintic, stomachic, digestive, appetizer, constipating, diuretic, antipyretic, stimulant, expectorant and anodyne; seeds also used in vomiting, dyspepsia, anorexia, colic, flatulence, diarrhoea, dysentery, chronic conjunctivitis, cephalalgia, epistaxis, erysipelas, strangury, dropsy, scrofula, helminthiasis, haemorrhoids, intermittent fevers, hyperdipsia, gout, rheumatism and giddiness; leaf paste used in allergic inflammations.

Foeniculum vulgare Mill.

Vernacular Name: Fennel (English); Saunf (Hindi).

Description: An aromatic, perennial herb 1-2 m tall; leaves alternate, 3 or 4 times pinnate, the ultimate leaflets very numerous, filiform, very elongated, the superior leaves with sheaths longer than the blade; umbels compound, large, long-pedunculate, nearly, regular; flowers yellow, not involucrate, July-October, calyx with 5 very slight teth. Petals 5, entire; tips involute; stamens 5, ovary 2-celled; stylopodium large, conical; fruit avoid, 6 mm long by 2 mm in diameter, greenish; glabrous mericarp compressed dorsally, semicylindrical, with 5 prominent, nearly regular ribs; seeds somewhat concave with longitudinal furrows; fruit tastes sweet and pungent and smell of anis seed.

Distribution: Commonly cultivated India; distributed all over the world.

Parts Used: Fruits.

Phytochemistry: Fennel oil (trans-anethole, fenchone, limonene, methyl chavicol, pinene, α-phellandrene, myrcene, and cis-ocimene.

Uses: Used in abdominal pain, stomachache, dysmenorrhoea, hernia pain; used to confer flavour and aroma to soup, meat dishes, sauces, bakery and confectionery products, liqueurs, and pickles.

Petroselinum crispum (Mill.) Nyman ex A.W. Hill

Vernacular Name: Parsley (English).

Description: A hardy, aromatic, biennial herb, sometimes lasting upto 4 years, producing a rosette finely divided radical leaves in the first year and a flowering stack upto 100 cm high in the second; leaves 2 or 3 pinnate, ultimate segment cuneate-ovate or some time linear, oblong, inside, dentate or entire, often much crisped, upper cauline leaves often ternate; flower yellow or yellowish-green in compound umbels fruit 2-3 mm long, crescent shaped, conspicuously ridge; consisting of two mericarp.

Distribution: Native to the central Mediterranean region (southern Italy, Greece, Algeria, and Tunisia), naturalized elsewhere in Europe, and widely cultivated as a herb, a spice, and a vegetable.

Parts Used: Leaves.

Phytochemistry: Leaf oil contains 1,3,8-p-menthatriene, β-phellandrene, myristicin and myrcene; seeds contain myristicin, α-and β-pinene, m-metntha-1-8-diene, elemecin, β-farnesene, and sabinene; luteolin, apigenin, folic acid, vitamin K, vitamin C, and vitamin A.

Uses: Fresh and dried leaves are used for garnishing and seasoning in sour, stews, salads, sauces, meat, and poultry; oil used as an exclusive flavouring in sauces, pickles, meat additives, and various spaces; highly nutritious.

Apocynaceae

Allamanda cathartica L.

Vernacular Name: Golden trumpet, common trumpetvine, yellow allamanda (English)

Description: Vigorous climbing plant, can sprawl to a height of 20 feet; leathery evergreen leaves arranged in whorls of 4 at intervals along the stem and bright yellow flowers; stems twine to 6 m or considerably more and contain milky sap; lance-shaped leaves are 6 cm long and 3 cm wide; golden-yellow funnel-shaped flowers up to around 9 cm long and 6 cm in diameter are held in groups of 12 on short branches at the ends of shoots or in the leaf axils; prickly seed pods contain winged seeds; clusters of particularly striking golden-yellow flowers, which contrast with the shiny dark green leaves to make a lush plant for the conservatory.

Distribution: Native to tropical South America, from Peru and Colombia eastwards to French Guiana and Brazil. It has also been recorded in Central America as far north as Honduras. It is widely distributed elsewhere in the tropics.

Part Used: Whole plant.

Phytochemistry: Leaves and stems contain 9,12,15-octadecatrienoic acid and its ethyl ester, n-hexadecanoic acid, 3-O-methyl-d-glucose, 2-furancarboxaldehyde 5-(hydroxymethyl); the plant also contains various phenolic compounds.

Uses: Leaves, roots and flowers have anti-tumour activity; leaf and root decoctions used as a laxative and emetic in traditional medicine in a number of tropical countries, but large doses are toxic.

Asclepias curassavica L.

Vernacular Name: Kaktundi (Hindi); Bharadwaji Raktapuspa, Dugdhaksupa, Vanapicula (Sanskrit); Wild Ipeccacunanha, Blood flower (English).

Description: An erect, simple or branched, perennial herb, 80-120 cm tall; leaves 7.5-10.0 × 1.3-2.5 cm, opposite, lanceolate or oblong-lanceolate, acute at the base, acuminate at apex; flower orange, numerous, borne in umbelliform cymes; follicles small, solitary, erect, ca 7.5 cm long; seeds with a thickened margin, ca 0.5 cm long ovoid.

Distribution: Widespread, ranging from southern North America through Central America and into South America; sporadic in moist places throughout up to 1500 m altitude in India and Mauritius.

Part Used: Whole plant.

Chemical Constituents: Cardiac glycosides (asclepin, calotropin, uzarin and their free genins, calactin, coroglucigenin and uzarigenin); oleanolic acid, β- sitosterol, and glycosides of asclepin.

Uses: Used in phthisis; considered to be emetic, styptic, purgative and cardiotonic; root used as an astringent and a remedy in piles and gonorrhoea; powder of root mixed with equal quantity of Acorus root is given in chronic ulcer; root decoction is used against cancer; juice of leaves is antidysenteric, anthelmintic, and is used against haemorrhages and gonorrhoea, useful externally in piles and in abdominal pain.

Catharanthus roseus (L.) G.Don

Vernacular Name: Madagascar perwinkle, Red perwinkle, Pink periwinkle, Rosy-flowered Indian perwinkle, Cape perwinkle, Old maid (English); Sadabahar (Hindi).

Description: Perennial herb, 30-80 cm tall; stems pinkish-red, much-branched; leaves opposite, obovate, glabrous on both sides, dark shining; flowers pink or white in the axil of the leaves; follicle cylindrical, narrow, slightly arched-recurved in pairs; seeds numerous, tiny, blackish-brown.

Distribution: Common wild plant in coastal areas and is also cultivated.

Part Used: Leaves and root.

Phytochemistry: Leaves contain alkaloids (serpentine, ajmaline, ajmaliine, catharanthine, catharanthinole, vindoline, vindolinine, vincaleucoblastine, leurosidine, vincristine).

Uses: Leaves are useful in oliguria, haematuria, diabetes and menstrual disorders; roots and the leaves in the form of a decoction are effective for hypertension; purified alkaloids extract of leaves is effective in treating leukaemia whereas root extract used to induce cerebrovascular dilatation and for hypertension.

***Calotropis gigantea* (L.) Dryand.**

Vernacular Name: Madar (Hindi); Crown (English)

Description: Shrubs 1-5 m tall; petiole 1-4 mm; leaf blade obovate-oblong or oblong, 7-30 × 3-15 cm, base cordate, apex obtuse, cottony tomentose when young, frequently glabrescent and glaucous green; lateral veins 4-8 pairs; cymes umbel-like, with fine woolly hairs; peduncle robust, 5-12 cm; pedicel thick, 2-5 cm; calyx almost flat, 1.2-1.5 cm in diam; flower buds cylindric; corolla usually purplish or lilac with paler greenish base, 2.5-3.5 cm in diam., fleshy, glabrous; lobes ovate, 1-1.5 × 0.6-1 cm, spreading or reflexed, margin revolute; corona shorter than gynostegium; follicles obliquely elliptic to oblong-lanceolate in outline, 5-10 × 2.5-4 cm, both ends incurved; seeds broadly ovate, 5-7 × 3-4 mm; coma 2.5-4 cm; flowering almost throughout the year.

Distribution: Distributed in China, India, Indonesia, Laos, Mauritius, Malaysia, Myanmar, Nepal, Pakistan, Sri Lanka, Thailand, Vietnam; tropical Africa.

Phytochemistry: Leaves, stem and flowers contain uscharin, calotoxin, calactin, and calotropin.

Uses: Bark used as a medicine for the treatment of neurodermatitis and syphilis; leaves used as a poultice; the juice is used in making a yellow dye and in tanning; fine fibre obtained from the stems.

***Calotropis procera* (Aiton) Dryand.**

Vernacular Name: Arkah, Aditya, Mandare (Sanskrit), Gigantic swallow-wort, Madar milk-weed (Eglish); Madar (Hindi).

Description: Large, hard much-branched milky shrub, very pale in colour; branches, leaves and inflorescence covered with loose soft whit wool; leaves opposite, subsessile, ovate, cordate base; flowers beautiful lilac, rosy or purple, tinted in umbellate lateral cymes; fruits follicles, green, with abundant with coma.

Distribution: Native to North Africa, Tropical Africa, Western Asia, South Asia, and Indochina; distributed throughout Africa and Asia.

Part Used: Whole plant.

Phytochemistry: Cardiac glycosides, calotropin, uscharin, calotoxin, calactin and uscharidin; gigantin.

Uses: The plant is good tonic, expectorant, depurative and anthelmintic; root bark is febrifuge, anthelmintic, depurative, expectorant and laxative, and is useful in cutaneous diseases, intestinal worms, cough, ascites and anasarca; powdered root promotes gastric secretion and is useful in asthma, bronchitis and dyspepsia; leaves are useful in the treatment of paralysis, arthralgia, swellings and intermittent fevers; flowers are bitter, digestive, astringent, stomachic, anthelmintic and tonic, also useful in asthma, catarrh, anorexia, inflammations and tumours; large doses as purgative and emetic.

***Carissa carandas* L.**

Vernacular Name: Karamadah, Avighnah (Sanskrit), Karaunda, Jasmine flowered carrisa (English); Karaunda (Hindi)

Description: Large dichotomously branched evergreen shrub with short stem and strong thorns in pairs; bark light grey, scaly; leaves simple, opposite, elliptic or obovate, shortly mucronate, glabrous, shining and coriaceous; flowers white, in pubescent terminal corymbose cymes; fruit ellipsoid or globose, purplish-black when ripe enclosing two or more seeds.

Distribution: Naturally grows in the Himalayas at elevations of 300-1800 meters, in the Siwalik Hills, the Western Ghats and in Nepal and Afghanistan. It is also grown in Mauritius and Sri Lanka lowland rain forests.

Parts Used: Roots, fruits.

Phytochemistry: Contain carissone, carindone, carinol and β-sitosterol.

Uses: Roots are anthelmintic, stomachic and antiscorbutic, and are useful in stomach disorder, intestinal worms, scabies and pruritus; unripe fruit is sour, astringent, bitter thermogenic, constipating, anaphrodisiac, appetizer and antipyretic, and is useful in hyperdipsia, diarrhoea, anorexia, and intermittent fever; ripe fruit is sweet cooling, appetizer and antiscorbutic, and is useful in anorexia, burning sensation, skin diseases scabies and pruritus.

***Gomphocarpus fruticosus* (L.) W.T.Aiton (Syn. *Asclepias fruticosa* L.)**

Vernacular Name: Milkweed, Cape cotton, Wild cotton (English)

Description: Herbaceous, perennial, spindly shrub, often with watery or milky sap; ovoid fruits (follicles) of a clear, pale green colour, covered with prickles and narrowing into a curved tip, and almost looks like a tiny swan; fibrous, multi-stemmed, growing up to 1.5-2 m high; evergreen and hardy; light brown stems branch higher up to form the crown; when cut, the stem exudes milky latex; leaves are pale to mid-green, long, narrow and opposite; attractive, creamy yellow flowers carried in pendulous clusters; fruit, borne in autumn, is an inflated, light brown, papery follicle, covered with bristle-like hairs and containing dark seeds; seeds and floss seeds are attached from cotton-like, silky hairs that aid in their dispersal.

Distribution: Native to South Africa; distributed in Asia and New Zealand.

Part Used: Whole plant

Phytochemistry: Leaves contains cardenolides, uzarigenin, gomphotin, desglucouzarin, etc.

Uses: The plant as a whole is poisonous to livestock; leaves are used as snuff and as a sedative in the treatment of headache and tuberculosis; roots are used to relieve stomach pain and general aches in the body.

***Holarrhena pubescens* Wall.ex G.Don**

Vernacular Name: Indrajau (Hindi); Conessi (English); Kalinga, Kutaja (Sanskrit)

Description: Deciduous shrubs or tree, to 8 m high, with rough, brown bark; leaves opposite, broadly ovate-elliptic, 10-45 x 5-10 cm, glabrescent or pubescent; petioles short; flowers white or cream coloured, 1.5-2.5 cm across, in terminal, corymbose cymes; calyx deeply 5-lobed; lobes lanceolate, acute; corolla tube cylindrical, swollen at base; lobes oblong, spreading; stamens 5; anthers subsessile; carpels 2, distinct, style short; stigma 2-lobed; follicles 2, slightly curved, cylindric, 15-40 × 0.5-1 cm, usually marked with specks; seeds liner, with pale brown coma.

Distribution: It is native to central and southern Africa, the Indian Subcontinent, Indochina, Mauritius and China.

Part Used: Seeds, leaves, bark and fruits.

Chemical Constituents: Seeds contain steroidal alkaloids (conessine, conessimine, isoconessimine, conarrhimine, conimin); stem bark contain kurchessine, holadysenterine.

Uses: Bark is Ayurvedic medicine for diarrhoea, piles, skin diseases and biliousness; bark also used externally in case of skin troubles, urinary troubles; the plant is useful in rheumatoid arthritis and osteoarthritis.

***Nerium oleander* L.**

Vernacular Name: Karavirah (Sanskrit), Indian oleander, Sweet-scented oleander (English), Kaner, Karber, Karuvira (Hindi).

Description: Large glabrous evergreen shrub with milky latex; leaves three in a whorl, shortly stalked, linear dark green and shiny above; flowers tubular with five lobes, red or pink in the wild, but maybe white, cream, yellow or purple in cultivars; some are scented; fruits composed of a pair of follicles that split along one side to release the seeds; seeds are oblong, with a plume of hairs at one end.

Distribution: Cultivated throughout India, China, Mauritius and other Asian countries.

Part Used: Roots, leaves.

Phytochemistry: Roots and seeds contain cardioactive glycoside (neriodorin, neriodorein and karabin); digitosigenin-β-D-diginoside, uzarigenin-β-D-diginoside, digitoxigenin-β-D-glucosido-β-D-diginoside, digitoxigenin-β-D-glucosido-β-D-digitaloside, digitoxigenin-β-D-glucosido-β-D-glucosido-monoacetly-β-D-digitaloside, digitoxigenin-β-

D-digitaloside, uzarigenin-β-D-glucoside-β-D-gluco-side-β-D-diginoside, uzarigenin-β-D-glucosido-β-D-diginoside, uzarigenin-β-D-glucosido-β-D-diginoside.

Uses: The roots are bitter, acrid, astringent, anthelmintic, thermogenic, aphrodisiac stomachic, febrifuge and diuretic. They are useful in cardiac asthma, surgery, renal and vesical calculi, chronic stomachalgia, arthralgia, leprosy, pruritus and ulcers; root bark effective for ringworm; leaves are powerful repellent and are used for scabies, haemorrhoids.

Tylophora indica (Burm. f.) Merr.

Vernacular Name: Emetic swallow-wory, Indan or Country ipecacuanha (English); Jangli pikvam, Antamul (Hindi); Lataksiri (Sanskrit).

Description: Slenderical, much-branched, tough laticiferous climber with long fleshy, knotty root; leaves simple, opposite, somewhat fleshy, ovate to orbicular, cordate, often apiculate, glabrous, acute or acuminate, more or less pubescent beneath; flowers in umbels, greenish-yellow outside, purplish within, pedicels filiform with a number of filiform hairy bracts at their base; fruits fusiform, divaricate, glabrous, follicles tapering to a fine point at the apex; seeds ovate with long come.

Distribution: Throughout India, in hedges and open forests up to 900 m.

Parts Used: Roots, leaves.

Phytochemistry: Contains alkaloids (tylophorine, tylophorinine, desmethyltylophorine, desmethyltylophorinine); flavonoid (kaempferol); root contains tylophorinidine, γ-fagarin; leaves contain d-septicine, d-isotylocrebrine, triterpenoids α and β-amyrine, β-sitosterol, stigmasterol, campesterol, phenylalanine and quercetin.

Uses: Roots and leaves are purgative, expectorant, vulnerary, diaphoretic, stomachic and antiviral; useful in asthma, bronchitis, whooping cough, dysentery, diarrhoea, hydrophobia, wounds, ulcers, dyspepsia, flatulence, haemorrhoids, gout, cancerous tumours and leukaemia.

Wrightia arborea (Dennst.) Mabb. (Syn. W. tomentosa Roem. & Schult.)

Vernacular Name: Woolly Dyeing Rosebay (English); Daira, Dharauli, Dudhi (Hindi); Kutaja (Sanskrit).

Description: Small deciduous trees; to 12 m high; bark 8.5 mm, pale grey, corky; blaze greenish-yellow, latex yellow; branchlets tomentose. Leaves simple, opposite, estipulate; petiole 3-7 mm, stout, tomentose; lamina 6-15 × 3-6.5 cm, elliptic, elliptic-ovate, ovate or obovate; base acute; apex acuminate; margin entire, tomentose on both sides, chartaceous; lateral nerves 9-13 pairs, tomentose, prominent, pinnate; intercostae reticulate; flowers bisexual, greenish-yellow, in terminal or pseudoaxillary dense shortly peduncled tomentose cymes; pedicel 1 cm long, stout, tomentose; calyx short, lobes 5, 4 mm long, broadly ovate, obtuse, pubescent without, glandular inside; corolla salver shaped, lobes 15 × 10 mm, obovate, puberulous, throat with 1-2 series of erect, entire or lobed orange corona scales, tube 5 mm long, shorter than the lobes; stamens 5, inserted on the mouth of the tube, anthers 8 mm long, exserted, sagitate, spurred at base, hairy; disc absent; carpels 2, 2 mm, connate; ovules many; style 3 mm, filiform; stigma ovoid with a basal ring usually toothed; fruit of 2 connate follicular mericarps, ca. 16 × 2 cm, cylindrical, subterete, green, sparsely white tubercled, rough, apically beaked; seeds linear, comose at one end.

Distribution: Found in India, Malesia, Mauritius.

Parts Used: Bark, leaves, aerial parts.

Phytochemistry: Conessine dehydrate, holarrhine, kurchisine, konkurchine, β-sitosterol, lupeol, α-amyrine and reducing sugars in the bark; leaves contain alkaloids, wrightiadione, phenolics and tannins.

Uses: Bark is used in irregular menstruation and renal complaints including kidney stone; various parts of the plant having antibacterial, antipyretic, anti-inflammatory, antidiabetic, antioxidant and anticancer activities.

Araceae

Colocasia esculenta (L.) Schott

Vernacular Name: Alupam, Alukam, Kachu (Sanskrit); Taro, Cocoyam (English); Arvi (Hindi).

Description: A tuberous perennial with a group of underground farinaceous corms consisting of a central large one and surrounding once of varying sixes; leaves with sheathing leaf base and erect petiole upto 1.2 m long bearing a thinly coriaceous peltate-ovate, cordate lamina; spadix shorter than petiole and much shorter than the spathe, appendix much shorter than the inflorescence.

Distribution: Native to Malaysia; cultivated in wet tropical India, Pakistan, Mauritius, Egypt, Indonesia, Australia, Turkey.

Part Used: Leaves, corms.

Phytochemistry: Dihydroxysterols, 14α-methyl-5α-cholesta-9, 24-diene-3β-, 7α-diol and 14α-methyl-24-methylene-5α-cholesta-9, 24-diene-3α-diol, β-sitosterol, stigmasterol, tetracos-20-en-1, 18-diol: 25-methyl tiracont-10-one: octacos-10en-1, 12-diol: pentatriacont-1, 7 tien-12-ol, 25-methyl-tritriacont-21-en-1, 11-trion, non-acosane and cyaniding-3-glucoside.

Uses: Leaf juice is styptic, stimulant and rubefacient, useful in internal haemorrhages, otalgia, otorrhoea, adenitis and buboes; corm juice is a laxative, demulcent and anodyne, useful in somatalgia, alopecia areata, haemorrhoids and congestion of the portal system.

Arecaceae

Acanthophoenix rubra (Bory) H.Wendl.

Vernacular Name: Barbel palm, Red palm, Mascarene Islands cabbage palm (English)

Description: A palm with a spiny stem; striking, red leaf sheaves bearing long, sharp spines, found at the top of the trunk; when matures spines fall away and the sheaves become gradually browner; leaves impressively large, reaching up to 3 metres long, and are composed of numerous pointed, paired leaflets, which project from the bristly leaf midrib; mature plants have around 10 leaves in total, which radiate from the crown in a spiral arrangement; during flowering it develops conspicuous, 50 centimetre-long, inflorescences that hang below the leaves; comprise a multitude of small, creamy-white flowers suspended from a central stem; fruits are roundish drupes, up to one centimetre long, which contain a single seed.

Distribution: Endemic to Mauritius, Rodrigues, and La Reunion that is prized for its edible palm hearts.

Part Used: Roots, leaves

Phytochemistry: Flavonoids, polyphenol oxidase, etc.

Uses: A decoction of the roots is used as a diuretic; palm heart (apical bud) edible.

Borassus flabellifer L. (Borassus flabelliformis L.)

Vernacular Name: Talah, Taladrumah (Sanskrit); Palmyra palm, Brab tree (English), Tar, Tal, Tad (Hindi).

Description: A tall palm attaining a height of about 30 m, with a black stem with a crown of leaves at the top; leaves 0.9-1.5 m in diameter, palmately fan-shaped, rigidly coriaceous, cleft into many lobes, petiole edges with hard horny spinescent serratures; flowers unisexual, male spadix branched, female spadix simple; fruits large, subglobose drupes, on the greatly enlarged perianth.

Distribution: Hotter part of India, Mauritius; wild as well as cultivated.

Part Used: Roots, leaves, inflorescences, fruits.

Phytochemistry: It contains minerals; iron, moisture, protein, sucrose, glucose, thiamine, riboflavin, nicotinic acid and vitamin-C.

Uses: Roots are cooling, restorative, diuretic, stimulant and antiphlogistic; useful in hyperdipsia, burning sensation, strangury and inflammation; juice of the leaf stalks and young roots is good for gastric catarrh and hiccough; ash obtained by burning the inflorescence is a good antacid and antiperiodic, and is useful in hyperacidity, heartburn, bilious fever, splenomegaly, hepatomegaly and skin diseases; fruit are sweet, cooling, laxative, sedative, anthelmintic, depurative, styptic, stomachic, aphrodisiac and tonic; useful in hyperdipsia, burning sensation, dyspepsia, flatulence, colic, constipation, intestinal worms, leprosy, skin diseases, haemorrhages, haemoptysis.

Cocos nucifera L.

Vernacular Name: Narikela (Sanskrit); Coconut (English); Nariyal (Hindi).

Description: Monopodial tree with long, narrow often curved trunk upto 30 m tall; foliage leaves all confined to apex of trunk each leaf paripinnate and up to or exceeding 6 m in length; flowers and fruit borne in drooping clusters arising from between the lower leaf petioles; fruit green brown, or reddish fibrous drupe up to or exceeding 30 cm in length.

Distribution: Common along the seashore to moderate elevations in inland areas, most abundant near human settlements; India and Mauritius.

Part Used: Fruit, stem, leaves, root.

Phytochemistry: Saccharose, myoinositol, scyllo-inositol, sorbitol, diphenylurea, alipahatic alcohols, ketones, leucoanthocyanins, 2-propyleneglycol, glycerol, sucrose, glucose, stachyose, bongrek acid xylan, glucosan, aliphatic fatty acids, polyphenols, ferricopnin, cocositol, mono and sesquiterpenes, α and β-amyrin, campesterol, cycloartenol, β-sitosterol, squalene, stigmasterol, α-tocopherol, and alkaloids (ligustrazine and 2,3,5-grimethylpyrazine).

Uses: Stemuseful in weakness after childbirth; juice from the midrib at the lower base of the leaf is used in treating maternal postpartum illness; coconut milk is used to treat fish poisoning; root used in treating stomachache and blood in the urine; oil from the kernel is rubbed onto stiff joints; oil is also is used to treat rheumatism and back pains or as an ointment to maintain smooth, soft skin; oil mixed with turmeric, in used to treat sick both newborn and mother; oil is used as an emetic and purgative; juice from a green coconut is to women who have difficult pregnancies; juice from the fruit is taken to treat kidney problems; coconut is said to have vermicide properties; haemorrhaging is stopped with the use of the dry, spongy kernel; diarrhoea and dysentery are treated with the plant.

Calamus rotang L.

Vernacular Name: Vetrah (Sanskrit); Common rattan, Rattan (English); Bet, Bemt (Hindi).

Description: A scandent slender shrub with cylindrical uniformly thick yellowish-white stems armed with short flat spines on leaf sheaths, glabrous, hollowed below, ochrea short, truncate, margined with small straight or recurved spines having laterally compressed bases, rachis near the base flat and smooth above, channelled at the sides to receive the leaflets, trigonous upwards armed below along the middle, leaflets very numerous, equidistant, 1-ribbed, the rib bearing scattered bristles; flowers unisexual, male spadix very long, flagelliferous, spathe elongate tubular, female flowers scattered along the slender branches of the spadix; fruits seated on the minute perianth, subglobose, mucronate, scales many, pale yellow in vertical series.

Distribution: India, Mauritius and nearby countries up to 450 m elevation.

Parts Used: Whole plant.

Phytochemistry: Saponin, alkaloid, flavonoid.

Uses: Roots are astringent, acrid, bitter, cooling, expectorant, alexeteric, antidysenteric, hypotensive, depurative, anti-inflammatory, diuretic, febrifuge and tonic; useful in burning sensation, hyperdipsia, cough, bronchitis, dysentery, leprosy, skin diseases, erysipelas, inflammations, strangury, vesical calculi, chronic fever and general debility; leaves are acrid, bitter, cooling, astringent and laxative; useful in cough, skin diseases and pruritus; sprouts are acrid, sweet and thermogenic.

Lodoicea maldivica (J.F.Gmel.) Pers.

Vernacular Name: Arkaragah, Agnijarah, Abdhinarikelah (Sanskrit); Double coconut palm, Sea coconut palm (English); Darya-ka-nariyal, Dariya-I-nariyal (Hindi).

Description: A tall palm, 12-30 m high, with smooth annulated trunk bearing a crown of large fan-shaped leaves; the palm is dioecious and its fruits are large weighing 11-27 kg, one-seeded, deeply bilobed and bony; shell is thick and black.

Distribution: Seychelles Islands, India, Mauritius.

Parts Used: Husk, Endosperm.

Phytochemistry: Tannins, phenolics, terpenoids

Uses: Decoction of the fibrous husk as well as the kernel used in diabetes; kernel is sweet, acrid, refrigerant, digestive, carminative, febrifuge and tonic; cholera, hyperdipsia, oedema, acute diarrhoea, colic and also as an antidote in opium and aconite poisoning; good cardiotonic; water of the green fruit and its kernel is antacid.

Phoenix dactylifera L.

Vernacular Name: Date, Date palm (English); Khajoor (Hindi).

Description: Flowering plant, cultivated for its edible sweet fruit; trees typically reach about 21–23 m in height, growing singly or forming a clump with several stems from a single root system; leaves are 4–6 m long, with spines on the petiole, and pinnate, with about 150 leaflets; leaflets are 30 cm long and 2 cm wide; full span of the crown ranges from 6–10 m.

Distribution: Originated from lands around Iraq; widely cultivated and is naturalized in many tropical and subtropical regions worldwide.

Part Used: Fruits, seeds

Phytochemistry: Fruits contain Vitamin A, β-carotene, lutein zeaxanthin, thiamine, riboflavin, niacin, pantothenic acid, folate, ascorbic acid, Vitamin E and K; seeds contain lauric acid, oxalic acid.

Uses: Fruits are highly nutritional; seeds are soaked and ground up for animal feed; seed oil is suitable for use in soap and cosmetics; seeds are also burned to make charcoal for silversmiths, and can be strung in necklaces; used in the manner of coffee beans, or as an additive to coffee; date pits exhibit anti-genotoxic and reduce DNA damage induced by N-Nitroso-N-methylurea.

Asparagaceae

Agave americana L.

Vernacular Name: Agave, Century plant (English); Banskeora, Bara kanwar, kantala (Hindi); Kalakantala, Kantala (Sanskrit).

Description: Succulent perennial with large rosette of 30-60 fleshy, sharply toothed leaves that reach a height of 2 m; cluster of yellow flowers, growing to 7 cm across, bloom on a pole-like stem after 10 years or more.

Distribution: Agave is native to deserts of Central America; now growing as an ornamental plant in tropical and sub-tropical areas around the world.

Part Used: Leaves.

Phytochemistry: Leaves contain oestrogen-like isoflavonoids, alkaloids, coumarins, vitamins A, B and C; hecogenin, chlorogenin, rockogenin, figogenin and dehydrohecogenin.

Uses: Demulcent, laxative and antiseptic: sap is a soothing and restorative remedy for many digestive ailments; used to treat ulcers and inflammatory conditions affecting the stomach and intestines, protecting these parts from infection and irritation and encouraging healing; employed to treat a wide range of other conditions, including syphilis, tuberculosis, jaundice and liver disease.

Asparagus officinalis L.

Vernacular Name: Asparagus (English).

Description: Slender-stemmed perennial growing to 2 m; long fronds of delicate needle-like leaves and bell-shaped yellow-green flowers that produce small bright red berries; shoots grow into tender stems in spring; the root is gathered after the shoots have been cut.

Distribution: Native to temperate regions in Europe, North Africa and Asia; cultivated worldwide as a vegetable.

Part Used: Root, clump.

Phytochemistry: It contains steroidal glycosides, bitter glycosides, asparagines and flavonoids.

Used: Used as a diuretic; decoction of the root to improve urine flow and to treat kidney problems, jaundice and sciatica; root used for teeth ache.

Asparagus racemosus Willd.

Vernacular Name: Satavari, Abhiru (Sanskrit); Wild asparagus (English); Satavar, Satamuli (Hindi).

Description: An armed, climbing undershrub with woody terete stems and recurved or rarely straight spines, young stems very delicate, brittle and smooth; leaves reduced to minute chaffy scales and spines; cladodes triquetrous, curved in tufts or 2-6; flowers white, fragrant, in simple or branched racemes on the naked nodes of the main shoots or in the axils of the thorns; fruits globular or obscurely 3-lobed, pulpy berries, purplish-black when ripe, seeds with hard brittle testa; tuberous succulent roots are 30 cm to a meter or more in length, fascicled at the stem base, smooth, tapering at both ends.

Distribution: Throughout India, Mauritius; also cultivated in Pakistan, Bangladesh.

Part Used: Tuberous roots.

Phytochemistry: Plant contains saponins (Shatavarin I to IV); mucilage and starch.

Used: Roots are bittersweet, emollient, cooling, nervine tonic, constipating, galactagogue, ophthalmic, anodyne, aphrodisiac, diurctic, rejuvenating, carminative, appetizer, stomachic, antispasmodic and tonic; useful in nervous disorder, dyspepsia, diarrhoea, dysentery, tumours, inflammations; burning sensations, hyperdipsia, ophthalmopathy, nephropathy, hepatopathy, strangury, scalding of urine, throat infections, tuberculosis, cough, bronchitis, gleet, gonorrhoea, leucorrhoea, leprosy, epilepsy, fatigue, hyperacidity, colic, haemorrhoids, cardiac debility, hypertension, abortion agalactia and general debility.

Dracaena reflexa Lam.

Vernacular Name: Pleomele, Song of India (English).

Description: Evergreen herb, may reach a height of 4–5 m; much smaller, especially when grown as a houseplant; slow-growing and upright in habit, tending to an oval shape with an open crown; lanceolate leaves are simple, spirally arranged, 5–20 cm long and 1.5–5 cm broad at the base, with a parallel venation and entire margin; they grow in tight whorls and are a uniform dark green; flowers are small, clustered, and usually white, appearing in mid-winter.

Distribution: A tropical tree native to Madagascar, Mauritius, and other nearby islands of the Indian Ocean; widely grown as an ornamental plant and houseplant.

Parts Used: Leaves, bark.

Phytochemistry: Leaves and stems contain alkaloids, flavonoids and tannins.

Uses: It is used to cure malarial symptoms, poisoning, dysentery, diarrhoea, dysmenorrhea, and to be useful as an antipyretic and hemostatic agent; leaves and bark are mixed with parts of a number of other native plants and mixed into herbal teas.

***Yucca gloriosa* L.**

Vernacular Name: Adam's needle, Palm lily (English).

Description: An evergreen shrub; grow to heights above 5 m; caulescent, usually with several stems arising from the base, the base thickening in adult specimens; long narrow leaves are straight and very stiff, growing to 30–50 cm long and 2-3.5 cm wide, dark green with entire margins, smooth, rarely finely denticulate, acuminate, with a sharp brown terminal spine; inflorescence is a panicle up to 2.5 m long, of bell-shaped white flowers, sometimes tinged purple or red; fruit is a leathery, elongate berry up to 8 cm long.

Distribution: Native to the beach scrub and sandy lowlands of the southeastern United States, from the Outer Banks of Virginia and North Carolina to Florida and Alabama; widely cultivated as an ornamental and reportedly has become established in the wild in Italy, Turkey, Mauritius, Réunion, Guam, the Northern Mariana Islands, Puerto Rico, Argentina and Uruguay.

Part Used:

Phytochemistry: Caudex and flower contain steroidal glycosides (smilagenin 3-O-β-d-glucopyranosyl-(1→2)-β-d-glucopyranoside, 3-O-β-d-glucopyranosyl-(1→2)-β-d-galactopyranoside, 3-O-β-d-glucopyranosyl- (1→2)-[β-d-glucopyranosyl-(1→3)]-β-d-glucopyranoside, 3-O-β-d-glucopyranosyl-(1→2)-[β-d-glucopyranosyl-(1→3)]-β-d-galactopyranoside, samogenin 3-O-β-d-glucopyranosyl-(1→2)- β-d-galactopyranoside, 3-O-β-D-glucopyranosyl-(1→2)-β-D-galactopy-ranosyl 5 β-(25R)-spirostan-3 β, 12 β-diol, 3-O-β-D-glucopyranosyl-(1→2)- [β-D-glucopyranosyl-(1→3)]-β-D-glucopyranosyl 5 β- (25R)-spirostan-3 β, 12 β-diol and 3-O-β-D-glucopyranosyl-(1→2)- β-D-galactopyranosyl 5 β-(25R)-spirostan-2 β,3 β,12 β-triol); rhizomes contain spirostane, furostane and cholestane glycosides; root contain phenolic constituents related to resveratrol.

Uses: Useful in pain relief; antioxidant and antiproliferative activities of roots and bark are justified.

Balsaminaceae

Impatiens balsamina L.

Vernacular Name: Dushpattrijati (Sanskrit); Garden balsam (English); Gulmendhi (Hindi).

Description: An annual herb, erect 30-90 cm high; stem fleshy, cylindrical, glabrous or pubescent, slightly branched; leaves alternate 5-15 cm long by 1.5-2.5cm wide, lanceolate, apex long acuminate, base cuneate and decurrent, margin serrate, petioate; flowers in summer, pink or white in colour, showy, pedicels 1-3, axillary, slender, pubescent, shorter than the leaves; sepals minute, ovate; standard small, orbicular, retuse, horned; wings broad, lateral lobes rounded, terminal sessile, very large; lip small, boat-shaped, mucronate; spur short or long, incurved; capsules tomentose; seed reticulate.

Distribution: Along watercourses and in the damp ground on field margins; commonly cultivated in China, India, Japan, Mauritius and Malaysia.

Parts Used: Seeds.

Phytochemistry: β-Amyrin, α-spinasterol, balsamin asterol, linoleic, parinaric, oleic, arachidic acids.

Uses: Amenorrhoea; dystocia, dysphagia; leaves and seeds edible; flowers mucilaginous and cooling used for lumbago and intercostals neuralgia; they improve circulation and relieve stasis; seeds contain a fatty oil used for cooking and as an illuminant; also suitable for use in the surface-coating industry.

Basellaceae

Basella alba L. (Syn. B. rubra L.)

Vernacular Name: Upodika (Sanskrit); Indian spinach (English); Poi, Lalbaclu (Hindi).

Description: Perennial succulent glabrous twining herb with white of red branches; leaves simple, alternate, broadly ovate, acute or acuminate, cordate at base, thick lamina narrowed into petiole; flowers white or red spikes, bracteoles longer than perianth; fruits red, white or black, globose, utricle enclosed in the perianth.

Distribution: Throughout India, cultivated; distributed to other countries including Mauritius.

Parts Used: Stems, leaves.

Phytochemistry: Iodine, fluorine, carotenoids, organic acids, vitamin-K.

Used: Stems and leaves are sweet, cooling emollient, aphrodisiac, laxative, haemostatic, appetizer, sedative, diuretic, demulcent, maturate and tonic; useful in burning sensation, constipation, flatulence, anorexia, haemorrhages, haemoptysis, sleeplessness, pruritus, leprosy, urticaria, ulcers, dysentery, gonorrhoea, balanitis, strangury, fatigue and general debility; used as laxative in children and pregnant women.

Bignoniaceae

Crescentia cujete L.

Vernacular Name: Calabash tree (English)

Description: A dicotyledonous plant with simple leaves, which are alternate or in fascicles (clusters) on short shoots; plant is a smooth, much-branched tree growing to a height of 4 to 5 meters; branches are arching with close-set clusters of leaves; leaves are alternate, often fascicled at the nodes, oblanceolate, 5 to 17 centimetres long, glossy at the upper surface, blunt at the tip and narrowed at the base; flowers develop from the buds that grow from the main trunk, yellowish and sometimes veined with purple, with a slightly foetid odour, occurring singly or in pairs at the leaf axils, stalked and about 6 centimeters long, and opens in the evening; calyx is about 2 centimetres long, and split into two lobes; fruit is short-stemmed, rounded, oval or oblong, green or purplish, 15 to 20 centimetres in diameter.

Distribution: Native to Central and South America; naturalized in India, Mauritius.

Parts Used: Fruit, bark, leaves

Phytochemistry: Fruit contains crescentic acid, tartaric acid, citric acid, tannic acid, cianhidric acid, β-sitosterol, stigmasterol, α and β amyrins, estearic acid, palmitic acid, quercetin, apigenin, acanthoside D, β-D-glucopransoyl benzoate, (R)-1-0-β-glucopyranosyl-1,3-octanediol, 6-O-p-hydroxybenzoyl-10-deoxyeucommiol, 6-O-benzoyl-10-deoxyeu-commiol, 6-O-benzoyl-dihydrocatalpolgenine, ningpogenin and 6-O-p-hydroxybenzoylaucubin; leaves have naphthoquinone, iridoid glycosides, aucubin, plumieride, and asperuloside; sarcocarp contains ningpogenin, 6-O-p-hydroxylbenzoyl-aucubin, 3,3′-bisdemethyl-pinoresinol, (22E,24R)-ergosta-7,22-dien-3β-ol, ergosta-4,6,8, 22-dien-3-one, cerevisterol, 5α,8α-epidiory-(22E,24R)-ergosta-6,22-dien-3β-ol, β-sitosterol, daucosterol, 3β,5α,9α-trihydroxyergosta-7,22-dien-6-one, ergosta-7,22-dien-3-one, sesquiterpene, 4-hydroxybenzonicacid, benzoic acid, p-hydroxybenzylethanol, p-hydroxybenzylalcohol, D-allitol, 5-hydroxymethyl-2-furancarboxa-ldehyde.

Uses: Fruit considered aperient, laxative, expectorant; considered anthelmintic, analgesic, anti-inflammatory, febrifuge; fruit used to make small vessels for serving or drinking.

Stereospermum tetragonum DC. (Syn. S. personatum (Hassk.) Chatterjee)

Vernacular Name: Trumpet flowers tree, Yellow snake tree (English); Paral, Patar, Padri (Hindi).

Description: Large deciduous tree 18-30 m in height with a straight trunk and fairly smooth yellowish-brown bark; leaves imparipinnate, large, clustered towards the end of the branchlets, leaflets 3-6 pairs, short-stalked, often unequal-sided at the bass, main nerves 8-10 pairs; flowers yellow, tinged with red, fragrant, in lax drooping terminal cymose panicles; fruits capsules cylindric, slender curved, somewhat rough with elevated white specks; seeds winged at each end, wings membranous.

Distribution: Throughout India and other Asian countries including in Mauritius.

Part Used: Whole plant.

Phytochemistry: Iridoid glycosides and naphthoquinones.

Uses: The roots are bitter, astringent, acrid, anodyne, appetizer, constipating, diuretic, lithontriptic, expectorant, cardiotonic, aphrodisiac, anti-inflammatory, antibacterial, febrifuge and tonic. They are useful in vitiated conditions of Vata, dyspepsia, diarrhoea, strangury, renal and vesical calculi, cough, asthma, hiccough, hyperdipsia, haemorrhoids, hyperacidity, vomiting, dropsy, fever and general debility. The leaves are useful in otalgia, odontalgia, rheumatalgia, malarial fever and wounds. The flowers are astringent, sweet, cooling, aphrodisiac, cardiotonic and tonic They are useful in burning sensation, vitiated conditions of pitta and vata, cardiopathy, hiccough and general debility. The fruits are astringent, sweet and cooling, and are useful in vitiated conditions of vata and pitta. The seeds are useful for external application in hemicranias.

Bixaceae

***Bixa orellana* L.**

Vernacular Name: Sinduri (Sanskrit); Annatto, Arnato (English); Sinduriya, Latkan (Hindi).

Description: A small evergreen tree; leaves large, cordate acuminate, glabrous on both surfaces; flowers white or pink in terminal panicles; fruits reddish-brown or green capsules, clothed with soft bristles; seeds trigonous covered with red pulp.

Distribution: Widely cultivated throughout India and Mauritius.

Part Used: Roots, bark, seeds.

Phytochemistry: Bixin.

Uses: Roots, bark and seeds are antiperiodic, antipyretic and astringent; useful in intermittent fever and gonorrhoea; pump surrounding the seed is a mosquito repellent; useful to treat dysentery; Annatto dye obtained from the pulp is used for colouring edible materials.

Boraginaceae

Cordia dichotoma G.Forst.

Vernacular Name: Slesmatakah, Bahuvarah (Sanskrit); Sebesten plum (English); Lasura (Hindi).

Description: A medium-sized tree with a short crooked trunk; leaves simple, entire or slightly dentate, elliptic-lanceolate to broad –ovate with a rounded or cordate base; flowers white, small in lax terminal or axillary cymes; fruits drupes yellowish brown, pink or nearly black when ripe with a viscid sweetish transparent pulp surrounding a central stony part.

Distribution: Throughout India and Mauritius in the Warmer regions.

Parts Used: Bark, leaves, fruits.

Phytochemistry: Sugars, flavonoids.

Uses: Bark useful in dyspepsia, fever, diarrhoea, burning sensation, helminthiasis, leprosy and skin disease; leaves are aphrodisiac, and useful in gonorrhoea and ophthalmodynia; fruits are sweet, cooling, emollient, anthelmintic, purgative, vulnerary, diuretic, expectorant, aphrodisiac, depurative and febrifuge, and useful in ulcers, leprosy skin disease, hyperdipsia, burning sensation, bronchitis, dry cough, pectoral disease, strangury, urethralgia urethritis, chronic fever, arthralgia, pharyngopathy, splenopathy and ringworm.

Brassicaceae

Brassica juncea (L.) Czern.

Vernacular Name: Indian mustard (English); Rayi (Hindi); Sarsapah, Rajika (Sanskrit).

Description: A glabrous annual herb with a few bristles at the bade upto 1.5 m in height; basal leaves long, broadly ovate, coarsely dentate, persistent middle leaves oblong, 8-

dentate, upper leaves broadly linear, entire; flowers yellow in racemes; fruits siliqua, breaking away from below upwards; seeds attached to the replum.

Distribution: Cultivated throughout India, Mauritius, Pakistan and Bangladesh.

Part Used: Seeds, oil.

Phytochemistry: Seeds contain sinigrin, gluconapin, sinapine base and a volatile isothiocyanate; essential oil.

Uses: Seeds are acrid, bitter, thermogenic, anodyne, anti-inflammatory, carminative, digestive, anthelmintic, aperients, sudorific and tonic; useful in dengue fever, abdominal colic, anorexia, dyspepsia, intestinal worms, flatulence, inflammations, morbid state of the cerebrospinal system, skin diseases, splenomegaly and persistent vomiting; mustard is used in large dose as an emetic in cases of poisoning and it will cause hyperdipsia, burning sensation.

Nasturtium officinale R.Br.

Vernacular Name: Watercress (English).

Description: A perennial much-branched, aquatic herb with creeping or floating stem; leaves pinnate, leaflets 7-11, sessile, ovate-oblong or sinuately lobed, obtuse; flowers white, in short racemes; fruit a siliqua, shortly cylindric; seeds minute, ovoid, muriculate.

Distribution: Found in temperate regions throughout the world, watercress thrives alongside or in fresh running water; commonly found in the wild in Asian countries.

Parts Used: Aerial parts.

Phytochemistry: Watercress contains vitamin A, B1, B2, C, E and minerals.

Uses: Used as a good detoxifying herb; useful for chronic diseases due to the presence of vitamin C and mineral; the plant is thought to stimulate the appetite and to relieve indigestion to help in cases of chronic bronchitis to be generally stimulating, and to act as a powerful diuretic.

Burseraceae

Canarium paniculatum (Lam.) Benth.ex Engl.

Vernacular Name: Colophon wood (English).

Description: Evergreen tree growing from 15 - 25 metres tall; bole can be up to 200 cm in diameter; tree is classified as Endangered in the IUCN Red List of Threatened Species (2011).

Distribution: It is endemic to Mauritius; also found in Africa.

Part Used: Leaves, stem bark.

Phytochemistry: Leaves contain (-)-epicatechin; bark has 3-O-galloylepigallocatechin; wood has methyl gallate ester, gallic acid and penta galloyl glucose; resin has methyl gallate ester as well as amyrin.

Uses: Leaf poultice and the resin are applied to body parts affected by rheumatism; leaf poultice is also applied on ulcerations; extracts of the stem, wood and bark have shown antibacterial activity; the tree is harvested from the wild for its timber and resin, which are used locally.

Cactaceae

Hylocereus undatus (Haw.) Britton & Rose

Vernacular Name: Pitahaya, Dragon fruit (English).

Description: Plants are scandent (climbing habit), creeping, sprawling or clambering, and branch profusely; there can be 4-7 of them, between 5 and 10 m or longer, with joints from 30–120 cm or longer, and 10–12 cm thick; with generally three ribs; margins are corneous (horn-like) with age, and undulate; areoles, that is, the small area bearing spines or hairs on a cactus, are 2 mm across with internodes 1–4 cm. Spines on the adult branches are 1-3, 2–4 mm long, being acicular (needle-like) to almost conical, and greyish brown to black in colour and spreading, with a deep green epidermis; scented, nocturnal flowers are 25–30 cm long, 15–17 cm wide with the pericarpel 2.5–5 cm long, about 2.5 cm thick, bracteoles ovate, acute, to 2.5 to less than 4 cm long; receptacle about 3 cm thick, bracteoles are linear-lanceolate, 3–8 cm long; outer tepals lanceolate-linear to linear, acuminate (tapering to a point), being 10–15 cm long, 10–15 mm wide and mucronate (ending in a short sharp point); colour is greenish-yellow or whitish, rarely rose-tinged; inner tepals are lanceolate (tapering to a point at the tip) to oblanceolate (i.e., more pointed at the base), up to 10–15 cm long about 40 mm wide at widest point, and mucronate, unbroken, sharp to acuminate (pointed), and white; stamens 5–10 cm long, are declinate, inserted in one continuous zone from throat to 35 mm above the pericarpel and cream. The style (bearing the stigma) to 17, they are 5-24.5 cm long, stout, 6–8 mm thick, cream, and up to 26 stigma lobes, they can be whole or sometimes split at the top, cream, about 25 mm long; nectar chambers are 30 mm long; fruit is oblong to oval, to 6–12 cm long, 4–9 cm thick, red with large bracteoles, with white pulp and are edible; seeds are black.

Distribution: Widely distributed through the tropics in cultivation.

Part Used: Whole plant.

Phytochemistry: Seed oil contains myristic acid, palmitic acid, stearic acid, palmitoleic acid, oleic acid, cis-vaccenic acid, linoleic acid, linolenic acid.

Uses: Fruits nutritional; fruit also converted into juice or wine, or used to flavour other beverages; flowers can be eaten or steeped like tea; fruits have antioxidant properties.

Opuntia dillenii (Ker Gawl.) Haw.

Vernacular Name: Erect prickly pear, Australian pest pear (English); Hathhathoria, Naghhana (Hindi).

Description: It is an erect or sprawling shrub up to 2 m in height, producing lemon yellow flowers in the spring and summer, followed by purplish-red fruits; stem segments not dis-articulating, green, flattened, narrowly elliptic or obovate, 10-25 × 7.5-15 cm, tuberculate, making margins appear scalloped between raised areoles, glabrous; areoles 3-5 per diagonal row across midstem segment, oval, 3-6.5 × 3.5 mm; wool dense, tan; spines 0-11 per areole, in nearly all areoles to only in some marginal areoles or absent, spreading in all directions, yellow, aging brown, straight or curving, the longest stout, oval in cross section, 12-40(-60) mm, not markedly barbed; glochids inconspicuous, few to many in crescent at adaxial edge of areole, yellow, ageing brown, often incurved, subequal to increasing in length toward adaxial edge of areole, to 4 mm; flowers: inner tepals light yellow throughout, 25-30 mm; filaments yellow; anthers yellow; style and stigma lobes yellowish; fruits purplish throughout, stipitate, ellipsoid or barrel-shaped, 40-60 × 25-30 mm, juicy, spineless; areoles 6-10. Seeds tan, subcircular, 4-5 × 4-4.5 mm, with a slightly irregular surface; girdle protruding to 1 mm. 2n = 44 (cultivated), 66.

Distribution: Mainly found in Southern North America, Central America, the Caribbean, and northern South America; distributed in Asian tropical areas.

Part Used: Fruits, leaves, flowers.

Phytochemistry: Fruit contains sugars, tannin, pectin, vitamin C, minerals, etc.

Uses: Fruits are refrigerant; leaves mashed up and applied as a poultice are said to allay heat and inflammation; ripe fruits useful in gonorrhoea; leaf made into a pulp is applied to the eyes in the case of ophthalmia; plant is bitter, hot, laxative; stomachic, carminative, antipyretic; cures biliousness, burning, leucoderma, urinary complaints, tumours, loss of consciousness, piles, inflammations, anaemia, ulcers and the enlargement of the spleen; flowers cure bronchitis and asthma; juice of the plant is heating, cures tumours and leucoderma.

Opuntia ficus-indica (L.) Mill. (Syn. O. vulgaris Mill.)

Vernacular Name: Prickly pear (English).

Description: Perennial cactus growing to 3 m; large spatula-shaped stems covered in clusters of spines, brilliant yellow flowers and roundish purple fruit.

Distribution: Prickle pear is native to Mexico and naturalize in semi-tropical regions around the world.

Parts Used: Flowers, fruit, stems.

Phytochemistry: Fruit contains mucilage, sugars, vitamin C and phenolics; flowers contain flavonoids.

Uses: Flowers are astringent and reduce bleeding, and used for the problem of the gastro-intestinal tract-particularly diarrhoea, colitis and irritable bowel syndrome; flowers are also taken to treat an enlarged prostate gland; fruit is nutritional.

Capparaceae

Crateva adansonii subsp. odora (Buch.-Ham.) Jacobs (C. roxburghii R.Br.)

Vernacular Name: Sacred garlic pear, Temple plant, Spider tree (English).

Description: A moderate-sized deciduous tree; bark grey, smooth horizontally wrinkled; leaves trifoliate; flowers white, or cream in many-flowered terminal corymbs; bark is grey, and the wood is yellowish-white, turning light-brown when old; leaves are clustered at the ends of branchlets, with a common petiole 5 to 10 centimetres long, at the summit of which are tree leaflets; leaflets are ovate-lanceolate or ovate, 7.5 to 12 centimetres long, 4 to 6 centimetres wide, and pointed at the base, with a rather slender point at the tip; flowers occur in terminal corymbs, are about 5 centimetres in diameter, greenish-yellow, and the stamens are purplish; petals are ovate or oblong, with the claw half as long as limb; fruit is ovoid or rounded, and 3 to 5 centimetres in diameters, with hard and rough rind; seeds are about 10 centimetres in length, numerous, kidney-shaped, and embedded in a yellow pulp.

Distribution: It is native to Japan, Australia, much of Southeast Asia and several South Pacific islands. It is grown elsewhere for fruit, especially in parts of the African continent.

Parts Used: Fruits, bark.

Phytochemistry: Bark yields ceryl alcohol, friedelin, lupeol, betulinic acid and diosgenin; fruit contains vitamin C.

Uses: It has anti-inflammatory, diuretic, lithontriptic, demulcent and tonic properties; useful in disorders of urinary organs, urinary tract infections, pain and burning micturition, renal and vesical calculi.

Crateva nurvala Buch.-Ham.

Vernacular Name: Varuna, Ashmarighan, ashmaghna (Sanskrit); Varuna (Hindi).

Description: A moderate-sized deciduous tree; bark grey, smooth horizontally wrinkled; leaves trifoliate; flowers white, or cream in many-flowered terminal corymbs; fruits multiple seeded, ovoid berry; seeds are embedded in yellow fleshy pulp.

Distribution: Growing widely in most of the part in south Asia.

Part Used: Roots, stem bark, leaves and flowers.

Phytochemistry: Root and stem bark contain triterpenoids (lupeol and varunol), tertiary and quaternary bases including choline; leaves yield flavonoids (rutin, quercetin and isoquercetin).

Uses: Urolithiasis and crystalluria, urinary tract infections; also used use as a cholagogue, antihelmintic and anti-ameodic in both intestinal and hepatic infestation; used in suppurative inflammations, in the small joints and in osteomyelitis; also used as an antipyretic.

Caprifoliaceae

Lonicera japonica Thunb.

Vernacular Name: Japanese honeysuckle (English).

Description: A twining evergreen plant, pubescent; stems brownish-red; leaves opposite, tomentose; flowers in axillary pedunculate pairs, often crowded at the end of branchlets, fragrant, white when blooming, then changing to yellow; berry globose, black when ripe.

Distribution: Grows wild in the mountains and is widely cultivated for ornament in India, Mauritius etc.

Parts Used: Whole plant.

Phytochemistry: Plant contains tannins, saponins, luteolin, inositol, carotenoid, cryptoxanthin.

Uses: Leaf-bearing stem is antibacterial and antiallergic; recommended in the treatment or boils, impetigo, urticaria, allergic rhinitis, fever, malaria, erythema, measles, diarrhoea, dysentery, syphilis, rheumatism and lichen tropicus.

Caryophyllaceae

Dianthus chinensis L.

Vernacular Name: Rainbow pink (English).

Description: Short-lived perennial herb; stems 30 cm tall, erect, internodes shorter than leaves; leaves opposite, often united at the base and forming a sheath about the stem;

the nodes usually swollen, basal leaves withering early; stem leaves 3-8 cm long, rarely to 0.5 cm wide, ciliate, sheath 0.5 cm long; inflorescence few-15 flowered, loosely clustered, epicalyx scales 4, abruptly contracted in to a long point; flowers solitary at terminals, 2 cm across, not fragrant, rosy-lilac with purplish eye; calyx tubular, 5-toothed, many-nerved, subtended by an epicalyx; petals 5, limb abruptly attenuated into an elongated claw, many-toothed, fimbriate with dark spot close to the tip; stamens 10; ovary 1-celled; styles 2; fruit a 4-valved capsuile, cylindrical, oblong-ovoid; seeds many.

Distribution: Native to northern China, Korea, Mongolia, and southeastern Russia, distributed to other parts of Asia.

Parts Used: Whole plant.

Phytochemistry: Eugenol, phenylethyl alcohol, melosides A and L, dianchinenosides A, B, C, D.

Uses: Diuretic, hypotensive, anthlmintic, increasing peristalsis of the intestine, antitumour, antioxidant activity; used in menostasis, gonorrhoea, cough, emmenagogue.

Casuarinaceae

Casuarina equisetifolia L.

Vernacular Name: Ironwood, She-oak, Beefwood tree (English); Jangli, Saru (Hindi).

Description: Tree to 25 high with drooping branches and needle-like branchlets; leaves highly reduced and scale-like giving the branchlets a pine-needle-like appearance; flowers anemophilous (wind-pollinated), male flowers born in spikes, each flower relatively inconspicuous, female flower brown in globose heads; fruit a globose woody aggregate (somewhat resembling a small pine cone) enclosing many small winged nuts.

Distribution: Native to Burma and Vietnam throughout Malaysia east to French Polynesia, New Caledonia, and Vanuatu, and south to Australia; introduced to the Southern United States, West Africa and Asia.

Part Used: Bark.

Phytochemistry: Ellagic acid, β-sitosterol, kaempferol and glycosides, quercetin, cupressuflivone, isoquerctrin, several common triterpenoids, trifolin, catechin and epicatechin, cholesterol, stigmasterol, campesterol, tannin, proantho-cyanidins juglanin, citrulline and amino acid, afzelin, casuarine, gallicin, catechol derivatives, gentisic acid, hydroquinone, rutin, trifolin.

Uses: Plant is used to treating nervous disorders, diarrhoea and gonorrhoea; used to treat coughs, ulcers, stomachaches and constipation; bark used for dysuria, menorrhoea, secondary amenorrhoea, throat infection, cough and stomachaches; infusion of the bark is used as a remedy for asthma, diabetes, mouth infections and urinary tract infections.

Celastraceae

Cassine orientalis (Jacq.) Kuntze. (Syn. Elaeodendron orientale Jacq.)

Vernacular Name: Indian olive wood, False olive (English).

Description: Trees reach up to 20 meters in height; fruits resemble small olives, from which the local name derives; juveniles have shiny narrow leaves with a bright red mid-rib and smooth edges; while adult leave are wide and oval with a serrated margin.

Distribution: Endemic to the Mascarene Islands of Mauritius, Reunion and Rodrigues.

Part Used: Leaves.

Phytochemistry: Flavan-3-ol derivatives, kaempferol, myricetin, quercetin, etc.

Uses: Leaves found active as anti-Chikungunya Virus; used against hypertension.

Cleomaceae

Cleome gynandra L. (Syn. Gynandropsis gynandra (L.) Briq.)

Vernacular Name: Karalia, Hulhul (Hindi); Putigandha, Tuparna (Sanskrit); Shona cabbage, African cabbage, spider wasp, cat's whiskers, stinkweed (English).

Description: Foetid, erect or ascending, 0.3-1 m tall, glandular-viscid herb; leaves predominantly 5-foliolate; petioles minutely distant-prickly; leaflets obovate or elliptic, 2-9 ×1-4.5 cm, lanceolate, acuminate or obtuse, outermost oblique; racemes subsequently elongate, subsequently elongate; flowers white; capsules 2-9 cm long, striate; seeds 1.5 mm in diameter, black.

Distribution: Native to Africa but has become widespread in many tropical and sub-tropical parts of the world.

Parts Used: Roots, leaves and seeds.

Chemical Constituents: β-Carotene, folic acid, ascorbic acid, calcium, vitamin E, iron, and oxalic acid.

Uses: Leaves act as rubefacient and vesicant; applied on boil and scorpion stings; juice of leaves is beneficial in otalgia and convulsions; leaves are used for muscular pain, headache and intestinal wounds; seeds are anthelmintic, sudorific, carminative and anti-spasmodic; also used in cough and piles; decoction of roots is useful in fever.

Clusiaceae

Calophyllum inophyllum L.

Vernacular Name: Nagachampa, Punnaga (Sanskrit); Sultanachampa (Hindi), Alexandrian laurel, Balltree, Beautyleaf, Indian doomba oiltree, Indian-laurel (English).

Description: Tree to 25 m tall with robust trunk which exudes white latex when bruised; leaves opposite, petiolate, thick and shiny with numerous parallel secondary veins; flowers borne in axillary cymes, flowers moderate-sized, white, fragrant, with variable number of perianth part and yellow anthers; fruit a purplish-black globoid-to-ovoid drupe when mature with a single seed.

Distribution: Native from East Africa, southern coastal India to Malesia, Mauritius and Australia.

Part Used: Leaves, fruits, bark.

Phytochemistry: Seed contains essential oil, amentoflavone, β-amyrin, apetaloid, pseudobrasilic acid, sixteen xanthones including buchanaxanthone, butyl citrate, calaustralin, calophylic acid, isocalophyllic acid, calo-pohylloid, calophynic acid, caloxanthones A-E, campesterol, canophyllal, canophylic acid canophyllol, canophyllum, epicatechin, cinnamic acid, costatolide, erucic acid, erythro-diol-3-acetate, euxanthone, epifriendelanol, frieedelin, inophenic acid, inophyllic acid, inophyllolide, 12-dihydro-inophyllolide, trans-inophyllolide, inophylloums, jacarebuin and derivatives, leucocyanidin, macluraxanthone, myricetin and glucoside, ponnalide, pyranoamentoflavone, quercetin, β-sitosterol, stigmasterol.

Uses: Leaves used to treat eye pain, sores and cut, skin rashes; oil from the fruit is rubbed onto joints to cure rheumatism; wounds are treated with fume from the bark; infusion of the leaves is ingested for diarrhoea.

Garcinia xanthochymus Hook.f. ex T.Anderson

Vernacular Name: Garcinia, False mangosteen (English)

Description: Tree with a rounded crown that grows from 4.5 to 7.6 m in height; leaves are large (15.4-30.5 cm in length) and leathery and are oblong to lance-shaped; small flowers (1.3 cm in diameter) are born in a dense cluster of 4-10 flowers and are greenish-white in colour; flowers are unisexual meaning that the ovule and pollen-producing structures are borne on separate flowers and in this species are borne on different trees; pollen-producing flowers have five petals and many stamens on which the pollen is produced that are arranged in bundles each containing five stamens; ovule producing flowers also have stamens but they are non-functional or sterile therefore they produce no pollen; bright yellow-orange fruit is almost round and is 5 cm to 8.9

cm in diameter; fleshy fruit usually contains 5 seeds that are surrounded by a yellow pulp that is edible.

Distribution: Native to northern India; distributed to nearby countries.

Parts Used: Fruits, leaves

Phytochemistry: Fruit contains xanthones, flavonoids and sugars.

Uses: Plant has antibacterial and antimalarial and anti-inflammatory activities.

Garcinia mangostana L.

Vernacular Name: Pursat mangosteen, Himalayan Mangosteen (English).

Description: Small tree of 7-8 m high with dense heavy branched crown; leaves are leathery; petioles are short and thick; flowers are 5 centimetres in diameter, 4-parted, bisexual, and borne singly or in pairs at the ends of the branchlets; seeds are large, flattened- and embedded in snowy-white or pinkish delicious pulp; fruit is the size of a small apple, purple-coloured, with a hard rind; inside there are typically five to seven seeds surrounded by a sweet, juicy cover (or aril).

Distribution: Native to South East Asia; cultivated in almost all tropical countries.

Part Used: Fruits.

Phytochemistry: Pericarp contains prenylated xanthones (α-, β-, and γ-mangostins, garcinone E, 8-deoxygartanin and gartanin), benzo-phenones, bioflavonoids and triterpenes.

Uses: Plant has anti-inflammatory properties; used in the treatment of skin infections and wounds; treatment of dysentery, different urinary disorders, cystitis and gonorrhoea.

Combretaceae

Combretum constrictum (Benth.) M.A.Lawson

Vernacular Name: Powderpuff Combretum, Thailand Powderpuff (English).

Description: Vining shrub or woody climber with red flowers; leaves are almost opposite, blade elliptic or oblong, 3-12 x 1.5-6.5 cm, pointed to rounded at the tip, usually hairless or almost so; leaf-stalk is 2-6 mm long, persisting as a blunt slightly curved spine; flowers are 5-merous, in short, head-like spikes; lower receptacle c. 3-4 mm long, subglabrous, upper 1.0-1.8 cm long, funnel-shaped at the apex, constricted at the middle for 4-7 mm, nearly hairless; sepals are reddish; petals 4-8 x 1.5-2.5 mm, hairy, filaments 1.5-2.5 cm long; fruit is 5-angled, sessile, c. 2.5 cm long, ellipsoid.

Distribution: Native to Somalia; distributed to tropical countries, Indian and Mauritius.

Part Used: Roots, flowers

Phytochemistry: Polyphenols, sugars, etc.

Uses: Roots are believed to be used as a remedy for treating worms in children; used in snakebites, aphrodisiac.

Combretum micranthum G.Don

Vernacular Name: Kinkeliba, Gambia, Bush tea (English).

Description: Compact deciduous shrub or small tree with long arching branches, 2–4(–10) m high or liana up to 20 m long; stem up to 10 cm in diameter; bark pale grey; slash orange to reddish; stems with bark peeling in long fibrous strips, red-brown. Leaves opposite or in whorls of 3, simple and entire; stipules absent; petiole up to 9 mm long; blade elliptical to ovate, 5–11 cm × 2–7 cm, apex short-acuminate, base rounded, with brown scales beneath, pinnately veined with 4–6 pairs of lateral veins; young leaves with many conspicuous scales, hairy on the midrib and often also on the main lateral nerves, margin ciliate, lateral veins prominent; inflorescence a spike 3–5 cm long in the axil of a leaf of the same season, rachis with hairs and scales; flowers bisexual, regular, 4-merous, sessile; receptacle consisting of 2 parts, lower part c. 2 mm long, upper part wider than deep, c. 2.5 mm in diameter; sepals triangular, tiny; petals free, obovate to spoon-shaped, c. 1.5 mm long, glabrous, cream-coloured; stamens 8, free, c. 3 mm long, filaments cream-coloured; ovary inferior, 1-celled, style up to 2.5 mm long, hairy to almost glabrous; fruit a 4-winged nut, broadly ovoid to nearly orbicular in outline, 1.2–1.5 cm in diameter, with 3–5 mm long stipe and c. 7 mm broad wings, reddish-brown, indehiscent, 1-seeded.

Distribution: Native to West Africa; found in other tropical areas.

Part Used: Leaves

Phytochemistry: Leaves contain orientin, homoorientin, myricetin-3-O-glucoside, myricetin-3-O-rutinoside, vitexin and isovitexin; gallic acid, rutin trihydrate, catechin, epicatechin, epigallocatechin and benzoic acid; stachydrine and hydrostachydrine, kinkeloids A–D; fatty acids (palmitic, oleic and linolenic acids), potassium nitrate, tannins, lupeol, α-amyrin, sorbitol, inositol, choline and β-sitosterol.

Uses: It is used for fatigue, liver ailments, headache, convalescence, blood disease, weight loss, cancer, sleep problems, liver disorders and malaria; tea made from the leaves as tonic drink and having detoxifying properties.

Commelinaceae

Commelina benghalensis L.

Vernacular Name: Benghal dayflower, Tropical spiderwort, Wandering Jew (English).

Description: Creeping, annual or perennial herb, stems creeping-ascending to 50 cm tall; branched, rooting at the nodes and growing in patch; roots white fibrous, fasciculate, also present at the nodes in contact with the soil; aerial stems are cylindrical, thick and lightly pubescent; underground stems are white, essentially fertile and thinner than aerial ones; simple alternate leaves, ovate to lanceolate, hairy, pale green, maximum 8 cm long and 5 cm wide with parallel veins; base narrow into a petiole; petioles of 0.6 cm long, membranous leaf sheath, cylindrical with long red hairs; pilosity variable, more important on the margin of the lamina (short one cell hairs and scales of 0.5 mm); bracts subtending 3 to 4 flowers (spathe) fused at the base, funnel-shaped, about 1.5 cm long and wide shortly pedunculate; external surfaces of the spathe with white hairs; first flower is long-peduncled, the others sessile; flowers lilac or blue, 2 lateral petals well visible and rounded and a central one very reduced; plant has white, burrowing rhizomes which can produce subterranean flowers and seeds; fruit apsule 3-lobed (2 anterior lobes dehiscent with 2 small seeds; 1 posterior lobe indehiscent with one big seed); 1 to 4 fruits by aerian spathe; only one by underground spathe; 5 seeds by capsule, dark brown, ellipsoidal in shape, trunqued, smaller 2.5 mm long, the biggest one 4 mm long.

Distribution: Native to tropical Asia and Africa. It has been widely introduced to areas outside its native range, including to the neotropics, Hawaii, the West Indies and to both coasts of North America.

Part Used: Whole plant.

Phytochemistry: Plant contains phlobatannins, carbohydrates, tannins, glycosides, volatile oils, resins, balsams, flavonoids and saponins.

Uses: Plant has diuretic, laxative, leprosy, febrifugal and anti-inflammatory effects; plant paste used to treat burns, and treat indigestion with a juice produced from the roots; paste used as a poultice.

Compositae

Ageratum conyzoides (L.) L.

Vernacular Name: Got wed, White weed, Appa grass (English).

Description: Annual herb, 30-50cm. high; stem erect hairy green or purple; leaves opposite, broadly ovate, crenate, coarsely hairy on both side, 3-nerved at the base; inflorescence is a terminal corymb of many small heads; flowers violet or white; achene minute, black 5 -echinate.

Distribution: Common weed widely distributed.

Part Used: Whole plant.

Phytochemistry: Essential oil (agratochromene, dimethoxy-ageratochromene, cadinene, caryophyllene); alkaloids and saponins.

Uses: Plant possesses anti-inflammatory and anti-allergic properties; juice from the fresh plant used to cure allergic rhinitis and sinusitis; also useful in post-partum uterine haemorrhage; decoction of the fresh plant makes the hair fragrant, soft and dandruff-free; prevents tetanus.

Ageratina riparia (Regel) R.M. King & H.Rob.

Vernacular Name: Mistflower, Creeping Croftonweed (English).

Description: A low growing, sprawling perennial herb that grows up to 40×60 cm high; stems produce roots at joints that touch the ground; serrated leaves on average reach 7.5 cm long and 2.5 cm wide and taper at each end; flowers are white with dense heads at the ends of the branches; buds from July until August, with full flowering occurring from August through October; seeds are black, slender, angular, 2 mm long, with fine white hairs at the tip; dispersed by wind and flowing water; mature plants can produce between 10000 and 100000 seeds per year.

Distribution: Native to Mexico, Cuba and Jamaica; distributed to Asia and Africa; introduced to southern Africa, tropical Asia, Australia, New Zealand, the Canary Islands, Madagascar, Mascarenes, Hawaii, and Peru.

Part Used: Whole plant.

Phytochemistry: Chromones, phenolics, alkaloids, etc.

Uses: Harvested to use as a tanning agent; having strong antifungal properties; poisonous for livestock; used as an ornamental plant.

Ambrosia tenuifolia Spreng.

Vernacular Name: Weakleaf bur ragweed (English).

Description: A perennial herb reaching heights between 30 cm and nearly two meters with bristly, fuzzy green to brown erect stems; multilobed fuzzy leaves have blades which can be nearly 16 cm long and are borne on petioles with lobed, winglike appendages; inflorescence has staminate (male) and pistillate (female) flower heads; pistillate heads yield one or two fruits which are burrs up to half a centimetre long and covered in short spines.

Distribution: Native to Northern Mexico and the southwestern United States from California east as far as Kansas, Oklahoma, and central Texas; naturalized in various other regions of Asia and Australia.

Part Used: Whole plant.

Phytochemistry: Sesquiterpene lactone, peruvin, ambrosic acid, cumanin, psilostachyin, psilostachyins B and C, dihydropsilostachyin C, altamisin and altamisic acid;

flavonoid, hispidulin.sesquiterpene lactone content. Lactones and lactone derivatives isolated were peruvin, ambrosic acid, cumanin, psilostachyin, psilostachyins B and C, dihydropsilostachyin C, altamisin and altamisic acidsesquiterpene lactone content. Lactones and lactone derivatives isolated were peruvin, ambrosic acid, cumanin, psilostachyin, psilostachyins B and C, dihydropsilostachyin C, altamisin and altamisic acid.

Uses: The plant showed trypanocidal and leishmanicidal activities.

Artemisia verlotiorum Lamotte (Syn. A. vulgaris var. aromatica Sacc.)

Vernacular Name: Chinese mugwort (English).

Description: Herb, perennial, 50-100 cm tall, pubescent, later glabrescent; leaves white gland-dotted adaxially and abaxially densely grey tomentose; basal and lowermost leaves shortly petiolate; leaf blade ovate, 1- or 2-pinnatisect; middle stem leaves ± sessile; leaf blade ovate or broadly ovate, 5-10 × 3-8 cm, 1-pinnatisect; segments 3 or 4 pairs, lanceolate or linear-lanceolate, 30-50 × 3-5 mm; uppermost leaves 3- or 5-lobed; leaflike bracts lanceolate or elliptic-lanceolate; capitula in somewhat broad panicles, ellipsoid or oblong-ellipsoid, 2-2.5 mm in diam., sessile; phyllaries glabrescent; marginal female florets 3-6; disk florets 8-18, bisexual; achenes obovoid or oblong.

Distribution: Found in Europe, North and South America and Oceania; Asia (Nepal, China, India, Indonesia, Mauritius, Japan, South Korea, Malaysia, Sri Lanka, Thailand); Africa.

Part Used: Whole plant.

Phytochemistry: Terpenoids, fatty acids, phenolic compounds.

Uses: Leaves showed anticancer activity.

Artemisia absinthium L.

Vernacular Name: Damara, Indhana, Aphasantin (Sanskrit); Wormwood, Abisinthe, Madarwood, Ajenjo (English); Vilayathi ajsanthin (Hindi).

Description: An aromatic and bitter shrubby plant; stem erect, angular and ribbed; leaves ovate to obovate, unequally 2-3 pinnatifidly cut into spreading linear or lanceolate-obtuse segments, hoary on both surfaces; flower heads heterogamous, numerous but hardly crowded, flowers, yellow, ray florets dilated below; outer involucral bracts oblong, hoary, narrowly scarious; inner orbicular, broadly scarious; receptacular hair long, straight; achenes elliptic-oblong.

Distribution: Native to Europe; distributed all around the Asian region.

Part Used: Whole plant.

Phytochemistry: Essential oil, artemisetin, absinthin, artabasin, myrcene, α-pinene, thujyl alcohol and thujyl acetate.

Uses: Used as a tonic, stomachic, febrifuge and anthelmintic; useful for inflammation of liver and menstrual disorders; tincture of the plant used for brain concussions; a remedy for enfeebled digestion and debility; flowers are vermifuge, tonic in intermittent fever; drug Afsanteen from the plant used in chronic fever, swellings and inflammation of liver; also used as tonic and stimulant.

Ayapana triplinervis (Vahl) R.M.King & H.Rob.

Vernacular Name: Ayapana, Water hemp, Pool root, White snakeroot (English).

Description: An ascending, slender perennial; leaves purple, long slender, subsessile, lanceolate, 3-nerved, acuminate, subentire, glabrous; inflorescence a lax, few-headed corymb, heads pedicellate, about 20-flowered; flowers are pale pink and the thin, hairless stem is reddish in colour.

Distribution: The plant is cultivated in Indonesia, the Philippines, Brazil; planted in Africa; cultivated in many other areas of the tropics and has become naturalized in some areas i.e., Mauritius.

Part Used: Whole plant.

Phytochemistry: Plant contains coumarins, heparin, cineol, α-phellandrene, α-terneol, ayapanin, ayapin, borneol, coumarin, sabinene, umbelliferone among many others.chemicals; leaves contain volatile ayapana oil; hexadecanoic acid, 2,6,10-trimethyl, 14-ethylene-14-pentadecne, bicyclo[4.1.0] heptane, 7-butyl, decanoic acid, 8-methyl-, methyl ester, 1-undecanol, 1-hexyl-1-nitrocyclohexane, 1,14-tetradecanediol, octadecanoic acid, 2-hydroxy- 1,3-propanediyl ester; leaves contain ayapanin and ayapin, stigmasterol, esculetin methylene ether, vitamin C and carotene.

Uses: It is stimulant, tonic in small doses and laxative when taken in quantity; hot infusion is emetic and diaphoretic; decoction of the leaves is antiseptic and haemostatic; useful against various kinds of haemorrhage and to clean foul ulcers; aqueous extract of the dried leaves is a cardiac stimulant; fresh leaves are used by the Marma for cut and stomachache; ethanolic extract of the aerial parts of the plant possess antibacterial and anti-fungal properties.

Bidens pilosa L.

Vernacular Name: Beggar-ticks, Black-jack, Broom stick, Devil's needles, Spanish needle (English).

Description: An annual forb of gracile habit, growing up to 1.8 m tall; leaves are oppositely arranged and pinnate in form with three to five dentate, ovate-to-lanceolate leaflets; petioles are slightly winged; flower at any time of the year, but in temperate regions, it blooms mainly in summer and autumn; flowers are borne in small heads on relatively long peduncles; heads bear about four or five broad white ray florets, surrounding many tubular yellow disc florets; fruits are slightly curved, stiff, rough black rods, tetragonal in cross-section, about 1 cm long, with typically two to three stiff, heavily barbed awns at their distal ends; infructescences from stellate spherical burrs about one to two centimetres in diameter; barbed awns catch onto fur or clothing, and can injure flesh.

Distribution: Native to the Americas; grows in Eurasia, Africa, Australia, and the Pacific Islands.

Part Used: Whole plant.

Phytochemistry: Polyacetylenes and flavonoids; chalcone okanin, ethyl caffeate, hydroxycinnamic acid.

Uses: Plant is applied on a fresh wound; showed anticancer activity.

***Calendula officinalis* L.**

Vernacular Name: Pot marigold, Ruddles, Common marigold, English marigold (English).

Description: A short-lived aromatic herbaceous perennial, growing to 80 cm tall, with sparsely branched lax or erect stems; leaves are oblong-lanceolate, 5–17 cm long, hairy on both sides, and with margins entire or occasionally waved or weakly toothed; inflorescences are yellow, comprising a thick capitulum or flowerhead 4–7 cm diameter surrounded by two rows of hairy bracts; in the wild plant they have a single ring of ray florets surrounding the central disc florets; disc florets are tubular and hermaphrodite, and generally of a more intense orange-yellow colour than the female, tridentate, peripheral ray florets; flowers may appear all year long where conditions are suitable; fruit is a thorny curved achene.

Distrlbution: Native to southern Europe; widely naturalised further north in Europe and elsewhere in warm temperate regions of the world.

Part Used: Flowers, leaves

Phytochemistry: Petals and pollen contain triterpenoid esters, carotenoids flavoxanthin and auroxanthin; leaves and stems contain lutein, zeaxanthin, and β-carotene; saponins, resins, and essential oils; flowers contain flavonol glycosides, triterpene oligoglycosides, oleanane-type triterpene glycosides, saponins, and sesquiterpene glucoside.

Uses: Plant has anti-viral, anti-genotoxic, and anti-inflammatory; flower oil protects the skin.

***Coreopsis tinctoria* Nutt.**

Vernacular Name: Plains coreopsis, Garden tickseed, Golden tickseed (English).

Description: Growing quickly; attain heights of 30–100 cm; leaves are pinnately-divided, glabrous and tending to thin at the top of the plant where numerous 2.5-to 4-cm flower heads sit atop slender stems; flower heads are brilliant yellow with maroon or brown disc florets of various sizes; flowering typically occurs in mid-summer; small, slender seeds germinate in fall or early spring.

Distribution: Common to Canada, Northeast Mexico, and much of the United States; widely cultivated and naturalized in China, India, Mauritius.

Part Used: Flowers.

Phytochemistry: Flowering tops contain chalcones, aurones, anthocyanins, flavanones, flavonols and phenylpropanoids.

Uses: Infusion of flowering tops used to control hyperglycemia.

***Cynara cardunculus* L.**

Vernacular Name: Artichoke thistle, Cardone, Cardoni, Carduni (English).

Description: Perennial herb growing to 1.5; large, thistle-like leaves grey-green above and woolly white beneath and very large purple-green flowerheads.

Distribution: Native to the Mediterranean region, artichoke thrives in rich loam in warm temperate climates.

Part Used: Flowers, leaves, root.

Phytochemistry: Plant contains sesquiterpene lactone, cynaropicrin and inulin; leaves contain cynarin.

Uses: It benefits the liver, protecting against toxins and infection; leaves are bitter and stimulate digestive secretions; useful for the treatment of gallbladder problems, nausea, indigestion, abdominal distension, and lower cholesterol level.

***Cynara scolymus* L.**

Vernacular Name: Globe artichoke (English).

Description: Plant grows to 1.4–2 m tall, with arching, deeply lobed, silvery, glaucous-green leaves 50–82 cm long; flowers develop in a large head from an edible bud about 8–15 cm diameter with numerous triangular scales; the individual florets are purple; edible portions of the buds consist primarily of the fleshy lower portions of the involucral bracts and the base; the mass of immature florets in the centre of the bud; inedible in older, larger flowers.

Distribution: Native to the Mediterranean region; both wild and cultivated varieties exist throughout the tropical region.

Part Used: Leaves, flowers

Phytochemistry: It contains apigenin, luteolin and cynarine.

Uses: Effective in cancer, cell oxidation and hypercholesterolaemia; leaves used as herbal tea.

Chamaemelum nobile (L.) All.

Vernacular Name: Roman chamomile (English).

Description: Aromatic perennial growing to 50 cm; feathery leaves and daisy-like flowerheads; daisy-like white flowers and procumbent stems; leaves are alternate, bipinnate, finely dissected, and downy to glabrous; solitary, terminal flowerheads, rising 20–30 cm above the ground, consist of prominent yellow disk flowers and silver-white ray flowers; flowering time is June and July, and its fragrance is sweet, crisp, fruity and herbaceous.

Distribution: Native to Western Europe; cultivated across Europe, and also in other temperate regions in Asia.

Part Used: Flowers.

Phytochemistry: It contains volatile oil, sesquiterpene lactones, flavonoids, coumarins and phenolic acids.

Uses: A remedy for the digestive system; used in nausea, vomiting, indigestion and loss of appetite; it is also sedative, antispasmodic and mildly analgesic, relieve colic, griping and other cramping pains.

Elephantopus scaber L.

Vernacular Name: Hastipadi, Gojihva (Sanskrit); Prickly leaves elephant's foot (English); Gobhi, Gojivha (Hindi).

Description: A rigid herb, usually perennial with short rootstock; leaves obovate-oblong, mostly radical formatting a spreading rosette on the ground, hairy on both surfaces, 12.5-20 cm long and 3.8-5.7 cm broad; flowers purple in heads, heads numerous, closely packed; fruits achenes, pappus of 4-6 bristles dilated at the base.

Distribution: Native to tropical Africa, Eastern Asia, Indian Subcontinent, Southeast Asia, and northern Australia; naturalized in tropical Africa and Latin America.

Parts Used: Roots, leaves, flowers.

Phytochemistry: Plant contains hydroxylated germacrianolides molephantin and molephantinin; phantomolin and its cis-epoxide, elephantin, elephantopin, deoxyelephantopin, iso-deoxyelephantopin, 11,13-dihydrodeoxyelephantopin, α-curcumene, β-amyrin, lupeol, epifriedelanol and stigmasterol.

Uses: Plant is bitter, acrid, astringent, antipyretic constipating, diuretic and tonic; decoction of the rots and leaves is gives in given dysuria, urethrorrhea, intermittent fever, diarrhoea and bronchitis; root decoction is specific for haemorrhoids and a paste made out of leaves is very specific for skin diseases; flowers are astringent, bitter, sweet, ophthalmic aphrodisiac and expectorant, and recommended for hepatopathy, ophthalmopathy, bronchitis, cough and swellings.

Launaea sarmentosa (Willd.) Sch.Bip. ex Kuntze

Vernacular Name: Beach Launaea (English).

Description: A glabrous perennial herb with a prostrate creeping stem; rootstock semi-woody; stems up to 1.4 m long, flagelliform, usually solitary, simple or sometimes branched, often rooting at the nodes, internodes 5–15 cm. long; leaves few to numerous, ± densely crowded at the stem base and also in fascicles at the nodes, the leaf fascicles often subtended by 2 broadly ovate-reniform foliaceous bracts; basal leaves subcoriaceous, up to 11-2.5 cm., spathulate to oblanceolate, apices rounded-obtuse, margins subentire to denticulate, rarely also sinuate-dentate to lobed, bases attenuate often subauriculate below; fascicle leaves similar but usually smaller; capitula solitary or few from the root crown and at the nodes; stalks 1–3 cm long, simple, bracteate; involucres 10–13 x 3–4 mm. in flowering capitula, up to 15 x 5 mm. long in fruiting capitula, narrowly oblong-cylindrical; phyllaries imbricate, increasing uniformly from 3–4 mm long and ovate on the outside to 13–15 mm. long and lanceolate on the inside, all eventually swollen and corky towards the base; Florets 12–20 per capitulum; corollas yellow, 15 mm. long, ligule c. 7 mm. long, lorate; achenes greyish-brown, 4–5 x 1.5 mm, subfusiform-cylindric, with 5 ± inflated spongy ribs; pappus copious of fine, white, minutely barbellate setae 9 mm long.

Distribution: Native to coastal areas in Africa, Madagascar, Seychelles, Mauritius, India, Sri Lanka and Southeast Asia; naturalized in Western Australia.

Part Used: Whole plant.

Phytochemistry: Root contains alkaloids, amino acids, carbohydrates, glycosides, tannin, and steroids.

Uses: Root is diuretic; the plant is antirheumatic, galactagogue; juice of the plant is used as a soporific for children; applied externally in the treatment of rheumatic affections.

Parthenium hysterophorus L.

Vernacular Name: Santa-Maria, Whitetop weed, Famine weed (English); Gajar Ghans (Hindi).

Description: A much-branched, short-lived, upright, aggressive ubiquitous annual herbaceous plant usually grows 2 to 4.5 ft. in height; stems and leaves are hairy; mature stems are greenish and longitudinally grooved, covered in small stiff hairs and become much-branched at maturity; leaves on the stem are alternate, and the upper leaves are entire to slightly lobed; lower leaves are relatively large and are deeply divided; leaves on the upper branches decrease in size and are also less divided than the lower leaves; leaves are light green and pubescent on both sides; flowers are small and whitish with tiny ray florets at each of the five distinct corners; flower-heads are arranged in clusters at the tips of the branches; colour changes to light brown when seeds are mature and about to shed; flowering can occur at any time of the year, but is most common during the rainy seasons; five small seeds are usually produced in each flower-head; seeds are dispersed by wind, water, animals, vehicles, tools and machinery and on clothing.

Distribution: Native to the American tropics; distributed throughout Asia; invasive species in India, Australia, and parts of Africa.

Part Used: Whole plant.

Phytochemistry: Allelopathic chemicals, parthenin, sesquiterpene lactones, hysterin, ambrosin, flavonoids (quercelagetin 3,7-dimethylether, 6-hydroxyl kaempferol 3-O-arabinoglucoside), fumaric acid, p-hydroxy benzoin and vanillic acid, caffeic acid, p-courmaric, p-anisic acid, chlorogenic acid, ferulic acid, and sitosterol.

Uses: The plant is a remedy for skin inflammation, rheumatic pain, diarrhoea, urinary tract infections, dysentery, malaria and neuralgia.

Senecio elegans L.

Vernacular Name: Redpurple ragwort, Purple groundsel, Wild cineraria, Purple ragwort (English).

Description: -An annual herb producing a single, erect, branching stem which grows to a maximum height around 60 cm; leaves have blades up to about 8 cm long which are deeply cut or divided into several toothed lobes; herbage is somewhat hairy and glandular, sticky to the touch; inflorescence bears flower heads lined with black-tipped phyllaries; contain many yellowish disc florets at the centre, each has usually 13 ray florets 1 to 1.5 cm long which can be almost any shade of red-purple.

Distribution: Native to southern Africa; growing wild in parts of New Zealand, Australia, the Azores, and on the Central Coast of California; found in a Mediterranean climate, often on the coast.

Part Used: Whole plant.

Phytochemistry: Flavonoids, tannins.

Uses: Plant is used as a diuretic.

Emilia sonchifolia (L.) DC. ex DC (Syn. Senecio sonchifolius (L.) Moench)

Vernacular Name: Sasasrutih (Sanskrit); Hirankhuri (Hindi).

Description: An annual herb, 30-40 cm in height, variously branched, sometimes procumbent and rooting near the nods; leaves simple, both radical and cauline, lyrate-

Pinnatifid with large terminal lobe, the basal leaf petioled, cauline acutely auricled; flowers purplish in lax corymbose heads, bracts almost equaling the corollas with scarious margins, pappus hairs with, soft, nearly equaling the corollas with scarious margins, pappus hairs white, soft nearly equaling the involucral bracts; fruits narrowly oblong, 5-ribbed brown achenes, scabrid on the ribs.

Distribution: Throughout India, Mauritius and nearby places.

Part Used: Whole plant.

Phytochemistry: Aerial parts contain pyrrolizidine alkaloids, senkirkine and doronine; simiaral, β-sitosterol, stigmasterol, palmitic and tracontannic acids.

Uses: Plant is astringent, sweet, thermogenic sudorific, antipyretic, ophthalmic, vulnerary and antiasthmatic; useful in infantile tympanitis, gastropathy, diarrhoea, otalgia, ophthalmia, nyctalopia, cuts and wounds, intermittent fever, pharyngodynia and asthma.

Sigesbeckia orientalis L.

Vernacular Name: Common St. Paul's wort (English).

Description: An annual herb about 50 cm tall; stem stiff, erect with branches horizontal below and dichotomously branched above; stem and branches tinged with purple, terete, pubescent; leaves opposite, 5-15.2 cm wide, triangular-ovate, acute or acuminate, deeply and irregular toothed, the uppermost leaves much smaller and nearly entire, all finely pubescent on both sides; base cuneate, running down wing-like into a somewhat obscure petiole; flowers in October Heads small, peduncled, in leafy panicles, flowers yellow, those of the ray red underneath; involucre bracts in 2 rows; achenes each enclosed in a boat-shaped bract, glabrous, slightly rough, black; leaves taste slightly bitter.

Distribution: Cambodia, China, India, Laos, Philippines, Mautitius, Sri Lanka, Vietnam.

Part Used: Whole plant.

Phytochemistry: Essential oil, salicylic acid and bitter glucoside.

Uses: Rheumatic arthritis, fatigue with fever, neurasthenia, ulcer, abscesses; boils.

Tanacetum parthenium (L.) Sch.Bip.

Vernacular Name: Feverfew (English).

Description: Plant is covered by flowers reminiscent of daisies and grows into a small bush up to around 46 cm high with citrus-scented leaves; spreads rapidly, and covers a wide area after a few years; leaves are variously pinnatifid with conspicuous flowers up to 20 mm across; outer florets are all ligulate and white; inner florets are yellow and tubular in lax corymbs.

Distribution: Originally from South-eastern Europe; commonly found in throughout Asia, Europe, Australia and North America.

Parts Used: Aerial parts.

Phytochemistry: Volatile oil, sesquiterpene lactones, sesquiterpenes.

Uses: It is used to reduce fever and to treat headaches, arthritis and digestive problems; promotes menstrual flow.

Xanthium strumarium L.

Vernacular Name: Arishta (Sanskrit); Ditch bur, Louse-bur, Broad cocklebur (English); Banokra, Gokhru, Chota-gokhru, Chota-dhatura, Adhasisi (Hindi).

Description: An erect, tall, coarse, unarmed weed with triangular, cordate or sub-orbicular, often three-lobed, irregularly toothed scabrid or hispid leaves; flowers numerous, white or green in terminal and axillary racemes with both male and female flowers on the same plant; male heads uppermost, globose, about 0.75 cm in diameter; female heads ovoid, 1.25 cm long with hooked bristles; achenes obovoid, thick, enclosed in the hardened, involucral bracts and clothed with strong hooked spines.

Distribution: Widely distributed throughout the world in the tropical and temperate region.

Part Used: Rots, leaves and fruits.

Phytochemistry: Aerial parts contain alkaloids, sesquiterpene lactones-xanthinin; stereoisomers-xanthinim and xanthatin; seeds have fatty acids.

Uses: It has diaphoretic properties; used for chronic malaria, leucorrhoea and urinary diseases; useful as a styptic and is considered a prophylactic against hydrophobia; root is bitter tonic, useful in cancer and scrofula; extract of the plant is used for ulcers, boils and abscesses; fruits are used as tonic, diuretic, diaphoretic, sedative, cooling and demulcent and are also given in small-pox; seeds are used resolving inflammatory swellings; herb is used in snake bite.

Convolvulaceae

Argyreia nervosa (Burm. f.) Bojer

Vernacular Name: Hawaiian baby Woodrose, Elephant creeper, Woolly morning glory (English); Adhoguda, vidhara) Sanskrit).

Description: A perennial climbing vine; large heart-shaped leaves and white trumpet-shaped flowers; large furry seeds grow in seed pods.

Distribution: Native to the Indian subcontinent and introduced to numerous areas worldwide, including Hawaii, Africa, and the Caribbean.

Part Used: Leaves, roots, flowers.

Phytochemistry: Seeds contain ergoline alkaloids, ergine, chanoclavine, elymoclavine; argyroside, (24R)-ergost-5-en-11-oxo-3β-ol-α-D-glucopyranoside; caffeic acid, myristoleic acid, palmitic acid, oleic acid, linoleic acid, etc.

Uses: Leaves used in ulcer; root and flower showed aphrodisiac activity.

Cuscuta chinensis Lam.

Vernacular Name: Chinese dooder (English).

Description: Leafless twining parasites; stems filiform, up to 1 m long, much-branched, fleshy, often forming dense yellow masses on some shrub or herbs amount the compositate or leguminouseae; flowers in summer, pentamerous, in cymes or dense racemes; capsule barely 2.5 mm in diameter, globose or ovoid, hyaline, divided into 2 loves by a deep furrow on tip. 2-4 seeded; seeds globose, 1-1.5 mm in diameter, with a greyish-brown or yellowish-brown surface.

Distribution: China, Japan, Korea, Mauritius, India.

Part Used: Seeds.

Phytochemistry: Flavonoids, phenolic acids, steroids, hydroquinone, volatile oil, lignans, polysaccharides, resin glycosides and fatty acids.

Uses: Used in tiredness in lower body and legs, frequent menstruation, vertigo, dizziness, poor vision, fetal distress.

Ipomoea batatas (L.) Lam.

Vernacular Name: Pindaluh, Raktaluh (Sanskrit); Sweet patato (English); Mitha alu, Sakkarkand (Hindi).

Description: A slender prostrate vine with tuberous roots which may be red or white in colour; leaves simple, alternate, ovate-cordate, entire, angular or deeply lobed; flowers purple, solitary or in cymose clusters, corolla widely tubular, funnel-shaped; fruits globose or ovoid capsules, seeds 2-4, flat, black and small.

Distribution: Native to the tropical regions in the Americas; cultivated all over the world.

Parts Used: Tubers.

Phytochemistry: Amylase, protease, invertase, catalase, laccase, arabinase, galactanase, polygalacturonase, peroxidases, monophe-nolase, catecholase, cytochrome C-oxidase, phosphorylase and phosphatases.

Uses: Roots are sweet, refrigerant, laxative, aphrodisiac diuretic and tonic; useful in burning sensation, hyperdipsia, constipation, strangury, renal and vesical calculi, diabetes and general weakness.

Ipomoea indica (Burm.) Merr.

Vernacular Name: Morning-glory, Oceanblue morning glory (English).

Description: Herbaceous twining sprawling vine; leaves alternate, long-petiolate, cordate with acuminate apex and round base, upto 17 cm long; flowers large and showy; trumpet-shaped, white to purple or blue, borne in small axillary clusters; fruit a subglobose capsule about 1 cm in diameter, usually with 4 large smooth brown seed.

Distribution: Native to tropical habitats throughout the world.

Parts Used: Roots and leaves.

Phytochemistry: Polyunsaturated fatty acids, saturated fatty acids, unsaturated fatty acids, tocopherol, tocotrienol and plastochromanol.

Uses: Plant is used as a laxative; paste made from the roots is applied as a poultice to backaches and sore muscles.

Ipomoea mauritiana Jacq.

Vernacular Name: Giant potato (English), Bhilaykand, Bhuyikohada (Hindi).

Description: A much-branched perennial climber with large tuberous roots and glabrous tough stems and branches; leaves palmately 5-7 lobed, ovate-lanceolate, acute or acuminate, glabrous, nerves prominent beneath; flowers purple, in pedunculate corymbose axillary panicles, lobes of the limb emarginated; fruits ovoid, 4-celled and 4-valved capsules, surrounded by enlarged fleshy sepals, seeds clothed with many long tawny cottony hairs.

Distribution: Found in West Africa, Australia's Northern Territory, naturalised in many parts of the world, including Asia.

Parts Used: Tubers.

Phytochemistry: Alkaloids, scopoletin, tannins, steroids and reducing sugars.

Uses: Roots are diuretic, depurative, stimulant, appetizer carminative, stomachic, anthelmintic, expectorant and tonic; useful in leprosy, skin diseases, anorexia, dyspepsia, colic, flatulence, helminthiasis, bronchitis, fever, burning sensation, nausea, vomiting, pharyngodynia, syphilis, gonorrhoea, inflammation, splenopathy, hepatopathy, menorrhagia and general debility.

Ipomoea nil (L.) Roth. (Syn. I. hederacea Jacq.)

Vernacular Name: Krsnabijah (Sanskrit); Morning glory, Pharbitis seeds (English); Kaladana, Jharmaric (Hindi).

Description: An extensively twining hairy herbaceous annual or perennial; leaves simple, ovate, cordate, more or less deeply 3-lobed, lobes ovate-acuminate, sparsely hairy;

flowers large, funnel-shaped, blue-tinged with pink, turning to red on again, in umbellate cymes; fruits subglobose or ovoid smooth capsules, seeds black, 4-6, nearly triquetrous, glabrous.

Distribution: Native to most of the tropical world; introduced widely.

Parts Used: Seeds.

Phytochemistry: Palmitic, stearic, arachidic, behenic, linolenic, linoleic, oleic, coprosterol, deep violet and red-violet flowers contain peonin chloride 3.5-diglucoside of peonidin chloride.

Uses: Seeds are anti-inflammatory, carminative and behenic; useful in constipation, verminosis, skin diseases, leucoderma, scabies, dyspepsia, flatulence bronchitis, gout, arthralgia, cephalalgia, hepatopathy, splenopathy and fever.

Costaceae

Cheilocostus speciosus (J.Koenig) C.D.Specht (Syn. Costus speciosus (J.Koenig) Sm.)

Vernacular Name: Keukand (Hindi), Crepe ginger (English).

Description: A tall and dramatic landscape plant with large dark green leaves arranged on the stalk in a spiral; grows to 10 ft tall in frost-free areas; flowers appear in late summer or early fall, and are quite unusual looking, form on red 4 in cone-shaped bracts, with several 2 in pure white crinkled flowers protruding from each cone; flowers look like crepe paper; after the flowers fade away, the attractive red cone-shaped bracts remain; large crepy object is not the petal, but the stamen - the three true petals of each flower are inconspicuous and are almost hidden by the bell-shaped stamen.

Distribution: Native to southeast Asia and surrounding regions, from India to China to Queensland; common on the Greater Sunda Islands in Indonesia; naturalized in Mauritius, Réunion, Fiji, Hawaii, Costa Rica, Belize, Melanesia, Micronesia, and the West Indies.

Part Used: Rhizomes.

Phytochemistry: Rhizomes contain diosgenin, sapogenin, tigogenin, steroids and alkaloids; diosgenin, 5α-stigmast-9(11)-en-3β-ol, sitosterol-β-D-glucoside, dioscin, prosapogenins A and B of dioscin, gracillin, quinones, α-tocopherol, tricontanoic acids, curcumin, tricontanol, aliphatic hydroxyl ketones, triterpenes, starch mucilage, oxa-acids, fatty acids, abscisic acid and corticosteroids; stems and rhizomes contains tigogenin, diosgenin, α-amyrin stearate, β-amyrin and lupeol.

Uses: Rhizome used to treat fever, rash, asthma, bronchitis, and intestinal worms; used to increase sexual attractiveness, treat kidney problems and other urinary problems.

Crassulaceae

Bryophyllum pinnatum (Lam.) Oken. (Syn. Kalanchoe pinnata (Lam.) Pers.)

Vernacular Name: Air plant, Cathedral bells, Life plant, Miracle leaf, Goethe plant (English).

Description: Perennial succulent herb, 40-60 cm high; stems tubular and glabrous, mottled with purple; leaves opposite, entire or 3-lobed, rarely 5-7 lobed; blade thick and fleshy; margins crenate; inflorescence in terminal paniculate cyme with long stalk; flowers pendulous, red of orange-red; the fruit of 4 follicles.

Distribution: Native to Madagascar, which is a popular houseplant; naturalized in tropical and subtropical areas.

Parts Used: Whole plant.

Phytochemistry: Plant contains bryophyllin and organic acids: citric, isocitric and malic; bryophillin A, bersaldegenin-3-acetate and bryophillin C.

Uses: Leaves possess antibacterial and demulcent properties; used in the treatment of burns, wounds, impetigo, ulcer, phlegmon and congestive ophthalmia; also considered as haemostatic and depurative.

Kalanchoe integra (Medik.) Kuntze. (Syn. K. spathulata DC.)

Vernacular Name: Bish-kobra, Rungru (Hindi).

Description: Perennial, glabrous herb, 30-90 cm high; leaves 6.5 x 1.5 cm, opposite fleshy, oblong-spathulate, crenate or entire, glaucous, tinged purple; lower leaves crowded, upper ones distant, sessile, oblong; flowers large, yellow, 1.0-2.5 cm long, in large terminal corymbs; follicles 4, ovoid-oblong, many-seeded, enclosed in dry persistent calyx and corolla.

Distribution: Found in tropical regions of Asia.

Part Used: Leaves.

Phytochemistry: Alkaloids, triterpenes, glycosides, flavonoids, cardenolides, steroids, bufadienolides and lipids; leaves contain bufadienolides, bryotoxin A-C, cardiac glycosides, digoxin and digitoxin.

Uses: Leaves are used as insecticidal, anticholerin, antiperiodic, tonic, purgative, and also used in wounds; the plant is poisonous to cattle.

Cucurbitaceae

Cucumis melo L.

Vernacular Name: Kharbuja (Hindi/ Sanskrit); Musk melon (English).

Description: An annual climbing or creeping herb with angular, scabrous steam; leaves simple, soft hairy, orbicular-reniform in outline, 5-angled or 5-lobed, scabrous on both surfaces; flowers unisexual, yellow.

Distribution: Cultivatedthroughout Asia.

Parts Used: Fruits, seeds.

Phytochemistry: Seeds contain isomultiflorenol, 24-methylene-sycloratenol, α-and β-amyrin, lupeol, and cycloartenol; volatile compounds (methyl-2-methylbutanoate-3-hexanal, 2-hexanal, 1,8-cineol, 1,5-octadien-3one, and ethyl-2methyl propane); green notes of musk melon are because of -2 and -3-hexenal,

Uses: Fruits useful in ophthalmia, urethrorrhea, dyspepsia, agalactia and general debility; seeds useful in urodynia, urochesia, ophthalmia, hepatopathy, nephropathy, bronchitis, chronic fever, hyperdipsia and burning sensation.

Cupressaceae

Juniperus bermudiana L.

Vernacular Name: Bermuda cedar, Bermuda juniper (English).

Description: An evergreen tree growing up to 15 m tall with a trunk up to 60 cm thick and thin bark that exfoliates in strips; foliage is produced in blue-green sprays, with the individual shoots 1.3–1.6 mm wide, four-sided (quadriform) in section; leaves are scale-like 1.5–2.5 mm long (up to 4 mm long on strong-growing shoots) and 1-1.5 mm broad, with an inconspicuous gland; they are arranged in opposite decussate pairs, occasionally decussate whorls of three; juvenile plants bear needle-like leaves 4–8 mm long; seed cones are irregularly globose to broad pyriform, 4–6 mm long and 5–8 mm broad, soft and berry-like, green at first, maturing bluish-purple about 8 months after pollination; they contain one or two seeds; male cones are 4–6 mm long, yellow, turning brown after pollen release in early spring.

Part Used: Stems.

Phytochemistry: Leaves contain deoxypodophyllotoxin and podophyllotoxin.

Distribution: Grown all over the world as an ornamental tree; roadside tree; agroforestry purpose.

Uses: Wood for construction, ship-building and furniture; ornamental; medicinal not well known.

Cycadaceae

Cycas circinalis L.

Vernacular Name: Hintalah (Sanskrit); False sago palm (English); Jangli madan must ka phul (Hindi).

Description: A slow-growing palm-like evergreen tree, reaching 5-12m tall; roots thick, dark brown, coral-like, arranged as a bead; trunk straight, cylindrical, unbranched, brown, with persistent leaf scars; leaves 1-3m long pinnately arranged; leaflets opposite, 60-100pairs, linear, 20-30 cm long, about 1 cm wide, single nerved, apex acuminate, margin entire, shiny, hairless; leaf-stalk 40-60 cm long, hairless, stout with few short stout slightly curved spines at base. Male and female flowers found in separate plants in the form of cones; male cones consist of numerous compactly arranged fleshy leaf-like structures called as microsporophylls, which produce thousands of powdery male spores; megasporophylls are specialized fleshy leaf-like structure, bearing the seeds, which are 2-4 in each megasporophyll, egg-shaped, hard, about 5 x 3 cm, hairless, smooth and brownish-yellow.

Distribution: Native to Sri Lankan; distributed to Mauritius and other tropical regions.

Parts Used: Bark, leaves, microsporophylls, seeds.

Phytochemistry: Leaflets contain bioflavonoids, tetrahydroisoginkgetin.

Uses: Seeds are used as an aphrodisiac as well as to improve the sperm production; bark and seeds are used as a poultice for sores and swellings; leaves relieve flatulence and vomiting; powdered endosperm is used to relieve burning sensation and general debility.

Dennstaedtiaceae

Sphenomeris chinensis (L.) Maxon

Vernacular Name: Lace fern (English).

Description: A fairly small evergreen fern with a short creeping rhizome; fronds are produced throughout the warm months and are 3 times pinnate giving them a very fine-textured look; frond length is variable depending on the exposure to the sun with long fronds growing in shader locations, anywhere from 10-100 cm long and 8-20 cm wide; pinnae grow in an alternating pattern, there often exists a basal pair of pinnae that are separate from the next pair further up the rachis by several centimetres; basal pinnae are the largest on the frond and the rest get progressively smaller as they approach the tip of the blade giving the frond a long triangular shape; tiny pinnules are wedge-shaped with the wide end at their terminus; reddish-brown stripe is completely scaleless and hairless and accounts for nearly half of the frond length on large fronds.

Distribution: Indigenous to Hawaii; other parts of the tropics and sub-tropics.

Part Used: Whole plant.

Phytochemistry: Tannins and phenolic compounds.

Uses: Plant used to treat various female ailments.

Dipterocarpaceae

Dipterocarpus alatus Roxb.ex G.Don

Vernacular Name: Garjan, Silgarjan (Hindi).

Description: A lofty evergreen tree with tall cylindric bole and comparatively small crown; leaves 9-21 cm long, elliptic, or ovate-elliptic, acute or shortly acuminate, coriaceous, nearly glabrous above and stellate-pubescent beneath; flowers subsessile, in simple, 3-7-flowered, axillary racemes, petals 1.5-2.5 cm long, stellately pubescent; fruit-belly 1.8-2.5 cm long, wings 7.5-12.5 mm long.

Distribution: Common in Thailand, Mauritius, Cambodia, Laos, Vietnam and the Philippines.

Part Used: Stem bark.

Phytochemistry: Tannins, resin; essential oil, 2 sesquiterpenes of eudesmane series.

Uses: Bark of a young tree is used against rheumatism and diseases of the liver, and to stimulate appetite in cattle; oleo-resin of the bark is applied externally in gonorrhoea and ulcerated wounds.

Elaeocarpaceae

Elaeocarpus serratus L.

Vernacular Name: Ceylon Olive (English).

Description: An ornamental medium-sized tree; bears smooth ovoid green fruits, size of about 2.5 cm long; recommended varieties having round and oval fruits; brown coloured seed inside the fruit.

Distribution: Found in the Indian Subcontinent, Indo-China and South East Asia, Sri Lanka and Mauritius.

Part Used: Fruits.

Phytochemistry: Fruits contain protein, starch, sugar and iron.

Uses: Fruits used to treat diarrhoea due to its constipating effect.

Erythroxylaceae

Erythroxylum laurifolium Lam.

Vernacular Name: Cafe marron petite (English).

Description: Heterophylle shrub from 2.5 to 7 m in height and trunk diameter of 10-15 cm; orthotropic axis and separate plagiotropic branches during the juvenile stage; flattened ultimate branches; leaves simple, alternate, long from 6 to 7 cm; heterophyllie strong enough (young leaves very narrow red petiole); blade elliptical, rounded tip, attenuated corner at the base; stipules intrapetiolaires, deciduous or evergreen scale-like after the fall of axillantes leaves, sometimes densely clustered at the ends of branches; inflorescences fascicles in the axils of the stipules or scales; group of flowers by 3 or 7, see 11 (very rarely solitary flowers); a booklet does not produce flowers all together but 2 or 3 times; the branches with many scales are very floriferous and form inflorescences composed of 10-12 leaflets; flowers hermaphrodite, solitary, white pedicel very long hail from 0.5 to 1.5 cm, 5 white petals fitted inside a fleshy ligule margin finely lobed; androecium consists of 10 United stamen tube at the base to a height of 1 mm; gynoecium composed of one ovary in three styles in longer styled flowers (3-3.5 mm) than styled flowers (1.2 mm) to stigma head; fruits drupes small like a pimento 1 to 2 cm long and red colour at maturity; seeds one per fruit.

Distribution: Native to Reunion Island and Mauritius; found in the low altitude in various Asian countries.

Part Used: Leaves.

Phytochemistry: Leaves contain proanthocyanidins, condensed tannins and flavonoids (quercitrin and ansafzelin).

Uses: Leaves has an antihypertensive effect; it showed angiotensin-converting enzyme (ACE) inhibitory.

Euphorbiaceae

Euphorbia thymifolia L.

Vernacular Name: Laghududhika (Sanskrit); Thyme-leaved spurge (English); Choti-dudhi (Hindi).

Description: Small annual herb with milky juice; stem prostrate, slender, purple and downy; leaves opposite, oval-shaped, obscurely serrulate, slightly pubescent beneath; inflorescence in a few-flowered axillary cyme; capsule hairy; seeds quadrangular and glabrous.

Distribution: Wildly grown in tropical countries; cultivated in gardens as ornamental.

Parts Used: Whole plant.

Phytochemistry: Plant contains 5,7,4-trithydroxy flavones, 7-glucoside; essential oil.

Uses: Plant has antibacterial and galactagogic properties; used in treating bacillary dysentery, boils, hypogalactia, galactophoritis, metrorrhagia and infantile lientery; a poultice of the pounded leaves is effective against dermatosis and wounds; the plant is also used as an insecticide.

Hypericaceae

Hypericum monogynum L.

Vernacular Name: Hypericum (English).

Description: Shrubs, 0.5-1.3 m tall, bushy or usually with branches lax, spreading; stems 2(-4)-lined and ancipitous when young, soon terete; internodes 1-5 cm, shorter than leaves; leaves with petiole 1.5-2 mm; blade oblong to elliptic, 2.5-5 × 1.2-2.3 cm, thickly papery, abaxially paler but not glaucous; laminar glands very small dots; abaxial glands absent; main lateral veins 2- or 3-paired, all or upper sometimes uniting with midvein branches to form ± arched intramarginal vein, tertiary reticulation very dense, not very conspicuous; base cuneate to subangustate, apex acute to rounded; inflorescence 1-15(-30)-flowered, from apical node, lax, shallowly round-topped, sometimes also from uppermost 1-3 nodes, rarely with 1 or 2 pairs of flowering branches; bracts caducous, linear-lanceolate, small; pedicels 0.8-2.8(-5) cm; flowers 3-6.5 cm in diam., stellate; buds ovoid, apex subacute to obtuse; sepals ± spreading, broadly to narrowly elliptic or oblanceolate, equal to unequal, 4.5-13 × 1.5-6 mm; laminar glands basally lines to streaks, dots toward apex, margin entire, apex acute to rounded; petals golden yellow to lemon yellow, triangular-obovate, 2-3.4 × 1-2 cm, 2.5-4.5 × as long as sepals; margin entire, eglandular; apiculus lateral, acute to rounded or obsolete; stamen fascicles each with 25-35 stamens, longest 1.8-3.2 cm, ca. as long as petals; ovary ovoid or ovoid-conic to subglobose, 2.5-5 × 2.5-3 mm; styles 1.2-2 cm, 3.5-5 × as long as ovary, united nearly to apices then outcurved or very rarely to half free; capsule broadly ovoid or rarely ovoid-conic to subglobose, 6-10 × 4-7 mm; seeds dark reddish brown, ca. 2 mm, narrowly carinate; testa linear-reticulate.

Distribution: Widely cultivated in South Africa, Asia, Australia, Central America, Europe, Mauritius, and the West Indies

Part Used: Whole plant.

Phytochemistry: Root contains polyprenylated tetraoxygenated xanthones; flowers contain polycyclic polyprenylated acylphloroglucinols; aerial parts contain 4-chlorobenzoic acid, quercitrin, astilbin, β-sitosterol, γ-sitosterol, friedelin and β-amyrin.

Uses: Plant is an alternative, antidote and astringent; used in the treatment of miasmatic diseases.

Harungana madagascariensis Lam. ex Poir. (Syn. Haronga madagascariensis (Lam. ex Poir.) Choisy)

Vernacular Name: Haronga (English).

Description: Small evergreen tree growing to 8 m. has black-doted leaves with a dark green upper surface and red-brown hairs underneath, and clusters of rust-coloured flowers.

Distribution: Native to Madagascar and East Africa; grows in tropical areas.

Parts Used: Leaves, bark.

Phytochemistry: Bark contains phenolic pigments, triterpenes, anthraquinones and tannins; leaves contain phenolic pigments, hypericin, flavonoids and tannins.

Uses: It stimulates bile secretion; used to treat indigestion and poor pancreatic function; diarrhoea and dysentery.

Juglandaceae

Juglans jamaicensis C. DC.

Vernacular Name: West Indian walnut; Nogal; Palo de nuez (English).

Description: Large tree reaches 25 m in height; compound leaves are each made up of several lance-shaped, toothed leaflets up to 9 centimetres long; bear male and female inflorescences, the male a catkin up to 11 centimetres long and the female an array of flowers at the end of a newly grown shoot; fruit is a drupe roughly 2 to 3 cm long with a black husk and a seed.

Distribution: Found in Cuba, the Dominican Republic, Haiti, and Puerto Rico; native to Jamaica; widely distributed.

Parts Used: Fruits, stem and bark.

Phytochemistry: Fruits and barks are rich in tannins; phenolic contents.

Uses: Seeds flesh inside raw or cooked is edible; stem used as timber.

Lamiaceae

Leonotis leonurus (L.) R.Br.

Vernacular Name: Lion's tail, Wild dagga (English).

Description: Shrub grows 1 to 2 m tall by 0.46 to 1.07 m wide; medium-dark green 5.1–10.2 cm long leaves are aromatic when crushed; tubular orange flowers in tiered whorls, typical to the mint family, that encircle the square stems; rise above the foliage mass during the summer season, with flowering continuing into winter in warmer climates.

Distribution: Native to South Africa and southern Africa; distributed to other tropical areas.

Part Used: Whole plant.

Phytochemistry: Alkaloid leonurine; flavonoids, terpenoids (labdane diterpenes, marrubiin).

Uses: Infusions made from flowers and seeds, leaves or stems are used to treat tuberculosis, jaundice, muscle cramps, high blood pressure, diabetes, viral hepatitis, dysentery, and diarrhea; leaves, roots and bark are used as an emetic for snakebites, bee and scorpion stings; fresh stem juice is used as an infusion drunk for blood impurity.

Leucas aspera (Willd.) Link

Vernacular Name: Thumbai (English).

Description: An annual herb or undershrub with heights of 15–60 cm; leaves opposite, subsessile or short petioled, linear or narrowly oblong- lanceolate, entire or distantly crenate, obtuse, narrowed at the base; stem quadrangular, much branched, hispid or scabrid and contains a wide stele; roots contains epidermal cells which are very narrow and closely packed together; cell walls of the epidermal cells are very thin, flattened and straight; infloresence verticillaster, flowers white, small, and directly attached to the base without a peduncle or stalk; flowers are held together in auxiliary whorls or dense terminals; flower complete, bisexual, irregular, zygomorphic, hypogynous, pentamerous, white; sepals 5, gomosepalous, 10 nerved, tubular, curved, 6-10 toothed, contracted at the mouth, glabrous below, ribbed and scabrid above; mouth oblique, produced on the upper side; teeth short, triangular, spinulose, ciliate, the upper one is the longest and 8 to 13 mm in length; petals 5, gamopetalous, bilabiate; tube annulate; lower lip 3 fid, spreding, mid lobe large; upper lip 2 fid, erect, concave, villous outside, white; fruit is 2.5 mm long, brown, smooth and oblong in shape.

Distribution: Found throughout India, India, the Philippines, Mauritius and Java.

Part Used: Whole plant.

Phytochemistry: Triterpenoids, oleanolic acid, ursolic acid, β-sitosterol, nicotine, sterols, glucoside, diterpenes, phenolic compounds (4-(24-hydroxy-1-oxo-5-n-propyltetracosanyl)-phenol).

Uses: Plant is used as an antipyretic and insecticide.

Ocimum basilicum L.

Vernacular Name: Munjariki, Surasa, Varvara (Sanskrit); Sweet basil, Common basil (English); Babui tulsi, Gulal tulsi, Kali tulsi (Hindi).

Description: An erect, almost glabrous herb, 30-90 cm high, leaves ovate-lanceolate, acuminate, toothed or entire, glabrous on both surfaces, glandular; flowers white or pale purple, in simple or much-branched racemes, often thysoid; nutlets ellipsoid, black, pitted.

Distribution: Throughout India, Mauritius, Sri Lanka and Bangladesh.

Parts Used: Leaves.

Phytochemistry: Essential oil (methyl cinnamate, linalool, methyl chavicol, ocimene, linalyl acetate, eugenol and trans-anethole).

Uses: Oil used as a flavouring agent in confectionery, baked goods, sauces, Ketchup, tomato pastes, pickles, fancy vinegar, spiced meats, sausages, and beverages; tincture of the herb finds use in liqueurs, as a flavour modifier.

Ocimum tenuiflorum L. (Syn. O. sanctum L.)

Vernacular Name: Surasah, Tulasi (Sanskrit); Holy basil, Sacred basil (English).

Description: An erect much-branched softly pubescent undershrub, 30-60 cm high with red or purple subquadrangular branches; leaves simple, opposite, elliptic, oblong, obtuse or acute, entire, serrate or dentate, pubescent on both sides, minutely gland-dotted, petioles slender, hairy; flowers purplish in elongate racemes in close whorls, stamens exserted, upper pair with a small bearded appendage at the base; fruits nutlets, smooth, not mucilaginous when wetted.

Distribution: Cultivated throughout Asia.

Parts Used: Whole plant.

Phytochemistry: Eugenol, eugenol methyl ether and carvacrol, methyl chavicol cineole and linalool.

Uses: Plant is useful in cardiopathy, haemopathy, leucoderma, asthma, bronchitis, catarrhal fever, otalgia, hepatopathy, vomiting, lumbago, hiccough, ophthalmia, gastropathy in children, genitourinary disorder, ringworm, verminosis and skin diseases.

Thymus mongolicus (Ronniger) Ronniger

Vernacular Name: Mongolian Thyme (English).

Description: Stems numerous, trailing to ascending; vegetative shoots pubescent; fertile branches (1.5-)2-10 cm, densely retrorse pilose or slightly spreading pilose below inflorescences, sparsely fine pilose at base, 2-4-leaved; petioles of lower stem leaves ca. 1/2 as long as blades; leaf blade ovate, 4-10 × 2-4.5 mm, glabrous, ± conspicuously glandular, base cuneate to attenuate, margin entire or 1- or 2-serrulate, apex obtuse to subacute; inflorescences capitate; floral leaves ciliate on lower 1/3 of margin; bractlets deciduous at base; calyx tubular-campanulate to narrowly campanulate, 4-4.5 mm, base pilose, apex subglabrous; teeth of upper lip less than 1/3 as long as lip, triangular, ciliate or glabrous; lower lip as long as to longer than upper; corolla purple-red, purple, or reddish, 6.5-8 mm, sparsely pubescent; tube elongated, 4-5 mm, slightly dilated upward; nutlets subglobose to ovoid, compressed.

Distribution: Native to temperate regions in Europe, North Africa and Asia.

Part Used: Whole plant.

Phytochemistry: Essential oil (linalool, carvacrol, and geranyl acetate).

Uses: Oil has antimicrobial activity.

Lauraceae

Cinnamomum cassia (L.) J.Presl

Vernacular Name: Camphor tree, Formosan wood (English).

Description: Evergreen tree about 15 m tall; trunk bark thick and grooved; leaves alternate, coriaceous, long-petiolate, shining on the upper side, 3-nerved at the base; inflorescence in axillary or terminal cymose panicle; flowers white; drupe ovoid, glabrous, purplish-brown when ripe.

Distribution: Distributed widely in South Asian countries.

Part Used: Bark.

Phytochemistry: Stem bark and twig bark contain essential oil; cinnamic aldehyde and coumarin.

Uses: Bark possesses antibacterial circulatory, respiratory, uterotonic and stomachic properties; also applied in the therapy of dyspepsia, colic, diarrhoea, cholera, dysentery, menstrual haematometra, coryza, influenza, cough, asthma, paresis and snake-bite.

Persea americana Mill. (Syn. P. gratissima C.F.Gaertn.)

Vernacular Name: Avocado (English).

Description: A large tree of 15-18 m high; leaves oblong or elliptic-lanceolate, upto 40 cm long; flowers small, greenish in compact panicles; fruit baccate, large 5-20 cm long, pear-shaped, ovate or spherical, yellow-green to maroon and purple, skin thin or woody; seed single, large, globe or pointed.

Distribution: Indigenous to Central America, avocado is widely cultivated for its fruit in tropical and sub-tropical areas, including Israel, Mauritius, Spain and South Africa.

Parts Used: Fruit, Seed and Leaves.

Phytochemistry: Leaves and bark contain volatile oil; flavonoids and tannins; fruit pulp contain unsaturated fats, protein, sesquiterpenes, vitamins A, B^1 and B^2.

Uses: Fruit highly nutritious; leaves and bark relive cough; promote menstrual flow; fruits pulp are emollients, carminative; rind eliminates worms.

Leguminosae

Barringtonia asiatica (L.) Kurz

Vernacular Name: Fish poison tree, Putat, Sea poison tree (English).

Description: Tree to 25 m tall with glossy alternate, petiolate, entire leaves, obovate, 12-40 cm long, 10-20cm broad; flowers large and showy, white, fruit a large fibrous drupe, shiny green, quadrangular, containing a large single seed; usually forms large spreading branches as well as a large, spreading buttress root system.

Distribution: Native to mangrove habitats on the tropical coasts and islands of the Indian Ocean including Mauritius and western Pacific Ocean from Zanzibar east to Taiwan, the Philippines, Fiji, New Caledonia, Solomon Islands, the Cook Islands, Wallis and Futuna and French Polynesia.

Parts Used: Seeds, leaves and bark.

Phytochemistry: Gallic acid, saponins (barrinin A), hydrocyanic acid, monosaccharides, triterpenoids (bartogenic acid, 19-epibartogenic acid and anhydrobartogenic acid).

Uses: Seeds powder mixed with coconut cream is used for burns; decoction of leaves is used to treat hernia; decoction of the bark is used to treat constipation and epilepsy; fruit or bark is used to treat yaws; seed to treat ringworm and the bark is used in tuberculosis.

Barringtonia racemosa (L.) Spreng.

Vernacular Name: Powder-puff tree (English).

Description: A mangrove tree; straight, unbranched stem that leads to a rounded crown and is usually 4-8 m tall; bark is greyish brown to pink with white blotches and raised dots and lines; branches are marked with leaf scars; leaves are alternate and carried in clusters at the ends of branches, are 180-320 x 55-145 mm, with petioles 5-12 mm long; midribs are prominent on the lower side of the leaf and the branching veins are visible on both sides; flowers are produced on hanging racemes up to 1 m long; buds are pinkish-red and split open to bring forth masses of delicate stamens in white sprays up to 35 mm wide, which are often tinged with pink; flowers give off a pungent, putrid yet faintly sweet odour in the morning, fruit are quadrangular, 65 x 40 mm; fruit contains a single seed surrounded by spongy, fibrous flesh that provides the buoyancy that allows the fruit to be carried off with the tide.

Distribution: Found in coastal swamp forests and on the edges of estuaries in the Indian Ocean, starting at the east coast of Mozambique and KwaZulu-Natal to Madagascar, India, Sri Lanka, Mauritius, Malaysia, Thailand, Laos, southern China, northern Australia, coastal Taiwan, the Ryukyu Islands and many Polynesian islands.

Part Used: Seeds, bark, stem, roots

Phytochemistry: Seeds, bark, stem, roots contain saponin; bark contains tannins.

Uses: Seeds, bark, wood and roots used to stun fish; plant extract is effective insecticides; fruit used to treat malaria; seeds are used as an antidote.

***Albizia versicolor* Oliv.**

Vernacular Name: Poison-pod albizia, Large-leaved false thorn (English).

Description: A deciduous tree with an open, spreading crown; usually growing from 5 - 18 metres tall, but sometimes smaller when growing in poor soils, or sometimes larger to 20 metres tall; bole is usually short and free of branches for up to 5 metres, occasionally for up to 12 metres; it is straight and cylindrical, up to 60cm, occasionally to 150cm in diameter.

Destribution: Tropical Africa; Indian subcontinent.

Part Used: Root, bark.

Phytochemistry: Bark contains kaempferol glycosides and triterpenes (lupeol, lupenone, betulinic acid and acacic acid lactone); tannins.

Uses: Root and bark are anthelmintic and purgative; decoction is used to treat anaemia, swollen glands and venereal diseases; bark maceration is used to relieve coughs, headache and sinusitis, and the bark powder is sniffed for the same purpose; bark decoction is applied externally to treat ophthalmia and skin rash.

***Albizia coriaria* Oliv.**

Vernacular Name: Mugavu (English).

Description: A deciduous tree with a heavily branched, spreading, dome-shaped crown; usually grows up to 35 metres tall; straight, cylindrical bole can be free of branches for up to 22 metres and up to 100 cm in diameter; usually unbuttressed, but can have small, thick buttresses up to 1.5 metres high; crown spreading, flat; bark rough, dark; young branchlets puberulous; leaf with 3-6 pairs of pinnae, with leaflets in 6-11 pairs, discolorous, the undersides paler; flowers: white, stamens red about, white below; pod dehiscent.

Distribution: Tropical Africa and Asia.

Part Used: Root and stem bark, leaves

Phytochemistry: Bark from roots and bole contains saponins and alkaloids

Uses: Bark is astringent and vermifuge; decoction is used to treat dysentery, bronchial affections and pain caused by fever; bark maceration is drunk and applied as a wash to treat jaundice; decoction of the bark is applied externally to treat sores, pimples and other skin complaints; maceration of the leaves is administered as an enema to induce abortion; leaf decoctions are used externally to treat headache, and as a wash or steam inhalation against fever (including malaria) and toothache; they are also applied as a wash to kill head lice.

***Cassia javanica* L.**

Vernacular Name: Java cassia, Pink shower, Apple blossom tree, Rainbow shower tree (English).

Description: A fast-growing, deciduous/ semi-deciduous tree; straight trunk that reaches heights of 25 – 40 m; leaves are paripinnate with 12 pairs of elliptical leaves; flowers range in colour from pale pink to crimson with yellow coloured stamens and are found in open clusters; fruit are housed in long cylindrical dark brown pods.

Distribution: Native to Southeast Asia; extensively grown in tropical areas worldwide as a garden tree.

Part Used: Whole plant.

Phytochemistry: Leaves contain flavones, sterols, hydrocarbons, anthraquinone, glycosides; seeds contain chrysophanol, physcion, 1,5 dihydroxy- 4,7 dimethoxy, 2 methyl antroquinone 3-O-α-L (-) rhamnopyranoside, 1,3,6,7,8-pentahydroxy-4-methoxy-2-methyl anthraquinone, chrysophanol and physcion; root bark contain quercetin, leucocyanidin-3-O-α-L-rhamnopyranoside and leucocya-nidin-3-O-α-L-(-) rhamnopyranoside.

Uses: Plant used in treating constipation, colic, chlorosis and urinary disorders; leaves are effective against herpes simplex; bark used as antidiabetic.

***Senna villosa* (Mill.) H.S.Irwin & Barneby (Syn. *Cassia villosa* Mill.)**

Vernacular Name: Senna (English).

Description: Coarse herb or weak shrub; stems and leaves stellate-pubescent; leaves alternate, once-pinnately compound, even-pinnate, with stalked glands between the leaflets; leaflets ovate-elliptic, base rounded, margin entire, apex acute; legume fruit flat, constricted between the seeds.

Distribution: South America, Mexico; tropical areas of Asia.

Part Used: Whole plant.

Phytochemistry: Leaves contain aliphatic esters, hexyl tetradecanoate, heptyl tetradecanoate and octyl tetradecanoate.

Uses: The plant extract showed antiparasitic activity against *Trypanosoma cruzi*; used topically to treat skin infections, pustules and eruptions and to heal wounds by scar formation.

***Crotalaria retusa* L.**

Vernacular Name: Sanapuspi, Sanaghantika (Sanskrit); Khunkhuniyan (Hindi).

Description: An erect, robust undershrub, 60-120 cm high, with glabrous striate branches; leaves oblanceolate-oblong, obtuse or retuse, usually glabrous above and silky pubescent beneath; flowers yellow, numerous, large showy in erect terminal racemes; fruit stalked pods, slightly broader upwards with 15-20 seeds per pod.

Distribution: Throughout Indian subcontinent.

Parts Used: Whole plant.

Phytochemistry: Seeds contain pyrrolizidine alkaloids, isatinecine, monocrotaline-N-oxide, retusamine and its N-oxide, retusine, spectabilime, acyclic-α-amino acids and vitexin-4'-O-galactoside.

Uses: Plant is useful in dyspepsia, fever, cardiac disorders and stomatopathy; leaves are used for curing ptyalism, diarrhoea, scabies and impetigo; seeds powdered and boiled with milk are said to be very useful for increasing body strength and life span; seeds are also useful in skin diseases, leprosy, flatulence and fever.

***Lablab purpureus* (L.) Sweet (Syn. *Dolichos lablab* L.)**

Vernacular Name: Nispavah (Sanskrit); Lablab bean, Bonavist bean, Indian butter bean (English); Sem, Bhatvas (Hindi).

Description: A tall, nearly glabrous twining perennial herb; leaves 10-20 cm long; flowers in racemes, white or pink, fruits white pods, green or purple margined; seeds white, yellow, purple or black.

Distribution: Native to Africa; cultivated throughout the tropical countries.

Parts Used: Leaves, seeds.

Phytochemistry: Plant contains alkaloids, trigonelline, and cyanogenic glycosides.

Uses: Leaves are given in colic; seed are useful in inflammation, strangury and agalactia.

***Dalbergia latifolia* Roxb.**

Vernacular Name: Indian rosewood (English); Kala-shisham, Vilayati shisham (Hindi).

Description: A single-stemmed deciduous tree with a dome-shaped crown of lush green foliage; tree can become 20-40 m tall, with a trunk diameter of 1.5-2 m; bark is grey, thin with irregular short cracks, peeling in fibrous longitudinal flakes; alternately arranged leaves are odd-pinnate with 5-7 unequal sized leaflets originating from the same rachis; leaflets are broadly blunt-tipped, dark green above and pale below; white flowers are borne in panicles, 0.5-1 cm long, in leaf axils; brown pods are oblong-lancelike, and pointed at both ends; contain 1-4 smooth brown seeds and do not open at maturity.

Distribution: East Asia, Indian subcontinent; Indonesia, Mauritius.

Part Used: Root and stem bark.

Phytochemistry: Plant contains flavonoids, isoflavonoid O-glycosides, dalbergione and dalbinol.

Uses: Bark is used in diarrhoea, indigestion and leprosy, and as a vermifuge.

Pongamia pinnata (L.) Pierre

Vernacular Name: Indian beech, Pongam oil tree, Karanj (English); Karanj, Karanja (Hindi).

Description: A medium-sized semi-evergreen glabrous tree with a short bole and spreading crown upto 18 m or more in height; bark greyish-green or brown, very often mottled with bark brown dots, specks, lines or streaks; leaves compound, leaflets 5-7, ovate acuminate or elliptic; flowers lilac or pinkish-white, fragrant in axillary racemes; fruits thick woody, smooth, compressed, with a short curved beak, seeds 1 or 2 per pod, reniform to nearly round, smooth or wrinkled.

Distribution: Naturally distributed in tropical and temperate Asia, from India to Japan to Thailand to Malesia to the north and north-eastern Australia to some Pacific islands; propagated and distributed further around the world in humid and subtropical environments from sea-level to 1200 m.

Parts Used: Root, bark, leaves, flowers.

Phytochemistry: Pongaglabrone, diketonepongamol, glabrin, karanjin, pongapin and kanjone.

Uses: Roots are good for cleaning foul ulcers, cleaning teeth, strengthening gums and gonorrhoea; root paste is used for to local application in scrofulous enlargements; fresh bark is anthelmintic and alexeteric and is useful in haemorrhoids, beriberi, ophthalmopathy, dermatopathy, vaginopathy and ulcers; leaves are digestive laxative and anthelmintic, and are good for flatulence, dyspepsia, diarrhoea, leprosy, gonorrhoea and cough; hot infusion of the leaves is good for rheumatalgia and for cleaning ulcers and wounds; flowers are useful in inflammations, otalgia lumbago, pectoral diseases, chronic fevers, hydrocele, haemorrhoids and anaemia; oil is recommended for scabies, ophthalmia, leprosy, haemorrhoids, ulcers, herpes and lumbago.

Loganiaceae

Strychnos spinosa Lam.

Vernacular Name: Natal orange, Spiny orange, Green monkey orange (English).

Description: Small to medium-sized, spiny deciduous tree with leaves turning yellow in autumn; canopy is flattish and irregular and the tree is heavily branched; produces small greenish-white flowers in dense heads at the ends of branches; fruit tend to appear only after good rains; smooth, hard fruit is large and green, ripening to yellow; inside are tightly packed seeds surrounded by a fleshy, edible covering.

Distribution: Indigenous to tropical and subtropical Africa; distributed to Asia.

Part Used: Bark, roots, leaves.

Phytochemistry: Bark contains strychnine and sarracenin

Uses: Plant is used for treating snakebite, increasing flow of breastmilk in lactating mothers, treatment of venereal disease and enhancing physical strength; roots, leaves and bark are used in the treatment of disorders of the male organs.

Lythraceae

Lagerstroemia speciosa (L.) Pers.

Vernacular Name: Giant crape-myrtle, Queen's crape-myrtle, Pride of India (English).

Description: A small to medium-sized tree growing to 20 m tall, with smooth, flaky bark; leaves are deciduous, oval to elliptic, 8–15 cm long and 3–7 cm broad, with an acute apex; flowers are produced in erect panicles 20–40 cm long, each flower with six white to purple petals 2–3.5 cm long; flowers in this plant blooms only once in a year at the peak of summer.

Distribution: Native to tropical southern Asia.

Part Used: Fruits, flowers.

Phytochemistry: Corosolic acid, lager-stroemin, flosin B, and reginin A.

Uses: Mature fruit is used in traditional medicine for reducing glucose in the blood.

Malvaceae

Adansonia digitata L.

Vernacular Name: Dead-rat tree, Monkey-bread tree, Upside-down tree (English).

Description: Tree can grow to 5–25 m in height; trunk tends to be bottle-shaped and can reach a diameter of 10–14 m; trunk is smooth and shiny and can range from being reddish-brown to grey; bark can feel cork-like; branches are thick and wide and very stout compared to the trunk; tree bears very large, heavy, white flowers, 12 cm across and open during the late afternoon to stay open for one night; flowers have 5 petals that are leathery and hairy on the inside; sepals are cup-shaped and 5-cleft; stamens are divided into multiple anthers and styles are 7-10 rayed; indehiscent fruit are large, egg-shaped capsules, filled with pulp that dries, hardens, and falls to pieces which look like chunks of powdery, dry bread; seed is hard, black and kidney-shaped.

Distribution: Native to the African continent; distributed to Asia.

Parts Used: Fruits, seeds, stem.

Phytochemistry: Fruit pulp has very high vitamin C, calcium, phosphorus, carbohydrates, fibres, potassium, proteins and lipids content; various part of the plant contains flavonoids, phytosterols, amino acids, fatty acids, vitamins and minerals.

Uses: Fruit pulp, seeds, leaves, flowers, roots, and bark are edible; fruit pulp used in seasoning as an appetizer and also used to make juices; the plant showed antimicrobial, anti-malarial, diarrhoea, anaemia, asthma, antiviral, anti-oxidant and anti-inflammatory activities The review summarizes the information on various aspects of traditional information, taxonomic description, medicinal properties and important nutritional value.

Gossypium arboreum L.

Vernacular Name: Cotton tree (English); Kapas (Hindi).

Description: A shrub attaining heights of 1-2 m; branches are covered with pubescence and are purple in colour; stipules are present at the leaf base and they are linear to lanceolate in shape and sometimes falcate; leaves are attached to the stem by a 1.5 to 10 cm petiole; blades are ovate to orbicular in shape and have five to seven lobes, making them superficially resemble a maple leaf; lobes are linear to lanceolate, and often a tooth is present in the sinus; glands are present along the midrib or occasionally on the adjacent nerves; leaves are glabrescent, meaning the pubescence is lost with age, but when it is present on young leaves, it is both stellate and simple; flowers are set on short pedicels; an epicalyx is present, which is a series of subtending bracts that resemble sepals; large, ovate segments are dentate, cordate at the base and acute at the apex; true calyx is small, measuring only about 5 mm in length; shape is cupular, and five subtle dentations are present; corolla is a pale yellow on colour, sometimes with a purple centre, and occasionally entirely purple, measures 3 to 4 cm in length; staminal tube bears the anthers and is 1.5 to 2 cm in length; fruit is a three- or four-celled capsule measuring 1.5 to 2.5 cm across, ovoid or oblong in shape and glabrous; surface is pitted and a beak is present at the terminal end; seeds within are globular and are covered in long white cotton.

Distribution: Native to India, Pakistan and other tropical and subtropical regions of the World.

Parts Used: Floss, seeds, roots.

Phytochemistry: Seed contains fatty acids; leaves contain tannins.

Uses: Juice of the root is used in the treatment of fevers.

***Grewia asiatica* L.**

Vernacular Name: Falsa Fruit (English); Phalsa (Hindi).

Description: A shrub or small tree growing to 8 m tall; leaves are broadly rounded, 5–18 cm long and broad, with a petiole 1–1.5 cm long; flowers are venkanna in cymes of several together, the individual flowers about 2 cm diameter, yellow, with five large (12 mm) sepals and five smaller (4–5 mm) petals; fruit is an edible drupe 5–12 mm diameter, purple to black when ripe.

Distribution: Native to southern Asia from Pakistan, India east to Cambodia, and widely cultivated in other tropical countries.

Part Used: Fruits, roots, leaves, stem bark.

Phytochemistry: Fruits contain sugars, carbohydrates, protein and vitamin A, B and C.

Uses: Fruit pulp used as an astringent, stomachic and cooling agent; root is used for rheumatism; stem bark is said to be used in refining sugar, for making ropes and its infusion is used as a demulcent; leaves are used as an application to pustular eruptions; infusion of the bark is used to treat diarrhoea, pain, rheumatism, and arthritis; bark relieves urinary troubles and relieves vaginal burning; leaves relieve all types of inflammations of the skin, including cuts, burning, boils, eczema, and pustular skin eruptions.

***Thespesia populnea* (L.) Sol. ex Corrêa. (Syn. *Hibiscus populneus* L.)**

Vernacular Name: Gardha-bhanda (Sanskrit); Thespesia (English); Parsipu, Porush, Paras-pipal, Gagjdanda (Hindi).

Description: Small tree to 12 m high; leaves, alternate petiolate, dark green, glossy, blades cordate, about 8-16 cm long on equally long petioles; flowers showy, yellow with a maroon to purple centre, borne single in the leaf axils; fruit a brown flattened-globose capsule enclosing a sticky yellow sap and about 10 hairy seed.

Distribution: Distributed throughout Asian countries.

Parts Used: Leaves, bark and stems.

Phytochemistry: Thespesin, gossypol, DL-gossypol, populinin, populetin, populneol, herbacetin, populnetin, glycosides of quercetin, gossypetin, epoxyleic acid, isoquercitrin, rutin, kaempferol 3-flucoside, lupenone, mansonones, myricyl alchohol, lipids, β-sitosterol, thespesone, thespone.

Uses: Decoction of leaves is used in treating cough, influenza, headache and relapses in illnesses; infusion of bark is used in treating dysentery, diabetes, gonorrhoea, yellow urine and thrush; bark used in indigestion, pelvic infection, appetite loss, ulcers and worms; stem is employed in treating breast cancer; decoction of bark and fruit is mixed with oil and use to treat scabies; bark is used to treat thrush; leaf and bark induction is used to treat eye injuries; inner bark is used to treat constipation and typhoid; crushed fruit is used in treatment of urinary tract problem and abdominal swelling.

Abelmoschus esculentus (L.) Moench. (Syn. Hibiscus esculentus L.)

Vernacular Name: Bandakaya (Sanskrit); Lady's finger, Okra (English); Bhindi (Hindi).

Description: A annual erect herb, 0.9-2.1 m in height, covered with hair throughout; leaves simple, alternate, palmately 3-5 lobed, coarsely toothed, scabrous; flowers large, yellow with a purple centre, stamina tube antheriferous throughout; fruit long, 6-8 ribbed, capsular; seeds many, rounded, striate.

Distribution: Cultivated throughout Asian countries.

Parts Used: Fruits, seeds and root.

Phytochemistry: Pods contain carotene, folic acid, thiamine, iboflavin, niacin, vitamin, oxalic, acid and the amino acids; fruit contains glucose, glucosamine and amino acids; seeds contain protein lipids, gossypol, thiamine, riboflavin and α-tocopherol; seed oil yields myristic, palmitic, palmitoleic, oleic, linolenic acids and tocopherols.

Uses: Fruits are nutritious, demulcent, emollient, diuretic; a mucilaginous preparation used as a plasma replacement; decoction of immature capsules as aodyne, demulcent, diuretic, anticatarrhal and useful in dysentery, ardor urinae, dysuria and gonorrhoea; infusion of roasted seeds is sudorific.

Meliaceae

Azadirachta indica A.Juss.

Vernacular Name: Nimbah, Prabhadrah (Sanskrit); Neem tree, Margoes tree, India lilac (English); Nim (Hindi).

Description: A medium to large-sized tree, 15-20 m in height with a clear bole of 7.0 m having grayish to dark grey tubercled bark; leaves compund imparipinnate, leaflets, subopposite, serrate, very oblique at base; flowers cream or yellowish-white in axillary panicles, stamina tubes conspicuous, cylindric, widening above, 9-10 lobed at the apex; fruits one-seeded drupes with woody endocarp greenish-yellow when ripe, seeds ellipsoid, cotyledons thick, fleshy and oily.

Distribution: Native to India and the Indian subcontinent including Nepal, Pakistan, Bangladesh, Mauritius and Sri Lanka; typically grown in tropical and semi-tropical regions.

Part Used: Bark, leaves, flowers, seeds.

Phytochemistry: Stem bark contains tannins; leaves contain nimbin, nimbinene, 6-desacetylnimbinene, nimbandiol, nimbolide, quercetin, β-sitosterol, n-hexacosanol and nonacosane; thiamine, niacin, carotene and amino acids (glutamic acid, tyrosine, aspartic acid alanine, proline, and glutamine); fruits contain gedunin, 7-deacetoxy-7α-hydroxy gedunin, azadiradioe, azadirone, 17β-hydroxy-azadiradione, 17-epiazadiradione and nimbiol.

Used: Bark is useful in malarial fever, wounds, ulcer, burning sensation, tumour, tubercular glands, anorexia, vomiting dyspepsia, intestinal worms, hepatopathy, cough, bronchitis, urorrhea, diabetes, inflammation, amenorrhoea, lumbago, haemorrhoids, otalgia, syphilis and fatigue; leaves are useful in leprosy, skin diseases, leucoderma, pruritus, ophthalmopathy, intestinal worms, dyspepsia, ulcers, tuberculosis, boils, eczema and malarial and intermittent fevers.

Melia azedarach L.

Vernacular Name: Mahanimba (Sanskrit); Margosa tree, China berry, Barbados lialc, Pride of India, Persian lialc, Beds tree, Hoop tree (English); Drek, Bakain (Hindi).

Description: A tree 10 cm high or more; bark pale-brown, lenticellate, longitudinally furrowed; leaves alternate, odd-pinnate; leaflets opposite, serrate, glabrous on both sides; inflorescence in axillary or terminal biparous cyme; flowers white-lilac; drupe ovate or ellipsoid-globose with 4 seeds, yellow when ripe.

Distribution: Native to Indomalaya and Australasia.

Parts Used: Stem, root bark.

Phytochemistry: Stem bark and root bark contain alkaloid (margosine), sterols and tannins; leaves contain paraisine and rutin; seeds contain fatty acids (stearic, palmitic, oleic and linoleic acids).

Uses: Root bark is used in treating ascariasis, oxyuriasis; vaginal injection of the decoction is effective for trichomonas infections.

Moraceae

Artocarpus heterophyllus Lam.

Vernacular Name: Panasah (Sanskrit); Jackfruit tree (English); Kathal (Hindi).

Description: A large monoecious evergreen tree, 18-25 m in height; bark black mottled with green, rough with warty excrescences; leaves simple, alternate, coriaceous, entire, dark, shiny, green above, penninerved, lateral nerves 7-8 pairs, stipules sheathing, leaving a scar on falling; male flowers crowded on cylindrical receptacles; female flowers crowded on blogose receptacles-both cauliflorous; multiple large, fleshy, globose or oblong covered with tubercles; seeds oval with a membranous testa, cotyledons unequal, tubercles; seeds oval, with a membranous testa, cotyledons unequal.

Distribution: Throughout Indian subcontinent.

Part Used: Fruits, seeds.

Phytochemistry: Fruit contains carbohydrates, essential amino acids, β-carotene cycloartenone, cycloartenol, β-sitosterol, butyrospermol, leucine, isoleucine, tyrosins and valine; seeds contain aurantiamide acetate; heartwood contains morin, dihydromrin, cyanomaclurin, artocarpin, isoartocarpin, cycloartocarpin, artocarpesin, cycloarto-carpesin, oxydihydroar-tocarpesin, artocarpetin, norartocarpetin, cycloartinone, artocarpanone, sucrose and ellagic acid; bark contains betulinic acid, cycloheterophyllin, isocycloheterophllin, cycloartenyl acetate, cycloartenol and cycloartenone; leaves contain cycloartenone, cycloartenol, β-sitosterol and tannins; root contains β-sitosterol, ursolic acid, cycloartenone, and artoflavanone.

Uses: Root has antidiarrhocal property; leaves are useful in fever, boils, wounds, skin diseases; unripe fruits are useful in dyspepsia and debility; ripe fruits are useful in ulcers; seeds are sweet, diuretic, aphrodisiac and constipating; wood is a nervine sedative, and useful in vonvulsions; latex is useful in dysopia, ophthalmitis and pharyngitis.

Moringaceae

Moringa oleifera Lam.

Vernacular Name: Sigruh, Sobhanjanah (Sanskrit), Hores-radish tree, Drumstick tree (English); Sahijan, Mungana (Hindi).

Description: An unarmed middle-sized graceful tree with corky grey bark and easily breakable branches; leaves usually tripinnate, rachis slender, thickened and articulated at the base, leaflets elliptic or obovate, rounded at the apex, nerves obscures; flowers white in large puberulous axillary panicles; fruits pods, upto 45 cm long, pendulous, greenish, triangular, 9-ribbed, seeds trigonous, the angles winged.

Distribution: Throughout Indian subcontinent.

Part Used: Roots, bark, leaves, seeds.

Phytochemistry: Carotene, nicotinic acid, ascorbic acid, prolamin; essential amino acids (arginine, histidine, lysine, tryptophan, phenylalanine, methionine, threonine, leucine, isoleucine, and valine).

Uses: Roots are useful in dyspepsia, anorexia, verminosis, diarrhoea, colic flatulence, otalgia, paralysis inflammations, amenorrhoea dysmenorrhoea, fever, strangury, vesical and renal calculi, ascites, ophthalmopathy, cough, asthma bronchitis, pectoral diseases, splenomegaly, epilepsy, hysteria, cardiopathy, abscess and pharyngodynia; bark is useful in ascites, and ringworm; leaves useful in scurvy, wounds, tumours, inflammation and helminthiasis; seeds are useful in neuralgia, inflammations, intermittent fevers and ophthalmopathy.

Myrtaceae

Corymbia citriodora (Hook.) K.D.Hill & L.A.S.Johnson (Syn. Eucalyptus citriodora Hook.)

Vernacular Name: Lemon-scented gum, Blue-spotted gum, Lemon eucalyptus (English).

Description: A tall tree, growing to 35 m in height; smooth, pale, uniform or slightly mottled bark, white to coppery in summer, and a conspicuously narrow-leaved crown which smells strongly of lemons; pear-shaped buds are borne in clusters of three, formed in the corner of leaf and stem junctions, whilst fruit (capsules) are urn-shaped; bark is smooth for the entire height of the tree, often powdery, shedding in thin curling flakes; prefers lighter, slightly acidic loamy soils and occurs in dry sclerophyll forest and woodlands in hilly country; lignotuber.

Distribution: Naturalised in Western Australia; distributed to Asian and African countries.

Parts Used: Leaves, stem.

Phytochemistry: Essential oil (citronellal, limonene and linalool).

Uses: Oil is used in perfumery; a refined form of this oil is used in insect repellents, especially against mosquitoes; medicinally used for arthritis, bronchitis, catarrh, cold sores, colds, coughing, fever, flu, poor circulation, and sinusitis.

Eucalyptus globulus Labill. (Syn. E. gigantea Dehnh.)

Vernacular Name: Nilaniryasa, Tailaparnah (Sankrit); Blue gum, Eucalypt (English); Yukeliptas (Hindi).

Description: A lofty tree about 90 m in height with a clean straight bole and smooth bark, peeling off in long strips; leaves opposite on juvenile shoots, alternate in adult shoots, lanceolate, 20-25 cm long, broad, rather thick and curved; flowers large, white, 1-3 together in axils, operculum thick and warty; fruit a hardened capsule, dehiscing longitudinally at the mouth, seeds very small.

Distribution: Distributed in various Asian and African countries; America and New Zealand.

Parts Used: Oil.

Phytochemistry: Eucalyptin, 8-desmethyleucalyptin, sideroxylin, 8-desmethylsideroxylin, quercetin, quercitrin, quercetol and its 3-deglucoside, chrysin, rutin hyperoside, phloroglucin derivatives, n-tritriacontan-16-18-dione, 4-hydroxytritriaconate-16-18-dion, 16-hydroxy-18-tritriacintanone, 11,12-dehydroursolic lactone acetat, caffeic, ferulic, gallic, gentisic, protocatechuic, maslinic and oleanolic acids.

Uses: Oil is acrid, bitter, astringent, thermogenic, oleaginous, antiseptic, deodorant, stimulant, anodyne, carminative, digestive, anthelmintic, cardiotonic, diuretic, sudorific, expectorant, insect repellent, rubefacient and antipyretic; useful in halitosis, cephalalgia, tuberculosis, chronic cough, asthma, pneumonitis, bronchitis, pyorrhea, burns, dyspepsia, flatulence, threadworm infestation, cardiac debility, pharyngodynia, hoarseness, strangury, skin disease and chronic and intermittent fever.

Syzygium aromaticum (L.) Merr. & L.M.Perry

Vernacular Name: Clove tree, Cloves (English); Lavamg, Laung (Hindi).

Description: A pyramidal or conical evergreen tree usually up to 12 m in height with a single main stem bearing obliquely oriented branches; leaves simple, lanceolate, gland-dotted, fragrant; flower buds greenish to pink, clustered at the ends of the branches, highly aromatic; fruits fleshy, dark pink drupes; seeds oblong, grooved on one side.

Distribution: Cultivated throughout the Indian subcontinent.

Parts Used: Flower buds (cloves), oil.

Phytochemistry: Protein, carbohydrates, tannin, oleanolic acid, eugenol acetate, caryophyllene and eugenol.

Uses: Clove are useful in halitosis, odontalgia, ophthalmopathy, flatulence, colic, gastropathy, anorexia, cough, asthma, skin diseases, helminthiasis, agalactia, impurity of breast milk, strangury, fever, cephalalgia, neuralgia, lumbago, nostalgia, dental, caries, hyperacidity, vomiting, dipsia, hepatopathy, general debility and tuberculosis;

oil is useful in catarrh, cough, bronchitis, gastrohelcosis, flatulence, colic, skin diseases, dyspepsia, vomiting, odontalgia, dental caries and cephalalgia.

Nyctaginaceae

Mirabilis jalapa L.

Vernacular Name: Marvel of Peru, Four o'clock flower (English); Dophariya, Gulbas (Hindi).

Description: Plant erect and spreading, 2-3 ft tall and just as wide; numerous branches and opposite, pointed leaves 2-4 in long; fragrant flowers are borne singly or in clusters, and can be red, magenta, pink, yellow or white, sometimes with more than one color on the same plant; plants continue to produce new flowers from late spring untill fall; large, black carrot-shaped tubers that can be a foot or more long; flowers are trumpet-shaped, about an inch across at the end and about two inches long, open in the evening and wilt the next morning, leafy, shrublike, multi-branched perennials which bloom throughout summer.

Distribution: Distributed in Asian and African countries.

Part Used: Whole plant.

Phytochemistry: Flowers contain betaxanthins (indicaxanthin, vulgaxanthin-I, miraxanthin-I, -II, -III, -IV, V and -VI); roots contain rotenoids (mirabijalone A, B, C and D, 9-O-methyl-4-hydroxyboeravinone B, boeravinone C and F, and 1,2,3,4-tetrahydro-1-methylisoquinoline-7,8-diol); seed contains fatty acid (8-hydroxyoctadeca-cis-11,14-dienoic acid); aerial parts contain β-sitosterol, stigmasterol, ursolic acid, oleanolic acid and brassicasterol.

Uses: Plant used as a diuretic, purgative, and for vulnerary (wound healing) purposes; root is aphrodisiac as well as diuretic and purgative; used in the treatment of dropsy; leaves are used to reduce inflammation; decoction of leaves used to treat abscesses; leaf juice used to treat wounds; powdered seed used as cosmetic and a dye.

Nymphaeaceae

Nymphaea nouchali Burm.f. (Syn. N. stellata Willd.)

Vernacular Name: Kumudam (Sanskrit); Indian water-lily (English); Kanval, Kokka (Hindi).

Description: A large perennial aquatic herb with short, erect, roundish, tuberous rhizomes; leaves floating, peltate, sharply sinuate-toothed, flowers large, floating, solitary, variable in colour from pure white to deep red; fruits spongy may seeded berries, seeds minute, greyish-black when dry with longitudinal striations.

Distribution: Native to southern and eastern parts of Asia, Sri Lanka, Bangladesh and Mauritius.

Part Used: Rhizomes, flowers, seeds.

Phytochemistry: Nymphaeine and Nymphalin.

Uses: Rhizome useful in diarrhoea, dysentery, dipsia, and general debility; flowers are astringent and cardiotonic; seeds are useful in dipsia, diarrhoea and dermatopathy.

Oleaceae

Jasminum angustifolium (L.) Willd.

Vernacular Name: Asphoa, Vanamallika (Sanskrit), Wild jasmine (English); Banmallika (Hindi).

Description: A wiry small scandent shrub with glabrous stem, pubescent branches; leaves simple, small, ovate, acute rounded at base, glabrous; flowers on long slender pedicles, more often in three at the ends of short lateral twigs, white corolla tube about 1.6 cm, lobes 7 or 8, equaling the tube; fruits, didymous berries, ripe carpels broadly ovoid, seed one in each carpel.

Distribution: Native to India, Mauritius, Sri Lanka and the Andaman Islands.

Parts Used: Roots, leaves.

Phytochemistry: Kaempferol-3-O-α-L-rhamnopyranosyl (1-3)-[α-L-rhamnopyranosyl (1-6)]-β-D-galactopyranoside, kaempferol-3-O-rutinoside, 7-ketologanin, oleoside-11-methyl ester, 7-glucosyl-11-methyl oleoside, ligstroside, oleuropein.

Uses: Roots are useful for external application in ringworm and herpes, and are recommended for ophthalmopathy, ulcerative stomatitis, leprosy, pruritus and wounds; leaves are used as an emetic in cases of poisoning.

Papaveraceae

Argemone mexicana L.

Vernacular Name: Svamaksiri, Brahmadanti (Sanskrit); Mexican poppy, Yellow Mexican poppy, Prickle poppy (English); Satyanasi, Bhatbhamt, Dhattura (Hindi).

Description: A strong, branched, prickly annual, 60-90 cm in height with yellow latex; leaves simple, sessile, spiny, semi-amplexcaul, sinuate-pinnatified, variegated with white, spinous, veins white; flowers large, bright yellow, terminal of short leafy branches; fruits prickle capsules, oblong-ovoid, opening by 4-6 valves; seeds numerous.

Distribution: Native to Mexico; widely naturalized in many parts of the world.

Part Used: Whole plant.

Phytochemistry: Allocryptopine, berberine, cheilanthifoline, chelerythrine, coptisine, cryptopine, dihydrosanguinarine, norchelerythrine, norsanguinarine, oxyhydrastinine, protopine, sanguinarine, β-scoiulerine methrohydroxide, α-and β-stylopine methohydroxide, 6-acetonyl-dihydrosanguinarine, α- tetrahydropalmatine methrohydroxide, maxicanol and mexicanic acid; flowers contain isorhamntin, isorhamnetin-3-glucoside and isorhamnetin-3, 7- diglucoside.

Uses: Roots are useful in guinea-worm infestation, skin disease, leprosy, pruritus, blennorrhagia, inflammations, all types of poisoning, constipation, flatulence, colic, malarial fever and vesicular calculus; leaves are useful in cough, wounds, ulcers and skin diseases; seeds are useful in cough, asthma, pertussis, skin diseases, leprosy, ulcer, wounds odontalgia, dental caries, constipation, rheumatialgia, colic and flatulence; latex is useful in dropsy, jaundice, skin disease, leprosy, blisters, indolent ulcers, conjunctivitis, inflammations, burning sensation and malarial fever; oil is useful in indolent ulcers, wounds, leprosy, skin diseases, constipation, flatulence, colic and rheumatalgia.

Fumaria officinalis subsp. wirtgenii (Koch) Arcang.

Vernacular Name: Fumitory, Drug fumitory, Earth smoke (English).

Description: An herbaceous annual plant, which grows weakly erect and scrambling, with stalks about 10 to 50 cm long; pink 7 to 9 mm flowers appear from April to October in the northern hemisphere; two lipped and spurred, with sepals running a quarter the length of the petals.

Distribution: Native to Europe and North Africa; grows in Asia, North America and Australia.

Parts Used: Flowers.

Phytochemistry: Flowers contain isoquinoline alkaloids (protopine and allocryptopine).

Uses: Plant has a stimulant action on the liver and gallbladder and is chiefly used to treat skin conditions such as eczema; also diuretic and mildly laxative.

Pinaceae

Pinus halepensis Mill.

Vernacular Name: Aleppo pine (English).

Description: A small to medium-sized tree, 15–25 m tall, with a trunk diameter up to 60 cm (24 in), exceptionally up to 1 m; bark is orange-red, thick, and deeply fissured at the base of the trunk, and thin and flaky in the upper crown; leaves are very slender, 6–12 cm long, distinctly yellowish-green, and produced in pairs (rarely a few in threes); cones are narrow conic, 5–12 cm long and 2–3 cm broad at the base when closed, green at first, ripening glossy red-brown when 24 months old, open slowly over the next few years, a process quickened if they are exposed to heat such as in forest fires; cones open 5–8 cm wide to allow the seeds to disperse; seeds are 5–6 mm long, with a 20-mm wing, and are wind-dispersed.

Distribution: Native to the Mediterranean region; distributed to Morocco, Algeria and Spain north to southern France, Italy, Croatia, Montenegro, Albania, and east to Greece, Malta and northern Tunisia, Lebanon, southern Turkey, Jordan, Israel, and Palestinian territories.

Part Used: Leaves, seeds.

Phytochemistry: Resin, terpenes and fatty acids.

Uses: Oil used in nasal discharge, stroke, lethargy, depression, pleurisy, haemorrhages, external wounds, respiratory tract diseases, tooth problems and wounds; diabetes prevention and treatment and also as an aphrodisiac agent.

Poaceae

Cymbopogon citratus (DC.) Stapf

Vernacular Name: Bhustrina (Sanskrit); Lemon grass, Ginger grass, Citronella grass, Sere grass (English); Gandhatrina (Hindi).

Description: Perennial, densely tufted, aromatic grass; stems underground, short, whitish or pale-violet. Leaves in dense clusters, linear amplexicaul, rough-margined, glaucous green on both sides; inflorescence in many-branched panicles without stalks; all part of the grass are lemon-flavoured.

Distribution: Cultivated throughout the world.

Parts Used: Whole Plant.

Phytochemistry: Plant contains essential oil (critral, limonene, isopulegol, citronellic acid, geranium acid α-camphorene).

Uses: Plant possesses antiseptic, antifebrile and stomachic properties; used in treating coryza, influenza and pyrexia; also used against dyspepsia and vomiting and as a carminative; externally to treat eczema.

Pteridaceae

Adiantum capillus-veneris L.

Vernacular Name: Maidenhair Fern (English); Hansraj, Mubaraka, Pursha (Hindi).

Description: Fern with delicate founds growing to 30 cm in height; fronds arising in clusters from creeping rhizomes 20-70 cm tall, with very delicate, light green fronds much subdivided into pinnae 5-10 mm long and broad; the frond rachis is black and wiry..

Distribution: Native to Europe and North America; distributed to Asia and Africa.

Parts Used: Aerial parts.

Phytochemistry: Plant contains flavonoids, terpenoids, tannins and mucilage; volatile oil, capillerin, tanning material, mucin, gallic acid, sugars and heterosides of kaempferol, quercetol and luteolol.

Uses: It is used in cough, bronchitis, catarrh, sore throat and chronic nasal catarrh.

Rhizophoraceae

Bruguiera gymnorhiza (L.) Lam.

Vernacular Name: Black mangrove (English).

Description: Tree to 20 m tall with a buttressed trunk and pneumato-phores; leaves opposite, petiolate, ovate and glossy with leathery texture; flowers borne singly, moderate-sized, with conspicuous basal cupule (calyx) that from a persistent crown-like structure surrounding the petals and ovary; fruit an elongate brown, cigar-like drupe which germinates while still in the tree; flowers and fruit available throughout the year.

Distribution: Africa and Asia.

Part Used: Bark, root.

Phytochemistry: Bark contains D-glucose, arabinose, tannins, bruguierol and isobruguierol; leaves contain β-sitosterol, campesterol, stigmasterol, and 28-isofuccosterol; α-amyrin, β-amyrin, lupeol, oleanolic acid, ursolic acid, taraxerol, gymnorhizol; ellagic acid and derivatives.

Uses: Bark used in syphilis and cancer; the root is used to restore lost appetite and used to treat diabetes.

Rubiaceae

Ixora coccinea L.

Vernacular Name: Paranti, Raktaka, Bandhuka (Sanskrit); Jungle-flame ixora, Sacred ixora (English); Rangan, Rukmini, Rajana (Hindi).

Description: A large glabrous shrub with few branches; leaves simple, opposite decussate, elliptic, ovate or obovate sessile or nearly so, main nerves 8-12 pairs, coriaceous, olive green when dry; flowers scarlet, in dense sessile or short peduncled corymbiform cymes; fruits globose, fleshy, smooth, crowned with the persistent calyx, purple when ripe, seeds concave on the ventral surface.

Distribution: Throughout Indian subcontinent.

Parts Used: Roots, leaves, flowers.

Phytochemistry: Roots contain aromatic oil, tannin, fatty acids; root bark contains octadecadienoic acid, mannitol and myristic acid.

Uses: Roots are useful in hiccough, fever, gonorrhea, anorexia, diarrhoea, dysentery, cephalalgia, sores, chronic ulcers and skin diseases; leaves are useful in diarrhoea; flowers are useful in dysentery, dysmenorrhoea, leucorrhoea, haemoptysis, catarrhal bronchitis, ophthalmopathy, sores and ulcer.

Salicaceae

Flacourtia indica (Burm. f.) Merr.

Vernacular Name: Aghori (Sanskrit); Madagascar plum (English); Kancu, Bilangra, Kataiya (Hindi).

Description: A small deciduous thorny shrub; leaves simple, elliptic, acute or acuminate, coarsely crenate or serrate, glabrous; flowers greenish-yellow, in short simple or branched tomentose racemes; fruits globose, dark purple drupes with juicy pulp and hard endocarp, seeds 8-16.

Distribution: Throughout Indian subcontinent.

Parts Used: Roots, leaves fruits.

Phytochemistry: Arabinose, mannose, glucose, phenolic glucoside ester, flacourtin, steroid, ramnoside, β-sitosterol and its β-D-gluco-pyranoside.

Uses: Roots are useful in poisonous bites, skin diseases; leaves are useful in pruritus and scabies; fruits are useful in strangury, jaundice, gastropathy and splenomegaly.

Sapindaceae

Cardiospermum halicacabum L.

Vernacular Name: Sakralata, Indravalli (Sanskrit); Ballon vine, Heart pea (English); Kanphuti, Kapalphoti (Hindi).

Description: A pubescent or nearly glabrous annual or perennial with slender branches climbing by means of tendrillar hook; leaves ternately bicompound, leaflets acuminate at the apex; flowers white, small; fruits membranous, depressed, pyriform capsule winged at the angles; seeds black with a large white-shaped aril.

Distribution: Throughout India and nearby countries.

Part Used: Roots, leaves, seeds.

Phytochemistry: Alkaloids, 1-triacountanol, n-pentacosane, n-triacontane, n-triacontanol, n-non- acosane, pelargonidin-3-galactoside, glucocappasalin, β-sitosterol and phthalic acid.

Uses: Roots are useful in strangury, fever, arthritis, amenorrhoea, lumbago and neuropathy; leaves are good for arthritis, otalgia and ophthalmodynia; seeds are good for arthritis and fever; the plant has sedative action on the central nervous system.

Litchi chinensis Sonn.

Vernacular Name: Lychee, Litchi (Hindi/ English).

Description: An evergreen tree reaching 10–28 m tall, the lychee bears fleshy fruits that are up to 5 cm long and 4 cm wide, approximately 20 g; outside of the fruit is covered by a pink-red, roughly textured rind that is inedible but easily removed to expose a layer of sweet, translucent white flesh.

Distribution: A tropical and subtropical fruit tree native to China; cultivated in many parts of the world.

Part Used: Fruits.

Phytochemistry: Polyphenols, flavan-3-ol monomers and dimers; cyanidin-3-glucoside and oligonol

Uses: Fruits edible; seeds known to have antidiabetic properties.

Sapindus mukorossi Gaertn.

Vernacular Name: Soapnut tree (English); Ritta (Hindi); Phenilah, Aristakah (Sanskrit).

Description: A medium-sized deciduous tree up to 20 m in height with grey smooth bark, peeling off in scales; leaves pinnate, leaflets 2-3 pairs, terminal pair being the largest; flowers white, polygamous, male flowers many, bisexual flowers few, all in the sane pubescent panicle; fruits fleshy drupes, the pulp becoming a saponaceous wrinkled rind on drying; seeds black.

Distribution: Throughout Indian subcontinent.

Part Used: Roots, bark, fruits.

Phytochemistry: Saponins (sapindoside A and sapindoside B); nuts contain kaempferol, quercetin and β-sitosterol; fruits contain saponin emarginatoside.

Uses: Roots and bark are expectorant and demulcent; roots are good for hemicranias, hysteria and epilepsy; decoction of the bark is good for cattle suffering from ulcers due to worm infestation after calving; fruits are good for asthma, diarrhoea, cholera, lumbago, verminosis and gastralgia due to dyspepsia.

Sapotaceae

Mimusops elengi L.

Vernacular Name: Bakulah (Sanskrit); Bullet-wood tree (English); Bakul, Maulsiri (Hindi).

Description: A large evergreen tree with dark grey fissured bark and dense spreading crown; leaves oblong, glabrous leathery with way margins; flowers white fragrant, axillary, solitary or fascicled; fruits ovoid or ellipsoid berries, seed 1-2 per fruit, ovoid, compressed greyish brown, shiny.

Distribution: Throughout India subcontinent.

Parts Used: Bark, flowers, fruits seeds.

Phytochemistry: Saponin, rhamnose, arabinose, glucose, tannin, palmitic, linoleic and oleic acids.

Uses: Bark, flowers and fruits are acrid, astringent, cooling and anthelmintic; bark is used as a gargle for odontopathy, ulitis and ulemorrhagia; stems are used as toothbrushes; useful in urethrorrhea, cystorrhoea, diarrhoea and dysentery; flowers are sued for preparing a lotion for wounds and ulcers; power of dried flowers is a brain tonic and is useful as a snuff to relieve cephalalgia; unripe fruit is used as a masticatory and will help to fix losses teeth; seeds are used for preparing suppositories in cases of constipation especially in children.

Solanaceae

Capsicum annuum L. (Capsicum frutescens L.)

Vernacular Name: Chilli (English); Lamirch (Hindi).

Description: A herbaceous, short-lived perennial, upto 1m in height; leaves oblong, glabrous, flowers solitary, rarely in pairs, white to bluish-white, berries green, maturing into yellow, thin, almost circular, having long placental connections.

Distribution: Cultivated throughout Asia and Africa.

Part Used: Fruits.

Phytochemistry: Fruits contain oleoresin (capasaicin); bell-pepper oil (2-methoxy-3-isobutylpyrazine, trans-β-ocimene, limonene, methyl salicylate, linalool, and hex-cis-3-enol); one-1-en-4-one, non-trans, trans-2, 5-din-4-one, 2-entyl-furan, benzaladehyde, 2-isobutyll-3-methoxy pyraxine, nona-trans, cis-2, 6-dienaland, deca-trans, trans-2, 4-dienal.

Uses: Green chillies are used fresh for garnishing or flavouring savoury dishes; used in cough, cold and low blood pressure.

Lycopersicon esculentum Mill.

Vernacular Name: Raktamaci (Sanskrit); Tomato (English); Tamatar (Hindi).

Description: An unarmed herbaceous spreading pubescent annual; leaves unevenly pinnate, curled, pubescent throughout, lobes acute or acuminate, veins prominent below; flowers yellow in extra axillary cymes; fruits globose, villous many on fleshy palcentum, flat, kidney-shaped.

Distribution: Cultivated in throughout the world.

Parts Used: Fruits.

Phytochemistry: Carotenoids, lycopene, carotene, xanthophylls and xanthophylls ester.

Uses: Fruits are useful in dyspepsia, anorexia, flatulence, constipation, asthma, bronchitis, dermatopathy, ulcerative stomatitis, hepatopathy, nephropathy and general debility.

Solanum americanum Mill. (Syn. S. nigrum L.)

Vernacular Name: Black night-shade (English); Makoy, Gurkkamai (Hindi); Kakmaci (Sanskrit).

Description: An erect, divaricately branched, unarmed, suffrutescent annual herb; leaves ovate or oblong, sinuate-toothed or lobed, glabrous; flowers 3-8 in extra-axillary drooping subumebellate cymes; fruits publish black or reddish berries; seeds many, discoid yellow, minutely pitted.

Distribution: Throughout Indian subcontinent.

Part Used: Whole plant.

Phytochemistry: Leaf contains riboflavin, nicotinic acid, vitamin C, β-carotene and citric acids; fruits contain glucose and fructose, vitamin C and β-carotene; immature green fruit of the plant contain steroidal glycol-alkaloids (solamargine, α and β-solanigrine).

Uses: Plant is useful in rheumatalgia, swellings, cough, asthma, bronchitis, wounds, ulcers, flatulence, dyspepsia, strangury, hepatomegaly, otalgia, hiccough, nasal catarrh, ophthalmopathy, vomiting, cardiopathy, leprosy, skin diseases, fever, splenomegaly, haemor-rhoids, plant depresses the central nervous system and reflexes of the spinal cord and has influence on cardiac activity and in the regulation of blood pressure; leaves are used as poultice for rheumatic and gouty joints and skin diseases; decoction of the berries and flowers is useful in cough, erysipelas, rat bite, bronchitis, pulmonary tuberculosis, fever, diarrhoea, ophthalmopathy and hydrophobia; seeds are useful in giddiness, dipsia, inflammations and skin diseases; root bark is useful in otopathy, ophthalmopathy, rhinopathy and hepatitis.

Urticaceae

Urtica dioica L.

Vernacular Name: Nettle (English); Bichhu booti (Hindi).

Description: A perennial growing to 1.5 m, with lance-shaped leaves and green flowers with yellow stamens in axillary cymes.

Distribution: Grows in temperate regions throughout the northern hemisphere, southern Africa, Asia, and Australia.

Parts Used: Root.

Phytochemistry: Flavonoids, amines, glucoquinone, minerals, sterols, and other phenolic compounds.

Uses: Diuretic, tonic, astringent, prevents haemorrhaging, anti-allergenic, increases breast-milk, production, reduces prostate enlargement.

Verbenaceae

Lantana aculeata L.

Vernacular Name: Caturangi, Vanacchedi (Sanskrit); Wild sage, Lantana weed (English); Caturag, Ghaneri (Hindi).

Description: A large scrambling evergreen, strong-smelling shrub with stout recurved prickles; leaves opposite, often rugose, scabrid on both side; flowers small, normally orange but often white to dark red, in heads which are prominently capitates; bracts conspicuous, persistent; fruits fleshy drupes, 5 mm in diameter, endocarp hard green when young and blue or black on ripening.

Distribution: A native of tropical America; naturalized in India, Pakistan, China and Mauritius.

Parts Used: Whole plant.

Phytochemistry: Catalase, amylase, invertase, lipase, tannase and glucosidase; tannins, sugars, glucoside; caryophyllene, l-α-phellandrene; aldehydes, alcohols, lantadene A, lantadene B, lancamarone and lantanine.

Uses: Plant is useful in tetanus, malaria, epilepsy and gastropathy; decoction of fresh roots is good to gargle for odontalgia and this is used by hill tribes for all type of dysentery; powdered leaves are used for cuts, wounds, ulcers and swellings; infusion of the leaves is good for bilious fever, eczema and eruption; fruits are useful in fistula, pustules, tumours and rheumatism.

Vitaceae

Cissus quadrangularis L.

Vernacular Name: Edible-stemmed vine, Adamant creeper, Bone-setter (English); Hadjod (Hindi); Asthisrnkhala, Vajravalli (Sanskrit).

Description: A tendril climber with stout fleshy jointed quadrangular stems, tendrils simple, long, slender and leaf-opposed, in addition to the normal roots, some aerial roots arising from the jointed nodes grow downward and strike the soil; leaves simple, broadly reniform, entire or toothed, rounded, truncate or cuneate at the base; flowers small, greenish in shortly peduncled cymes, petals 4, hooded at the apex; fruits ovoid or globose, red berries, seeds ellipsoid.

Distribution: Throughout the hotter parts of the Indian subcontinent.

Part Used: Whole plant.

Phytochemistry: Stem contains triterpenoids, onocer-7-ene-3α-, 21β-diol and onocer-7-ene-3β, 21α-diol; steroid; β-sitosterol, δ-amyrin and δ-amyrone.

Uses: Plant is useful in helminthiasis, anorexia, dyspepsia, colic, flatulence, skin diseases, leprosy, haemorrhages, haemoptysis, ophthalmopathy, otorrhoea, chronic ulcers, tumours, haemorrhoids, epilepsy, convulsions, spanomenorrhea, fractures, swellings; shorts are useful in colonopathy, scurvy, otorrhoea, asthma, burns and wounds, powdered roots as well as the stem paste are very specific for bone fractures.

Zingiberaceae

Curcuma pseudomontana J.Graham

Vernacular Name: Hill Turmeric (English); Kachura (Hindi).

Description: An erect herb, reaching 75 cm tall, with stout rootstock bearing small almond-like sub-globose tubers at the ends of the fibrous roots; tubers are fleshy and with inside, aromatic; leaves 3-5, oblong-lanceolate, 20-30 x 6-9 cm, base acute, apex acuminate, margin entire, hairless, shiny; leaf stalk and the sheath up to 20 cm long; flowering spikes seen in the center of the previously formed tuft of leaves, 10-25 cm long, bearing numerous compactly arranged flowers; flowering bracts conspicuous, inverted egg-shaped to lanceolate, 3-5 x 1.5-2cm, apex rounded to acute, hairless, green with a pink tip; non-flowering bracts (coma) oblong-lanceolate, conspicuous, purple below and pinkish-purple above; flowers bisexual, 2-4 in each fertile bract, bring yellow, about 3 cm long and 4 cm broad; capsules globose, splitting by 3-valves, smooth; seeds ovoid or oblong, usually covered with arils.

Distribution: Distributed in Indian subcontinent.

Parts Used: Rhizomes.

Phytochemistry: Quercetin, kaempferol, quercitrin; dihydroflavons, flavans, anthocyanidins, proanthocyanidins, chalcones, catechin and leucoanthocyanidins.

Uses: Tuber used to cure jaundice, body swellings; boiled tubers to increase lactation; tuber paste on the head for cooling effect.

Chapter 5

Common Medicinal Plants of Myanmar and Mauritius

Introduction

Apart from the plants given in previous chapters, there are many plants found in these areas which are common in both the countries. Moreover, their traditional importance in most of the areas of Myanmar and Mauritius is also the same. Hence, this chapter has been given separately to avoid repetition of the plants and their description.

Acanthaceae

Justicia adhatoda L. (Syn. Adhatoda vasica Nees)

Vernacular Name: Muyakri (Burmese); Vasa (Sanskrit); Adusa (Hindi); Malabarnut, Vasaka (English)

Description: Small evergreen and subherbacious brush, leaves are 10-16 cm in length, minutely pubescent and broadly lanceolate; inflorescence dense, short pedunculate, bractate and spike terminal; corolla large and white with lower lip streaked purple and 4-seeded small capsule.

Distribution: Mauritius, India, Sri Lanka, Burma and Malaysia up to 1000 m high from sea level.

Part Used: Whole plant

Phytochemistry: Leaves contain essential oil, vasicine, N-oxides of vasicine, vasicinone, deoxyvasicine, maiontone. Roots contain vasicinolons, vasicol, peganine and 2'-

glucosyl-oxychalcone. Flowers contain β-sitosterol-D-glucoside, kaempferol, kaempferol glycosides and quercetin.

Uses: Leaf extract used in bronchitis and asthma; relieves cough and breathlessness; Ayurveda prescription for bleeding due to idiopathic thrombocytopenic purpura, local bleeding due to peptic ulcer, piles, menorrhagia etc.; fresh leaf juice used in tuberculosis; leaf extract used for pyorrhea and bleeding gums.

Amaryllidaceae

Allium cepa L.

Vernacular Name: Kyet-Thun-U-Gyi (Burmese); Onion (English); Pyaj (Hindi).

Description: A biennial or perennial herb with aromatic fleshy underground bulb; leaves linear, hollow, cylindric and fleshy; flowers many, white in globular umbels, spathe composed of 2-3 reflexed walls; underground bulbs vary in size, colour, shape, firmness, the strength of flavor and keeping quality.

Distribution: Cultivated throughout the world.

Part Used: Bulbs.

Phytochemistry: Essential amino acids (arginine, histidine, lysine, tryptophan, phenylalanine, methionine, threonine, leucine, isoleucine); essential oil (protocatechuic acid, p-hydroxybenzoic acid, vanillic acid, caffeic acid, and o-and p-coumaric acids); citric, malic, abietic and oxalic acids; oligosaccharides.

Uses: Bulbs are antibacterial, aphrodisiac, emmenagogue, emollient expectorant, carminative, stomach ache, diuretic, anodyne and tonic; also useful in haemorrhoids, dysentery, flatulence, dyspepsia, colic, jaundice, splenopathy, hepatopathy, pneumonopathy, asthma, bronchitis, ophthalmia, vomiting, otalgia, pharyngodynia, malaria fever, lumbago, epilepsy, tumours, wounds paralysis, arthralgia, leucoderma and skin diseases.

Allium sativum L.

Vernacular Name: Kyet-Thunb-byu (Burmese); Garlic (English); Lahsan (Hindi).

Description: An annual glabrous, bulbous herb with a pungent odour; bulb short, consists of several smaller cloves and is surrounded by a thin, white or pinkish sheath; leaves flat and narrow, attenuate-acute at the apex; nerves parallel; flowers white or pink with bulbils in globose head covered with a large bract.

Distribution: Cultivated all over the world.

Part Used: Bulbs.

Phytochemistry: Essential oil (allicin, diallyl disulfide, allyl propyl disulfide and other sulfur compounds).

Uses: Bulbused as antibacterial, anti-inflammatory, amoebiasis, oxyuriasis and colitis; cures cough, bronchitis and pertussis; an aqueous solution of bulb active against coryza and influenza; also useful in hypercholesterolaemia and atherosclerosis; a poultice of pounded bulb used to treat boils, abscesses, phlegmons and centipede hits.

Anacardiaceae

Anacardium occidentale L.

Vernacular Name: Shitkaie, Thihothayet (Burmese); Vrkkabijah (Sanskrit); Cashewnut tree (English); Kaju (Hindi).

Description: A small tree with short, thick, crooked trunk; leaves obovate or ovate-oblong, hard glabrous, obtuse-retuse or rounded at tip, cuneate at base, 10-20 cm long; flowers small, yellow, with pink stripes borne in 15-25 cm long, terminal panicles, with both staminate and hermaphrodite flowers; fruit, a kidney-shaped nut, 2.5 cm long, borne and on 5.0-7.5 cm long, pyriform, fleshy receptacle, the cashew apple, having a thin bright yellow to scarlet skin and soft and juicy flesh. The nut is greyish-green with a hard, smooth and shining oleaginous pericarp enclosing a curved white kernel covered with a thin reddish-brown testa.

Distribution: Cultivated and wild especially on the sea coast.

Parts Used: Roots bark, leaves, fruits.

Phytochemistry: Amino acids (arginine, histidine, lysine, phenylalanine, methionine, threonine, tryptophan, leucine, isoleucine, valine, alanine, aspartic acid, cystine, glutamic acid, glycine, serine and tyrosine); catechin, epicatechin, ethylgallate, leucocyanidin, proanthocyanidin, leucodelphinidin and epiafazelchin.

Uses: Roots considered as purgative; combination of bark and inflorescence used in snakebit; bark and leaves are useful in odontalgia and ulitis; gum from the bark is recommended in leprosy, ringworm, corns and obstinate ulcers; fruits are anthelmintic, useful in skin diseases, dysentery, haemorrhoids and anorexia; used for preventing hair loss.

Mangifera indica L.

Vernacular Name: Bur-thyet (Burmese); Amrah (Sanskrit); Mango tree (English); Aam (Hindi).

Description: A large spreading evergreen tree upto 45 m in height with a heavy doom shaped crown, straight stout bole and thick rough dark grey bark; leaves simple, crowded at the ends of branches, linear-oblong or elliptic-lanceolate, acute, acuminate or sub-obtuse; flowers small, pungently odorous, reddish-white or yellowish-green in large many-flowered pubescent panicles, longer than the leaves; fruits large, fleshy drupes, green, orange, yellow or red in colour, seeds solitary, encased in a hard compressed fibrous endocarp.

Distribution: It is found in the wild in India, Bangladesh, Mauritius and Pakistan where it is indigenous and cultivated varieties have been introduced to other warm regions of the world; grow upto 1200 m.

Parts Used: All parts.

Phytochemistry: Ripened fruits containβ-carotene, xanthophylls; various part of the plant also contain 2-octene, α-and β-pinene, α-phelandrene, limonene, dipentene, nerol, geraniol, neryl acetate, citronellal, sesquiterpene alcohol named mangiferol and sesquiterpene ketones.

Uses: Roots and bark used as antisyphilitic, vulnerary, antiemetic, anti-inflammatory and constipating; useful in metrorrhagia, colonorrhagia, pneumorrhagia, leucorrhoea, syphilis, wounds, ulcer, vomiting, uteritis, diarrhoea, dysentery, diphtheria and rheumatism; effective for cough, hyperdipsia, burning sensation, and stomatopathy.

Apiaceae

Apium graveolens L.

Vernacular Name: Tayok-Nan-nan (Burmese); Ajmoda (Sanskrit); Celery (English); Ajmod (Hindi).

Description: An annual or biennial herb; stem upto 2-4 m. high, erect, branching; radical leaves pinnate with deeply lobed segments, cauline 3-partite; segments once or twice trifid, coarsely toothed at the apex; peduncle 6 mm or less; umbel rays 5-10, pedicels 6-16; flowers white, very small; fruit a cremocarp, each with 2 mericarps, 1-1.5 mm. long with narrow ridges, broad vittae.

Distribution: Cultivated in Asia, Europe and America.

Part Used: Fruits.

Phytochemistry: Choline, myristic acid, myristicis acid, myristoleic acid, stearic acid, apiin, isoquercitrin, apigenin, apigravin, apiumetin, apiumoside, bergapten, celereoside, rutaretin; α-eudesmol, β-eudesmil, santatol, 3-n-butyl phthalide, sedanenolide, d-limonene, selinene, oleoresin.

Uses: Used as carminative, stomachic, anodyne, nerve tonic, emmenagogue, stomach trouble; complaints of depression and sleeplessness; obstinate retention of urine, throbbing headache; rheumatic pain in muscles of neck and sacrum, dysmenorrhoea with short pains in both ovarian regions better by flexing legs, left ovarian pain better by lying on left side; sharp stitching pain in left iliac region going over to right side.

Centella asiatica (L.) Urb.

Vernacular Name: Indian pennywort (English); Brahmamanduki Bemgsag (Hindi); Mandukaparni (Sanskrit).

Description: Small trailing herb; stems slender, prostrate, rooting at the joints; leaves alternate or tufted at each node, orbicular, round or kidney-shaped, obviously crenate; inflorescence in single umble, bearing 1-5 small flowers, white or reddish, without stalks; fruit very small, compressed.

Distribution: Indigenous to the Indian subcontinent, Southeast Asia, and wetland regions of the Southeastern US.

Part Used: Whole plant.

Phytochemistry: Plant contains asiaticoside, madecassoside, brahmoside, bicycloelemene, centelloside, indocentelloside, oxyasiatiocoside, thankuniside, asiatic acid, betulinic acid, centellic acid, madecassic acid, centellose, kaempferol, hydrocotyline, phellandrene, vitamin C and linamarase.

Uses: Plant is used to treating varicose veins, chronic venous insufficiency, psoriasis, minor wounds, strangury, and to increase lactation.

Cuminum cyminum L.

Vernacular Name: ZiyaZiyarbyu (Burmese); Jiraka, Jira (Sanskrit); Cumin (English); Safed zeera (Hindi).

Description: An annual herb grows to 30-50 cm tall; slender, glabrous, branched stem that is 20-30 cm tall and has a diameter of 3-5 cm; each branch has two to three sub-branches; all the branches attain the same height; stem is coloured grey or dark green; leaves are 5-10 cm long, pinnate or bipinnate, with thread-like leaflets; flowers are small, white or pink, and borne in umbels; each umbel has five to seven umbellts; fruit is a lateral fusiform or ovoid achene 4-5 mm long, containing two mericarps with a single seed; seeds have eight ridges with oil canals, oblong in shape, longitudinally ridged, and yellow-brown in colour.

Distribution: Cultivated mainly in India and little in other Asian countries.

Part Used: Fruits.

Phytochemistry: Cumin oil (sminaldehyde, 1, 3-p-menthadien-7-al, 1, 4-p-menthadien-7-al).

Uses: Oil is used in curries and other culinary preparations for flavour. Confectionery, beverages, liqueurs and cordials; helps in digestion; in Ayurveda, dried cumin seeds are believed to have medicinal importance in mild digestive disorders, diarrhoea, dyspepsia, flatulence, morning sickness, colic, dyspeptic headache and bloating.

Daucus carota L.

Vernacular Name: Monla-u-wa (Burmese); Garjara, Narangavarnaka (Sanskrit); Carrot (English); Gajar (Hindi).

Description: Annual or biennial herb, grows between 30 and 60 cm tall, and is roughly hairy, with a stiff, solid stem; leaves are tripinnate, finely divided and lacy, and overall triangular in shape; flowers are small and dull white, clustered in flat, dense umbels; maybe pink in bud and may have a reddish flower in the centre of the umbel; lower bracts are three-forked or pinnate; fruits are oval and flattened, with short styles and hooked spines; dried umbels detach from the plant, becoming tumbleweeds.

Distribution: Cultivated in all over the world, mainly in Asian and African countries.

Parts Used: Seeds and roots.

Phytochemistry: Carrot seeds oil contains carotol, daucol, bisabolene, daucene oxide carota-1, 4-oxide; root oil contains terbinolene, β-and γ-bisabolene, γ-terpinene, caryophyllene, sabinene, limonene, 2-nonenal, octanal, 2-decenal, 2,4-decadinenal, 2-methoxy-3-sec-butylpyrazine and 2-methoxy-3-iso-propylpyrazin; oil from aerial parts contain trans-asarone and β-bisabolene.

Uses: Carrot seeds oil used in spice blends, sauces, cured meats, seasonings, etc. and also for flavouring alcoholic beverages; roots are highly nutritious.

Apocynaceae

Alstonia scholaris (L.) R. Br.

Vernacular Name: Kyese-Kalwalettok (Burmese); Saptaparna (Sanskrit); Devil tree, Shaitan wood (English); Chattiyan, Saittan ka jat (Hindi).

Description: A large evergreen tree upto 3.0 m tall, bark grayish brown, rough, lenticellate abounding in bitter, milky latex; leaves 4-7 in a whori, coriaceous, elliptic-oblong, pale beneath; flowers small, greenish white, numerous in umbellate panicles, corolla tube short, very strongly scented; fruits follicles, 30-60 cm long; seeds papillose with brownish hair at each end.

Distribution: Throughout India and Mauritius in deciduous and evergreen forests.

Part Used: Bark, leaves, milky exudates.

Phytochemistry: Bark contains alkaloids (ditaine, echitenine, echitamine and echitamidine) and triterpenes (α-amyrin and lupeol).

Uses: Bark used for digestive problems; good remedy as laxative, anthelmintic, febrifuge, antipyretic, depurative, galactagogue, stomachic, cardiotonic and tonic; useful in malarial fever, abdominal disorders, diarrhea, dysentery, dyspepsia, leprosy, skin diseases, pruritus, tumours, chronic, and foul ulcers, asthma, bronchitis, cardiopathy, helminthiasis, agalactia and debility; leaves and milk exudate are good for ulcers with foul discharge.

Cascabela thevetia (L.) Lippold (Syn. Thevetia peruviana (Pers.) K.Schum.)

Vernacular Name: Bur-set-hna-ya-thi (Burmese); Yellow oleander, Lucky nut (English).

Description: Evergreen tropical shrub or small tree; leaves are willow-like, linear-lanceolate, and glossy green in color, covered in waxy coating to reduce water loss (typical of oleanders); stem is green turning silver/grey as it ages; flowers bloom from summer to fall, long funnel-shaped sometimes-fragrant yellow (less commonly apricot, sometimes white) flowers are in few-flowered terminal clusters; fruit is deep red-black in colour encasing a large seed that bears some resemblance to a Chinese lucky nut.

Distribution: Native to central and southern Mexico and Central America, and cultivated widely as an ornamental.

Parts Used: Stems, flowers and leaves.

Phytochemistry: Cardenolides (thevetin A and thevetin B); peruvoside, neriifolin, thevetoxin and ruvoside.

Uses: The plant's toxins are used in biological pest control; seed oil used to make paint with antifungal, antibacterial and anti-termite properties.

Plumeria rubra L. (Syn. P. acuminate Ait; P. acutifolia Poir.)

Vernacular Name: Bur-toyoksaga (Burmese); Temple tree, pagoa tree, frangipani plant, (English); Golainchi, (Hindi); Kshirachamp, (Sanskrit).

Description: Medium-sized tree of 5-6m high; branches fleshy, brittle; leaves alternate, usually crowded in thick terminal cluster, enter and attenuate at both ends; inflorescence in terminal cyme; flowers white with a yellow centre of reddish-pink, fragrant; follicles divaricate, with numerous winged seeds; all parts of the plant yield a milky latex.

Distribution: Originally native to Mexico, Central America, Colombia and Venezuela, it has been widely cultivated in subtropical and tropical climates worldwide.

Part Used: Bark, latex and flowers.

Phytochemistry: The stem-bark contains plumierides and fulvoplumierin.

Uses: Flowers are used in treating cough, constipation, acute enteritis, dysentery and haemophilia; decoction of the stem bark is used as hydropsia and purgative; flowers and bark used for the treatment of fever, bacillary dysentery, and pertussis.

Wrightia antidysenterica (L.) R.Br. (Syn. Holarrhena antidysenterica (L.) Wall.)

Vernacular Name: Bur-lettok-gyi (Burmese); Kutaja (Sanskrit); Snowflake, Milky Way, White angel (English); Kurchi (Hindi).

Description: A perennial, small and compact semi-deciduous shrub, reaching 1.2-2 meters in height, with a spread of about 1.5 meter; unfolding from buds to flowers; moderate grower with several short and divaricate branches that turn chocolaty brown as it ages, and adorned with dark green, ovate and acuminate leaves (2.5-6cm long) that are oppositely arranged; pure white tubular 5-petaled flowers with yellow centres appear in corymb-like cymes at the end of branches; showy 2.5-3.5cm star-shaped flowers all year round.

Distribution: Native to Sri Lanka, distributed to India, Mauritius and nearby places.

Part Used: Root and stem bark

Phytochemistry: Stem bark contains alkaloids (conessine, conessimine, kurchine, conamine, conimine, conessidine, conarrhimine, holarrhimine, holarrhine and kurchicine); steroidal alkaloids (regholarrhenine A, B, C, O, E and F).

Uses: Bark has anti-microbial and anti-inflammatory properties; juice from its bark is used for the relief of mouth sores and its leaves are used for the treatment of several skin disorders, psoriasis and another non-specific dermatitis; useful in treating gut motility disorders; juice of the plant is a potent ingredient for a mixture of wall plaster.

Arecaceae

Areca catechu L.

Vernacular Name: Kunkunllee (Burmese); Gubak, Poogiphalam, Tantusara (Sanskrit); Betal-nut palm, Areca-nut palm, Betel palm, Areca palm, Penang palm, catechu palm (English); Supari (Hindi).

Description: Slender palm about 10 m tall; stem erect, surmounted by a crown of pinnate leaves; petiole broadly expanded at the base; inflorescence in spadix encased in a spathe; flowers yellowish-white in much-branched raceme, which bears both male and female flowers; fruit ovoid; pericarp hard and fibrous; kernel (seed) brown.

Distribution: Grown and cultivated in India, Mauritius.

Part Used: Pericarps and kernels.

Phytochemistry: Kernels contain tannin, catechin, lipids (laurin, olein and myristin); glucosides; alkaloids (arecoline, arecaidine, arecaine, guvacine and guvacoline).

Uses: Pericarp is effective in the treatment of flatulence, oedema, dysuria and hyperemesis of pregnancy; the kernel is used to treat diarrhoea and dysentery; cure malaria.

Phoenix sylvestris (L.) Roxb.

Vernacular Name: Bur-sun-ba-lun (Burmese); Silver date palm, Sugar date palm, Wild date palm (English); Khajur, Khajuri (Hindi).

Description: Trees range from 4 to 15 m in height and 40 cm in diameter; leaves are 3 m long, gently recurved, on 1 m petioles with acanthophylls near the base; leaf crown grows to 10 m wide and 7.5 to 10 m tall containing up to 100 leaves; inflorescence grows to 1 metre with white, unisexual flowers forming to a large, pendent infructescence; single-seeded fruit ripens to a purple-red colour.

Distribution: Growing in plains and scrubland to 1300 m; native to southern Pakistan, most of India, Sri Lanka, Nepal, Bhutan, Burma, and Bangladesh; naturalized in Mauritius, the Chagos Archipelago, Puerto Rico and the Leeward Islands.

Part Used: Fruits, leaves.

Phytochemistry: Fruits contain ascorbic acid, free amino acids, mainly alanine; sugars, tannins, mucilage, vitamins A, B and D; salts and iron in an assimilable form; seeds contain a fatty oil containing palmitic, oleic and linoleic acids; juice of the tree contains reducing sugars, aspartic acid, nicotinamide, nicotinic acid, isoniazid and thiamine.

Uses: Fruit is cooling, oleaginous, cardiotonic, fattening, constipative, good in heart complaints, abdominal complaints, fevers, vomiting and loss of consciousness; juice obtained from the tree is considered to be a cooling beverage; roots are used to stop toothache; fruit pounded and mixed with almonds, quince seeds, pistachio nuts and sugar, form a restorative remedy; central tender part of the plant is used in gonorrhoea.

Bignoniaceae

Oroxylum indicum (L.) Kurz

Vernacular Name: Bur-waphyu-wa-new (Burmese); Syonakah, Tuntukah (Sanskrit); Indian trumpet tree (English); Sonapatha, Syona, Arul, Pharkhat (Hindi).

Description: Small to medium-sized deciduous tree upto 12 m in height with soft light brown bark with corky lenticels; leave very large, 90-180 cm long, 2-3 pinnate with 5 or more pairs of primary pinnate, rachis very soft, cylindric, swollen at the junction of the branches, leaflets 2-4 pairs ovate or elliptic, acuminate, glabrous; flowers lurid to reddish-purple outside, dull or pale pinkish-yellow within, numerous in large erect racemes; fruits flat capsules, upto 1m long tapering to both ends.

Distribution: Throughout India in deciduous forests, Mauritius in moist areas.

Part Used: Roots, leaves fruits, seeds.

Phytochemistry: Stem and root barks contain oroxylin-A, baicalein, chrysin; bark contains alkaloid, charya; seeds contain baicalein and tetuin.

Uses: Roots are sweet, astringent, bitter, acrid, refrigerant, and anti-inflammatory, anodyne, aphrodisiac, expectorant, appetizing, carminative, digestive anthelmintic, constipating, diaphoretic, diuretic, antiarthritic, febrifuge, and tonic; useful in inflammation, dropsy, sprains, neuralgia, hiccough, cough, asthma, bronchitis, anorexia, dyspepsia, flatulence, colic, helminthiasis, diarrhoea, dysentery, strangury, vomiting, leucoderma, wounds rheumatoid arthritis and fever; leaves are stomachic and anodyne, and are useful in stomachalgia, flatulence, cephalalgia, ulcers, splenomegaly and are also useful in cough, bronchitis, dyspepsia, flatulence, colic and leucoderma; mature fruits are acrid, sweet, anthelmintic and stomachic, useful in pharyngodynia, cardiac disorder, helminthiasis, gastropathy, bronchitis and haemorrhoids; seeds are purgative.

Tecoma stans (L.) Juss.ex Kunth

Vernacular Name: Bur-pha-yaung-ban (Burmese); Yellow trumpetbush, Yellow bells, Yellow elder, Ginger-thomas (English).

Description: A large shrub or small, much-branched, tree usually growing 1.5 to 5 m tall; younger stems are smooth, hairless (glabrous) and greenish in colour, slightly four-angled in cross-section (quadrangular) and turn pale brown or reddish-brown in colour as they age; bark on the main stem is light brown to pale grey in colour, furrowed, and relatively rough in texture; leaves (10-25 cm long) are arranged in pairs (oppositely arranged) and are borne on slender stalks (petioles) 1-9 cm long, compound (pinnate) with 3-13 leaflets, but usually have 3-7 leaflets; leaflets are narrowly egg-shaped in outline with broad end at base (ovate-lanceolate) or elongated in shape (25-100 mm long and 8-30 mm wide) and have a pointed tip (acuminate apex), margins are irregularly and sharply toothed (serrated) and they are mostly hairless (glabrous), though a few hairs may be present on their undersides near the midrib; showy flowers are borne in several-flowered clusters (5-15 cm long) that are first produced at the ends of the branches (in terminal clusters), and then later in the leaf forks (axils) near the tips of the branches (in axillary clusters), bright yellow in colour, tubular in shape, and borne on short stalks (pedicels) that are somewhat curved or twisted; flower tube (corolla tube) is 30-50 mm long and has five rounded lobes that are 8-30 mm long; several faint reddish lines in the throat of the flower, which is slightly ridged and hairy; flowering may occur throughout the year; fruit are large, elongated (linear), and somewhat flattened (compressed) capsules (10-30 cm long and 5-20 mm wide), turn from green to brown in colour as they mature and finally split open to release numerous papery seeds; are very flat, oblong in shape (7-8 mm long and about 4 mm wide), and have a transparent wing at each end (the size of entire seed including the wings is about 20×6 mm).

Distribution: Native to the Americas; distributed to India and Mauritius; East Africa.

Part Used: Leaves, roots.

Phytochemistry: Alkaloids (tecomanine, tecostanine), lapachol, sugars (glucose, fructose, sucrose and xylose), triterpenoids (ursolic and oleanolic acids and α-amyrine), β-sitosterol and phenolics (chlorogenic, caffeic, vanillic, o-cumaric and sinapicnacids).

Uses: Leaf infusion can be taken orally for diabetes and stomach pains; a strong leaf and root decoction is taken orally as a diuretic, to treat syphilis or for intestinal worms; the plant is desirable fodder when it grows in fields grazed by livestock.

Boraginaceae

Heliotropium indicum L.

Vernacular Name: Bur-sin-let-maunggalay (Burmese); Vrscikali, Hathi-sunda (Sanskrit); Indian turnsole (English); Siriyari, Hattasura, Hatta-juri (Hindi).

Description: A coarse, somewhat succulent foetid herbaceous annual, 15-60 cm height with densely hirsute ascending branches; leaves simple, alternate or subopposite, ovate, obtuse, narrowed or cordate at base, often unequal upto 10cm long and 7.5-10 cm broad, hispid-pubescent, veins and veinlets conspicuous on the lower side; flowers pale violet, many, 2-ranked, in extra axillary, bristly scorpioid cymes; fruits nutlets combined in pairs, separating later, beaked, each lobe 4-ribbed.

Distribution: Various Asian countries including India and Mauritius.

Parts Used: Whole plant.

Phytochemistry: Tannin and alkaloids.

Uses: Plant is diuretic; useful for local application for ulcers, sores, wound, skin affections, stings of insects, ulitis and rheumatism; root useful in cough, bronchitis, fever, ringworm, rheumatism and ophthalmopathy; leaves are useful in fever, urticaria, ulcer, wounds, localized inflammation, gonorrhoea, ringworm, rheumatism, erysipelas and pharyngodynai; flowers are emmenagogue in small doses and abortifacient in large does; seeds are stomachic.

Brassicaceae

Lepidium sativum L.

Vernacular Name: Bur-samon-ni (Burmese); Asalika, Asalic (Sanskrit); Garden cress, Water crass (English); Chandrasur, Halim, Halo (Hindi).

Description: An erect herbaceous glabrous annual upto 45 cm in height; leaves entire or variously lobed or pinnatisect, the upper sessile; flowers white, small, in long racemes; fruits obvate, small pods, notched at the apex with two seeds per pod.

Distribution: Cultivated throughout India, Pakistan, Bangladesh, and Mauritius as a culinary vegetable.

Parts Used: Roots, leaves, seeds.

Phytochemistry: Leaves contain protein, fat, carbohydrate, mineral matter, calcium, and phosphorus, trace elements; iron nickel cobalt and iodine; vitamins; vitamin A, thiamine, riboflavin, niacin and ascorbic acid; seeds contain glucotropaeloin, sinapin, sinapic acid, mucilaginous matter and uric acid; seed oil contains palmitic, stearic, arachidic, behenic, lignoceric, oleic, and linolenic acids; usaponifiable mater contain β-sitosterol and α-tocopherol.

Uses: Roots useful in secondary syphilis and tenesmus; leaves are stimulant, diuretic and antibacterial and useful in scorbutic diseases and hepatopathy; seeds bitter, thermogenic, depurative, rubefacient, galactagogue, emmenagogue, tonic, aphrodisiac, ophthalmic and diuretic, used as poultices for sprains and in leprosy, skin diseases, dysentery, diarrhoea, splenomegaly, dyspepsia, lumbago, ophthalmopathy, leucorrhoea, scurvy, seminal weakness, asthma, cough, hiccough, haemorrhoids.

Raphanus raphanistrum sub sp. sativus (L.) Domin. (Syn. R. sativus L.)

Vernacular Name: Bur-mon-la-u (Burmese); Moolaka (English); Muli (Hindi).

Description: Annual herb; small root annual, growing rapidly; stem 0.2-0.6 cm high, flexuose, fistular, branched in the upper part, glabrous or hispid; lower leaves on hairy petioles 3cm long; lyrate-pinnatifid 3-4 rarely 5 jugate, terminal, terminal lobe, suborbicular or shortly ovate, rounded at the apex, with the margin crenate, and more or less confluent ith lateral lobes; leaves roughly pilose, bright green; flowers scented-sepals 6.5-10 mm long oblong, something brown red; petals 1.5-2.2 cm long, blade obovate subemarginate at the apex white or lilac with yellow or purple veins; roots white cylindrical and fleshy with tiny rootlets from the surface.

Distribution: Cultivated all over the world in temperate and warm countries.

Part Used: Leaves roots, seeds.

Phytochemistry: Seeds contain Raphanin, glycosinolates, enzymes, trace elements, acids, aldehydes, anthocyanin, pectin and arabinogalactan proteins; leaves contain vitamin C.

Uses: Used in indigestion, oedema, skin infections; Ayurvedic medicine as a diuretic, carminative and a digestive; flowers are cholagogue; juice of fresh leaves is also used as a diuretic, a laxative and in flatulence.

Calophyllaceae

Mesua ferrea L.

Vernacular Name: Gangaw, thanocobrasaffron (Burmese); Nagapuspah, Nagakesarah (Sanskrit); Mesua, Iron-wood tree (English).

Description: A medium-sized to large glabrous, evergreen tree, 18-30 m in height and reddish-brown bark which peels off in thin flakes; leaves simple, opposite, thick lanceolate, coriaceous, covered with waxy bloom underneath, red when young, acute or acuminate, nerves.

Distribution: Throughout India, Pakistan, Bangladesh, Mauritius and other nearby areas.

Parts Used: Flowers.

Phytochemistry: Palmitic, stearic, oleic, linoleic acids, palmitostearo- olein, dipalmito-olein distearo-olein, stearo, diolein, palmito-diolein, linoleodiolein, triolein.

Uses: Flowers are astringent, bitter, acrid, mildly heating, anodyne, sudorific, digestive, carminative, constipating, anthelmintic, diuretic, alexipharmic, expectorant, stomachic, haemostatic, aphrodisiac, febrifuge and cardiotonic; useful in asthma cough, hiccough, halitosis, leprosy, scabies, dermatopathy, pruritus, pharyngodynia, vomiting, dysentery, haemorrhoids, ulcer burning sensation of the feet; dipsia, impotency, leucorrhoea, haemoptysis, strangury, cephalalgia, fever and cardiac debility; seed oil is used in skin diseases.

Cannabaceae

Cannabis sativa L.

Vernacular Name: Say-gyauk, Bhang (Burmese); Bhanga, Vijaya (Sanskrit); True hemp, Soft hemp, Indian hemp (English); Bhang (Hindi).

Description: A large aromatic resinous annual herb 1.2-4.8 m in height with erect angular stem; female plant usually taller than the male; leaves palmately divided, lower 3-8 foliate with long petioles, upper 3-1 foliate palling into bracts; male flowers in short drooping panicles, females in short axillary crowded spikes; fruits achenes seeds black, flattened.

Distribution: Throughout India, Pakistan, Bangladesh, and Mauritius.

Parts Used: Leaves, flowering shoots.

Phytochemistry: It contains tetrahydrocannabinol, cannabidiol, cannadinol, cannabichromene, cannabicitran, cannabicyclol, cannabigerol, cannabielsion, stereo isomers of cannabitriol, cannabinodiol, cannabicumaronone, cannabiglendol and cannabitetrol.

Uses: Leaves are bitter, astringent, tonic, aphrodisiac, antidiarrhoeic, intoxicating, stomachic, analgesic and abortifacient; used in convulsions, otalgia, abdominal disorders, diarrhoea, stomatalgia and haematorrhoea; excessive used causes dyspepsia, cough, melancholy, dropsy, hyperpyrexia and insanity; bark is tonic, and is useful in inflammations, haemorrhoids and hydrocele; seeds are carminative, astringent, aphrodisiac, antiemetic and anti-inflammatory.

Caricaceae

Carica papaya L.

Vernacular Name: Thinbaw (Burmese); Brahmairandah, Erandakarkati (Sanskrit); Papaw tree, Papaya (English); Pappita (Hindi).

Description: A small, soft- wooded, fast-growing, short-lived laticiferous tree upto 8.0 m in height with a straight cylindric stem bearing characteristic leaf-scare throughout and with a tuft of leaves at the top; leaves deeply lobed, palm-like with characteristically long, hollow petiole; flowers unisexual, white or yellowish-white, rarely bisexual, males in long drooping panicles; females in short clusters; fruits one-chambered, succulent, indehiscent, spherical or cylindrical; seeds many, yellowish-brown, ash-coloured or black.

Distribution: Cultivated throughout Asia, Africa.

Parts Used: Fruits, leaves.

Phytochemistry: Cryptoxantin, β-carotene, cryptoflavin, violaxanthin, phytoene, phytofluene, cis-β-carotene, ζ-carotene, γ-carotene, 5,6-monoepoxy-β-carotene, mutatochorme, aurochrome, cis-violaxanthin, antheraxanthin, chrysanthemaxanthin, and neoxanthin.

Uses: Fruits are bitter, acrid, thermogenic, anodyne aphrodisiac, stomachic, appetizer, digestive, carminative, anthelmintic, styptic, anti-inflammatory, antifungal, demulcent and diuretic; useful in cough, bronchitis, stomachalgia, dyspepsia, anorexia, intestinal worms, haemoptysis, haemorrhoids, inflammations, splenomegaly, ringworm, skin diseases, psoriasis, urinary calculus and injures of the urinary tract; latex is anthelmintic, anodyne, laxative, digestive, emmenagogue, galactagogue, depurative, sudorific, febrifuge and tonic; useful in roundworm infestation, stomachalgia, dyspepsia, constipation, amenorrhoea, dysmenorrhoea, agalactia, skin diseases, leprosy, haemorrhoids, fever, stranger and general debility.

Cleomaceae

Cleome viscosa L.

Vernacular Name: Hingala (Burmese); Wild mustard, Dog mustard, Sticky cleome (English); Hulhul, Hurhur (Hindi); Pasugandha (Sanskrit).

Description: Annual stick herb with strong penetrating odour and clothed with glandular and simple hairs; leaves 3-5 foliate, gradually becoming shorter upwards; flowers yellow in lax raceme; fruits capsules, compressed, hairy throughout; seeds brownish black when ripe.

Distribution: A weed throughout the plains in India, Mauritius.

Part Used: Whole plant.

Phytochemistry: Aerial parts contain a macrocyclic diterpene, 20-oxocembra-3,7,11,15-tetra-en-19-oic and a bicyclic diterpene, cleomeolide; seeds contain coumarino-lignans, cleomiscosin, A, B, C and D.

Uses: Plant is acrid, thermogenic, anthelmintic and sudorific; roots are stimulant, antiscorbutic and vermifuge; leaf juice is digestive and is good for otalgia; seeds are anthelmintic, carminative, constipating, febrifuge and cardiac stimulant and are useful in fever, diarrhoea, worm infestations, cardiac disorders and dyspepsia.

Combretaceae

Combretum indicum (L.) DeFilipps (Syn. Quisqualis indica var. villosa C.B.Clarke)

Vernacular Name: Bur-Dawe-hnaing (Burmese); Chinese honeysuckle, Rangoon creeper (English); Madhumalti (Hindi).

Description: A ligneous vine that can reach from 2.5 to 8 m; leaves are elliptical with an acuminate tip and a rounded base, grow from 7 to 15 cm and their arrangement is opposite; flowers are fragrant and tubular and their colour varies from white to pink to red; 30 to 35 mm long fruit is ellipsoidal and has five prominent wings; fruit tastes like almonds when mature.

Distribution: It found in the Philippines, India and Malaysia. It has since been cultivated and naturalized in tropical areas including Mauritius.

Part Used: Root, leaves, seeds, fruits.

Phytochemistry: Seed contains quisqualic,, choline and shikimic acids; triterpenoids, sterols, alkaloids, glycosides, flavonoids, tannins, phenols.

Uses: Decoctions of the root, seed or fruit used as antihelmintic to expel parasitic worms or for alleviating diarrhoea; fruit decoction used for gargling, combat nephritis; leaves used to relieve pain, fever; roots used to treat rheumatism.

Terminalia arjuna (Roxb. ex DC.) Wight & Arn.

Vernacular Name: Bur-phauk-kyant (Burmese); Arjun (English/ Hindi); Arjunah, Kakubhah (Sanskrit).

Description: Large evergreen tree with buttressed trunk and spreading crown with drooping branches, bark smooth, grey outside and flesh coloured inside, flaking off in large flat pieces; leaves simple, subopposite, oblong or elliptic, coriaceous, crenulate, pale dull green above, pale brown beneath, often unequal sided, nerves 10-15 pairs, reticulate; flowers white in panicles of spikes with linear bracteoles; fruits ovoid or oblong with 5-7 short, hard angles or wings, the lines on the wings oblique and curving upwards.

Distribution: Throughout Asia including India, Mauritius, etc.

Parts Used: Bark.

Phytochemistry: Arjunolic acid, tomentosic acid, β-sitosterol, ellagic acid, saponin and leucodelphinidin; bark contain a crystalline compound arjunine, a lactone, arjunetin, essential oil, tannin pyrocatachol, large quantities of calcium salts and traces of aluminium and magnesium salts, reducing sugars and colouring matter.

Uses: Bark is astringent, sweet, acrid, cooling, aphrodisiac, demulcent, cardiotonic, styptic, anti dysenteric, urinary astringent, expectorant, alexiteric, lithontriptic and tonic; useful in fractures, ulcer, urethrorrhea, leucorrhoea, diabetes, vitiated conditions of pitta, anaemia, cardiopathy, hyperhidrosis, fatigue, asthma, bronchitis, tumours, otalgia, dysentery, inflammations, internal and external haemorrhages, cirrhosis of the liver and hypertension.

Terminalia bellirica (Gaertn.) Roxb.

Vernacular Name: Bur-thit-sein (Burmese); Belliric myrobalan (English); Bahenda, Bahaira (Hindi); Vibhitakah, Aksah (Sanskrit).

Description: A large deciduous buttressed tree, 20-30 m in height thick brownish-grey back having shallow longitudinal fissures; leaves simple, alternate, long-petioled, crowded about the extremities of the branches, broadly elliptic, margin entire, main nerves 6-8 pairs, midrib prominent on the both surfaces; flowers pale greenish-yellow with an offensive odour, in axillary spikes, longer than the petioles but shorter than the leaves; fruits ovoid grey drupes, obscurely 5 angled, narrowed into a very short stalk.

Distribution: Throughout India, Mauritius.

Parts Used: Bark, fruits.

Phytochemistry: Fruits contain tannins, β-sitosterol, gallic acid, ellagic acid, ethyl galate, galloyl glucose and chebulagic acid; heartwood and bark contain ellagic acid; seed-coat contains gallic acid, bellericanin.

Uses: Bark is useful in anaemia and leucoderma; fruits are astringent, acrid, sweet, thermogenic, anti-inflammatory, anodyne, styptic, narcotic, digestive, anthelmintic, aperients, expectorant, ophthalmic, antipyretic, antiemetic and rejuvenating; useful in cough, bronchitis, pharyngitis, insomnia, dropsy, dyspepsia, flatulence, dipsia, vomiting, haemorrhages, ophthalmopathy, strangury, splenomegaly, cephalalgia, skin diseases, leprosy, fevers, ulcer and general debility; mature and dry fruit is useful in diarrhoea and dysentery; oil obtained from the seeds is useful in dyspepsia, skin diseases, leucoderma and greyness of hair; fruit effective in high blood pressure.

Terminalia catappa L.

Vernacular Name: Bur-badan (Burmese); Deshi-badam (Hindi); Almond (English).

Description: Large, spreading tree to 30 m tall; leaves mostly near ends of branches; leaves alternate, short-petiote, the blades obovate to 30 cm long, deciduous and turning orange to red before falling; flowers small, borne in densely packed spikes; fruit a reddish flattened ovoid drupe up to 6 cm long, with fibrous outer layer with a single edible seed within; flowers and fruit available throughout the year.

Distribution: Common along beaches, rocky coats, upper margins of mangrove swamps, low land clearings and secondary forests.

Parts Used: Bark, leaves.

Phytochemistry: Ellagic acid, corilagin, tannins, tannic acids, organic acids; palmitic, oleic, linoleic and myristic acids; essential oils, reducing sugars, flavonoids and amino acids.

Uses: Fluid from the bark is used to treat diabetes and as a tonic; juice of the leaves is ingested for coughs; infusion of the leaves is used to treat jaundice; leaves are used in indigestion; bark is used to treat mouth sores and stomachache; sores, pimples and fungal skin diseases are treated with the bark; leaves are used to treat wounds and burns; leaves are used to treat bronchitis and tuberculosis.

Compositae

Carthamus tinctorius L.

Vernacular Name: Su-ban, Wild-saffron (Burmese); Safflower, American saffron, False saffron (English).

Description: An annual, erect, branching herb 0.6-1.0m high, glabrous; stem whitish, striate; leaves sessile, somewhat clasping, broad-laneolate or lanceolate-oblong, attenuated at both ends, strongly dentate, slightly spinous; ventation pinnate and netted; inflorescence a broad corymb, heads 3-5 born on leafy peduncles; flowers orange-red, fruit a white achene, obovoid, 4-angled, truncate at the top with 4 bosses; no pappus.

Distribution: Cultivated in Cambodia, China, India, Mauritius, Laos, Middle East, Vietnam; United States, Mexico and Argentina.

Part Used: Flowers.

Phytochemistry: Seed oil contains oleic acid and linoleic acid.

Uses: Used in dysmennorhoea; amennorhoea, coronary heart diseases; angina pectoris, injuries; pains due to blood stasis.

Eclipta prostrata (L.) L. (Syn. Eclipta alba (L.) Hassk.)

Vernacular Name: Kycik-hman (Burmese); Eclipta (English).

Description: Erect, annual herb, 10-60 cm tall; leaves opposite, lanceolate, 3-10 cm long, 5-25 mm wide, acuminate, sessile or short-petiolate, serrulate, densey strigoes-pilose on both sides; heads terminal on stem and branches or axillary, 1 cm across, with pedicel 2-4.5 cm long; involucres campanulate, 5 mm long, 6-7 mm across; phyllaries 5-6, green oblong; ray flowers white, 2.5-3 mm long, 0.4 mm wide, fertile; disc flowers bisexual, fertile, white, 4-lobed; anthers very shortly bifid at base; achenes 2.8 mm long, 3-angled in ray flowers, compressed and 4-angled in disc florets.

Distribution: Korea, Japan, India, Mauritius and China.

Parts Used: Whole plant.

Phytochemistry: Eliptine, wedelactones, and thiophene-derivatives.

Uses: Antidote for snake venom, antihepatotoxic, anti-inflammatory; used in eruption, leucorrhoea, enterohaemorrhage.

Eupatorium triplinerve Vahl. (Syn. Eupatorium ayapana Vent.)

Vernacular Name: Bizat (Burmese); Ayaparnah, Ajaparnah (Sanskrit); Ayapana tea (English); Ayapana (Hindi).

Description: A profusely branching prostate undershrub upto 90-120 cm in height with a few ascending branches; leaves simple, opposite, subsessile, lanceolate, 3-nerved, acuminate, subentire, glabrous; flowers slaty blue, all tubular in few headed coryms; fruits achenes, truncate, 5-angled with uniseriate pappus hairs.

Distribution: Native of Brazil; found throughout the Indian subcontinent.

Parts Used: Whole plant.

Phytochemistry: Plant contains daphnetin dimethyl ether, hydrangetin, daphnetin-7-methyl ether, umbelliferone, daphenetin, ayapin and herniarin.

Uses: Plant is bitter, astringent, acrid, thermogenic, stimulant, digestive, carminative, alterant, emetic and purgative in large does, haemostatic, vulnerary, cardiotonic, antiscorbutic, alexeteric, sudorific, antipyretic, detergent, anti-inflammatory, expectorant and tonic; useful in gastropathy, dyspepsia, epistaxis, haematemesis, haemorrhages, haemoptysis, haematuria, menorrhagia, haemorrhoids, wounds, ulcer, ulcerative stomatitis, cardiac debility, skin eruption, scabies, pruritus, poison bites, intermittent fever especially yellow fever, inflammations, and asthma.

Lactuca sativa L.

Vernacular Name: Bur-Salat (Burmese); Lettuce (English).

Description: Plant is 15 to 30 cm tall; leaves are colourful, mainly in the green and red colour spectrums, with some variegated varieties; root system that includes a main taproot and smaller secondary roots; flowers becomes bitter and unsaleable, plants grown for consumption are rarely allowed to grow to maturity; flowers more quickly in hot temperatures, while freezing temperatures cause slower growth and sometimes damage to outer leaves; inflorescences are composed of multiple florets, each with a modified calyx called a pappus, a corolla of five petals fused into a ligule or strap, and the reproductive parts, include fused anthers that form a tube which surrounds a style and bipartite stigma; anthers shed pollen, the style elongates to allow the stigmas; ovaries form compressed, obovate dry fruits that do not open at maturity, measuring 3 to 4 mm long; fruits have 5–7 ribs on each side and are tipped by two rows of small white hairs.

Distribution: Native range spreads from the Mediterranean to Siberia; distributed to almost all areas of the world.

Part Used: Leaves.

Phytochemistry: It has vitamin K, vitamin A, provitamin A compound, β-carotene, folate and iron.

Uses: The plant is used for sexual potency; effective in smallpox, typhoid, pain, rheumatism, tension and nervousness, coughs and insanity.

Tagetes erecta L.

Vernacular Name: Bur-Kadipan-pan-wa (Burmese); Mexican marigold, Aztec marigold (English).

Description: An erect annual herb that grows up to a height of 180 cm; inflorescence is a solitary terminal head, up to 12 cm in diameter, bright yellow in wild types, lemon-yellow to deep brown-red in cultivated types; numerous cultivars differing in flower colour, flowerhead size and plant height.

Distribution: Native to Mexico, the Americas; cultivated ornamentally throughout the world.

Part Used: Whole plant.

Phytochemistry: Essential oil, fatty acids and phenolics.

Uses: Whole herb is anthelmintic, aromatic, digestive, diuretic, emmenagogue, sedative and stomachic; used internally in the treatment of indigestion, colic, severe constipation, coughs and dysentery; externally used to treat sores, ulcers, eczema. sore eyes and rheumatism; paste of the leaves is applied externally to treat boils, carbuncles and earaches; flowers are carminative, diuretic and vermifuge; decoction is used to treat colds, and mumps; applied externally to treat skin diseases, conjunctivitis and sore eyes; the root is laxative.

Tridax procumbens (L.) L.

Vernacular Name: Mywe-zoknegya (Burmese); Mexican Dausy, Coat Button (English).

Description: Annual or perennial, decumbent-ascending or erect, strangling, pubescent herb, 10-70 cm high; leaves ovate or lanceolate, 1.5-6 x 1-3.5 cm, coarsely serrate-dentate or lobed, cuneate at base, pubescent on both surfaces; petioles 0.5-2.5 cm long; heads solitary, radiate, 0.6-1.5 cm across, on long, hirsute peduncles; outer involucral bracts green, ovate-lanceolate, hirsute-hairy; inner ones greenish-white; ray florets pale-yellow or whitish, 2-3 lobed; disc florets yellow tubular; achenes obconical, pilose; pappus white, unequal, plumose.

Distribution: Native to the tropical Americas; introduced to tropical, subtropical, and mild temperate regions worldwide.

Part Used: Whole plant.

Phytochemistry: Aerial parts contain procumbenetin; alkyl esters, sterols, pentacyclic triterpenes, fatty acids, polysaccharides, 3,6-Dimethoxy-5,7,2',3',4'-pentahydroxyflavone 7-O-β- D-glucopyranoside (procum-betin).

Uses: Leaves cooked as a vegetable; used as an anticoagulant, hair tonic, antifungal and insect repellent, in bronchial catarrh, diarrhoea, dysentery, and wound healing.

Convolvulaceae

Cuscuta reflexa Roxb.

Vernacular Name: Shwe-new (Burmese); Akasavalli, Asparsa (Sanskrit); Dodder (English); Akasbela, Amarabela (Hindi).

Description: A slender much-branched leafless yellow intertwining parasite on shrubs and tree almost covering them, without contact with the soil; flowers sometimes solitary or in clusters of 2-4; fruit capsules, 6-8mm diameters, depressed, globose with 2-4 black seeds; filiform yellowish branches freely swing in the air to and fro; broken branches, when they happen to be in contact with a lining plant, will soon establish themselves on the host and spread rapidly and will ultimately destroy the host plant.

Distribution: Throughout South Asia.

Part Used: Whole plant.

Phytochemistry: Plant contains scoparone, melanettin, quercetin, hyperoside, quercetin 3-O-galactoside and 3-O-β-glucosode; 3,4-dica-feoylquinic and 3-caffeoylquinic acids.

Uses: It is bitter, sweet, astringent, expectorant, carminative, tonic, aphrodisiac, diaphoretic, anthelmintic and diuretic; useful in jaundice, myalgia, cough, bronchitis, strangury, cephalagia, fever and paralysis; seeds are carminative, depurative; applied as an anodyne.

Operculina turpethum (L.) Silva Manso. (Syn. Ipomoea turpethum (L.) R. Br.)

Vernacular Name: Bur-Kya-hin-net "Phyu" (Burmese); Trivrt (Sanskrit); Indian jalap, Turpeth (English); Nisoth, Nisotar, Pitohri (Hindi).

Description: A large stout perennial twiner with milky juice, winged, twisted, pubescent, twining stems and long slender fleshy much-branched roots; leaves simple, large, very variable in shape, subacute, mucronate, more or less pubescent on both sides, base cordate or truncate; flowers white, tubular-campanulate in few-flowered cymes; fruits globes capsules, enclosed in enlarged bitter imbricate sepals, seeds, seeds 4 or less, glabrous dull black.

Distribution: Endemic to India; distributed to nearby areas.

Parts Used: Roots.

Phytochemistry: Glycosidic resin, turpethin, α-turpetein and β-turpethein.

Uses: Roots are purgative, carminative, anthelmintic, axpectorant, antipyretic, hepatic, stimulant and hydragogue; useful in colic, constipation, dropsy, paralysis, myalgia, arthralgia, pectoralgia, bronchitis, obesity, helminthiasis, gastropathy, ascites, inflammations, intermittent fever, leucoderma, pruritus, ulcers, erysipelas, haemorrhoids, tumours, jaundice, consumption and ophthalmia.

Cucurbitaceae

Coccinia grandis (L.) Voigt (Syn. Coccinia indica Wight & Arn.)

Vernacular Name: Aung-me-bywe (Burmese); Bimbi, Bimbika (Sanskrit); Ivy ground (English); Bimb, Kanturi (Hindi).

Description: A perennial, much-branched handsome tendril climber, roots sometimes tuberous; leaves deltoid or subrotund, angled or lobed, bright green above and pale beneath, palmately 5-nerved from a cordate base with circular glands between the nerves; flowers white, large, unisexual; fruits ovoid or oblong or ellipsoid berries with white streaks, bright scarlet red when ripe; seeds ovoid, compressed, yellowish-grey.

Distribution: Throughout Asia, cultivated as well as wild.

Parts Used: Roots, leaves fruits.

Phytochemistry: Fruits contains taraxerone, sitosterol, tarazerol, 24-ethylcholest-5-en-3β-ol; aerial parts contain triterpene alcohols, acetates of cycloartenol, α-& β-amyrin, isomultiflorenol, lupeol and 24-methylenecy-cloartanol; caffeic acid, quercetin, kaempferol and β-sitosterol.

Uses: Roots useful in vomiting, burning sensation and uterine discharges; fruits used in leprosy, skin diseases, intermittent fever, agalactia, asthma, cough, bronchitis, consumption and jaundice; fruits and leaves are anti-inflammatory, anthelmintic, digestive, liver tonic, alexeteric, depurative, febrifuge, sudorific and expectorant; used in wounds ulcer, helminthiasis, dyspepsia, hepatopathy, jaundice, fever asthma, cough, diabetes, stomatitis and anaemia.

Momordica charantia L.

Vernacular Name: Kyet-hin-ga (Burmese); Sushavi (Sanskrit); Bitter ground, Balsam pear, Balsam apple (English), Karela (Hindi).

Description: Climbing or scrambling herbaceous vine with tendrils; leaves alternate, petiolate, the blade with 5-7 deep palmate lobes and quite variable in size; flowers unisexual, tubular, 5-lobed, moderate-sized, pale yellow to orangish; fruit a pepo with black seeds embedded in a reddish pulp.

Distribution: Cultivated all over the world.

Part Used: Fruits.

Phytochemistry: Vicine, mycose, steroidal glucoside, momorcharaside A, B, cucubitane triterpenoids, momordicines I and II, cycloeucalenol, spinasterol, stigmasterol, taraxerol, lophenol, momordicosides, diosgenin, thiocyanogen, 24-methylencycloartenol, phenyl propanoids, carotenoids, squalene, stigmastandien-3-β-O- glucoside.

Uses: Fruit used to treat leprosy and malignant ulcer to treat stomach worms, fever and phlegm, hypertension, dysentery and diabetes.

Cycadaceae

Cyperus rotundus L.

Vernacular Name: Myet-mon-hnying-u (Burmese); Mustaka, Musta (Sanskrit); Nut grass, Cocograss (English); Mutha, Motha (Hindi).

Description: Glabrous perennial grass, 20-30 cm high; rhizome tuberous, reddish-brown, aromatic, leaves long, very narrow, sheathed and one-nerved; inflorescence consisting of 4-6 spikes radiating from a slender, triangular stalk, purplish; fruit angular, greyish.

Distribution: Weed found everywhere in Vietnam, Mauritius and nearby areas.

Parts Used: Rhizomes.

Phytochemistry: Rhizomes contain essential oil (cyperene, cyperol, a-cyperone, cineol and L-α-pinene); starch.

Uses: Rhizomes used for the treatment of irregular menstruation, dysmenorrhoea, gastralgia, dyspepsia, diarrhoea and vomiting.

Elaeagnaceae

Elaeagnus latifolia L.

Vernacular Name: Hmangu (Burmese); Bastard oleaster (English).

Description: A deciduous shrub growing around 3 metres tall; larger leaves and fruit; flowers tubulose-campanulate.

Distribution: A plant of the subtropics and warm temperate zones, it is also found at higher elevations in the tropics.

Part Used: Stem bark.

Phytochemistry: Fruits contain vitamins A, C and E, flavonoids and essential fatty acids; carbohydrates, ascorbic acid, tannins, and phenolics.

Uses: Flowers are astringent and cardiac; fruit is an astringent and rich source of vitamins and minerals; capable of reducing the incidence of cancer.

Euphorbiaceae

Acalypha indica L.

Vernacular Name: Haritamanjari (Sanskrit); Indian acalypha (English); Khokli, Kuppi (Hindi).

Description: An erect, annual herb, 30-100cm in height, leaves ovate or rhombic-ovate, 2.5-7.5cm long with slender petioles; spikes slender, erect, up to 7.6 cm, terminal or axillary with mail flowers minute, clustered at the top, and female flowers with accrescent, board, leafy bract; capsules often one-seeded, concealed in the bract; seeds pale-brown, ovoid, acute and smooth.

Distribution: Throughout South Asia, Tropical Africa eastward to Sri Lanka, India, Burma, Timor, Philippines.

Part Used: Whole plant.

Phytochemistry: Plant contains kaempferol, tracetonamine; leaves and twigs contain acalyphine, acalypus, acalyphamide, quinine, sterols and cyanogenic glycoside.

Uses: Plant possesses diuretic, cathartic, expectorant, emetic, anthelmintic, anodyne and hypnotic properties; decoction of the herb is used as a cure tooth and earache; useful in bronchitis, pneumonia and asthma; root is cathartic; leaves used in scabies and in snake bite; fresh leaf juice useful in rheumatoid arthritis and skin affection; paste of leaves applied on burns; powder of leaves for bed-sores and wounds.

Euphorbia hirta L.

Vernacular Name: Kywè:kyaung:ming-sé (Burmese); Pill-bearing spurge, Asthma plant (English); Dudhi (Hindi).

Description: Erect annual or perennial plant growing to 50 cm, with pointed oval leaves and clusters of small flowers.

Distribution: Native to India and Australia; widespread throughout the tropics.

Parts Used: Aerial parts.

Phytochemistry: Spurge contains flavonoids, terpenoids, alkanes, phenolic acids, shikimic acid and choline.

Uses: Useful in bronchial asthma; pill-bearing spurge relaxes the bronchial tubes and eases breathing; mildly sedative and expectorant; also taken in intestinal amoebiasis.

Euphorbia tirucalli L.

Vernacular Name: Shazaung-letnyo (Burmese); Aveloz, Firestick plants, Indian tree spurge, Naked lady, Pencil tree, Milk bush (English); Saptala, Satala (Sanskrit).

Description: A many-branched, succulent plant, usually 3-5 m but may reach 10 m on occasion; bark of very old specimens is grey and rough with longitudinal dents and ridges that break up into very small fragments; branches are cylindrical, smooth and glabrous-green, 5-8 mm in diameter, forming brush-like masses that are the best known feature of this species; leaves are small and slender, up to 12 x 1.5 mm, rarely seen, as they fall very early; leaf scars on young twigs form conspicuous dents which contract until they are no more than grey dots on older branchlets and can even be seen on fairly thick green stems; thin twigs are pendant, pale green and occur opposite each other, alternate or in groups on the branchlets which gives a rather untidy and rounded appearance to the crowns; flowers are yellow, inconspicuous, and carried in clusters at the apex of the short branches or in the angles of branches; fruits are tripartite capsules (divided into three parts), about 12 mm in diameter, longitudinally very slightly lobed, short-stalked (8 mm), pale green, with a pink tinge and conspicuously pubescent (clothed with soft hairs); capsules dehisce while still on the tree; seeds are oval, about 4 x 3 mm, glabrous, smooth and dark brown with a white line around the small white caruncle (fleshy wart near the hilum of the seed).

Distribution: Widely distributed in Africa; introduced to many other tropical regions, Sri Lanka and Mauritius.

Part Used: Whole plant.

Phytochemistry: Plant latex contains triterpenes (euphol, tirucallol, glut-5-en-3-b -ol, cycloeuphordenol, euphorginol, aamyrin, lanosterol, cycloartenol); 12,20-dideoxyphorbol-13 isobutyrate, 12-deoxy-4 β-hydroxyphorbol-13-phynylacetate-20-acetate, euphol; stem bark contains glut-5-en-3β-oland cycloart-23-en-3β, 25-diolisolate; tirucalicine, cyclotirucanenol and euphorginol.

Uses: It is used for cancer, excrescence, tumours, warts, asthma, cough, earache, neuralgia, rheumatism, toothache.

Euphorbia neriifolia L.

Vernacular Name: Myin-na (Burmese); Hedge Euphorbia, Indian Spurge Tree, Oleander Spurge (English); Thor (Hindi).

Description: A succulent plant with twisting columnar branches; leaves come out in the late summer/ fall; flowers are small, yellow, like small buttons, alternating with the thorns, all along the stem, arranged spirally.

Distribution: East Asia to South China, India, Myanmar, Mauritius, Vietnam, Malaysia and New Guinea.

Part Used: Roots, leaves, bark.

Phytochemistry: Plant contains terpenes, anthocyanins, alcohols and steroids.

Uses: Latex is diuretic, purgative, rubefacient and vermifuge; used in the treatment of asthma; latex applied externally on haemorrhoids; used to relieve sore throat and cracked lips; to treat gonorrhoea; and also to remove warts and cutaneous eruptions; bark has been used as a strong purgative; root is considered as antiseptic; leaves are diuretic, antiasthmatic; also used in colds, stomach upset, fevers, coughs and diabetes.

Euphorbia nivulia Buch.-Ham.

Vernacular Name: Zizaung, shazaung (Burmese); Leafy Milk Hedge, Holy Milk Hedge, Dog's Tongue (English).

Description: An open-crowned succulent-branched spiny, glabrous, deciduous tree up to 9 m tall and 1 m in girth; bark is thick and rough; branches are terete, 2.5 cm thick, with paws of sharp stipular spines arising from spirally-arranged tubercles; alternately arranged leaves are stalkless or nearly so; leaf-blades obovate-spoon-shaped to inverted- lance-shaped, 10-25 x 3-8 cm, rounded at the tip, tapered to the base, lateral nerves 6-8 pairs, obscure, thick, fleshy, with the midrib prominent beneath; stipules are spiny, 5 mm long, paired on each spine- shield; cyathia arise in stalked cymes of 3 each, in leaf axils, towards the tips of the branches, reddish; stalks are 1 cm long; glands are transversely oblong, yellowish; fruits is triradiate, the lobes laterally compressed, 6 x 13 mm, smooth, glabrous, exsert on a pedicel, 1 cm long; styles is stout, united to above the middle, 2 mm long; seeds are ovoid- quadrangular, 4 mm long, smooth.

Distribution: Distributed to India, Burma, Pakistan and Mauritius.

Part Used: Stem, leaves.

Phytochemistry: Leaves and stem contain phenolics and terpenes (euphol and tirucallol) and proanthocyanidin.

Uses: Latex used for the treatment of cancer; showed antioxidant activity.

Jatropha curcas L.

Vernacular Name: Bur-kyetsu, gyisiyokyetsu, (Burmese); Dravanti (Sanskrit); Purging nut, Physic nut (English); Jungli errand, Pahari errand, Ratanjot (Hindi).

Description: A large deciduous soft wooded shrub, 3-4 m in height with sticky juice; leaves alternate, broadly ovate, cordate, 3-5 lobed, glabrous, base seven nerved, stipules zero, flowers yellowish-green in loose panicles of cymes; fruits ovoid, black, splitting into three, 2-valved cocci, seeds dull brownish-black.

Distribution: Throughout India, Mauritius and nearby areas.

Parts Used: Leaves, seeds.

Phytochemistry: Fatty acid (myristic, stearic, arachidic, oleic, and linoleic acid).

Uses: Leaves are galactagogue, rubefacient, suppurative and have insecticidal properties; useful in foul ulcers, tumours and scadies; latex is good for wounds and ulcer; seeds are powerful purgative, acrid sweet, aphrodisiac, thermogenic, digestive tonic, anthelmintic and depurative; useful in haemorrhoids, wounds, splenomegaly and skin diseases.

Manihot esculenta Crantz (Syn. M. utilissima Pohl)

Vernacular Name: Pilaw-pinau (Burmese); Darukandah, Kalpakandah (Sanskrit); Cassava, Manioc, Tapioca (English); Sakkarkand (Hindi).

Description: A shrub 2-5 m in height with stems and branches of varying colour, marked by leaf scars and clusters of tuberous roots; leaves palmate, pale green, 5-9 lobed, lobes lanceolate, acuminate, entire; flowers yellow or greenish-white, unisexual, in terminal cymes; fruits capsules containing 3 seeds.

Distribution: Cultivated throughout Asian countries.

Parts Used: Tubers, leaves.

Phytochemistry: Tuber contains albumin, globulin, glutelin, prolamine, essential amino acids (arginine, histidine, isoleucine, leucine, lysine, methionine, phenylalanine, threonine, tryptophan, and valine).

Uses: Juice of the tubers to treat constipation and indigestion; boiled tuber used in diarrhoea; fresh tubers used as a poultice on sores and boils; leaves used to treat fever; the stem is folded and rubbed across the eyes of people suffering from glaucoma.

Ricinus communis L.

Vernacular Name: Htaw-la-bal (Burmese); Castor Bean (English); Erandi (Hindi).

Description: Shrub or small tree to about 4 m high with conspicuous ring-like scars on the hollow stem; leaves alternate, long-petiolate, palmate with 7-11 lobes and serrated edge, 20-60 cm long and often tinged with red; terminal inflorescence is a narrow panicle; there are separate unisexual flowers, fruit a spiny subglobose schizocarp about 1.5 cm long splitting into three sections when mature, each section containing one mottled smooth brown seed in each of three sections flowers and fruit available throughout the year.

Distribution: Indigenous to the southeastern Mediterranean Basin, Eastern Africa, and India; widespread throughout tropical regions.

Part Used: Seeds, bark.

Phytochemistry: β-Amyrin, ricins, 5-dehydro-avenasterol, β-carotene, tannins, brasicasterol, campesterol, β-sitosterol, stigmasterol, lupeol, quercetin, casbene, chlorogenic acid, coumarin, ellagic acid, diethyleneglycol disulphide, hemaglutinin, quinic acid, indoleacetic acid, ricinine, N-demethylricinine, ricinus agglustinins, glycoproteins, vitamin B_6 and B_1, saponin, epicatechin hyperuside, kaempferol flycoside.

Uses: Seed oil is used as a purgative; infusion of the bark is used to treat skin inflammations and rashes; drink of the juice in water is taken to treat breast tumours and boils.

Lamiaceae

Clerodendrum indicum (L.) Kuntze (Syn. C. siphonanthus R.Br.)

Vernacular Name: Ngayanpadu (Burmese); Tubeflower, Turk's-Turban, Sky Rocket, Bowing Lady (English); Bharangi (Hindi).

Description: A semideciduous, perennial plant with stems that are sometimes herbaceous but more often become woody, with the plant adopting the habit of a shrub or a small tree; growing from an aggressively suckering, stoloniferous rootstock, the plant can be up to 3 metres tall, the stems usually very straight or arching, mostly unbranched and hollow.

Distribution: Southern China, India, Sri Lanka, Nepal, Bangladesh, Myanmar, Thailand, Mauritius, Cambodia, Laos, Vietnam, Malaysia, Indonesia.

Part Used: Whole plant.

Phytochemistry: Leaves contain flavonoids (scutellarein, hispidulin and their 7-O-glucuronides) and sterols; flowers contain β-sitosterol and triterpenoids; bark yields hexitol and sorbitol; flavone (pectolinarin) and diterpene (oncinotine).

Uses: Leaves are used as a bitter tonic and vermifuge; dried leaves are smoked like cigarettes to relieve asthma; juice of the tender parts of the plant is used as an external application for skin complaints; pounded root, combined with ginger, is considered useful in the treatment of asthma, coughs and other pulmonary complaints; resin extracted from the plant is used in the treatment of syphilitic rheumatism.

Gmelina arborea Roxb.

Vernacular Name: Gumbhar, Kumbhar, Shivparni (Hindi); Gambhari, Gandhari (Sanskrit); Malaya-bushbeech (English).

Description: Deciduous tree up to 15.5 m high greyish-brown bark, exfoliating in scrubby flakes; leaves 9.0-23.5 x 7-15.5 cm, opposite, broadly ovate-cordate, glandular at base, acuminate, entire, finely tomentose and glaucous beneath; flowers 2-2.5 cm across, reddish-brown or yellow, appearing with the leaves, arranged in tomentose panicles; drupes 1.5-2.5 cm, ovoid or pyriform, orange-yellow when ripe.

Distribution: Distributed in India, Mauritius, Pakistan, and Bangladesh, Myanmar.

Part Used: Whole plant.

Phytochemistry: Heartwood contains lignans (6"-bromo-isoarboreol, 4-hydroxysesamin, 4,8-dihydroxysesamin, gummadiol, 2-piperonyl-3- hydroxymethyl-4- (α-hydroxy-3,4-methylenedioxybenzyl)-4-hydroxytetrahydrofuran and 4-O-glucoside of 4-epigummadiol); arboreol or gmelanone; roots contains umbelliferone 7-apiosylglucoside; (+)-7′-O-ethyl arboreol, (+)-paulownin, (+)-gmelinol, (+)-epieudesmin and β-sitosterol.

Uses: Plant is used as antidote for snake-bite and scorpion sting; leaves are used as demulcent, antigonorrhoic, bechic for removing worm from ulcer; whole plants is used as astringent, bitter tonic, stomachic, digestive, cardiotonic, laxative, galactagogue, pulmonary and nervine tonic; improves, memory, overcome giddiness and heart disease, nervous disorders and piles; pulverized root is applied locally for gout; drupes are sweetish and bitter and are used an astringent refrigerant decoction for fevers and bilious affections; paste of leaves is applied to the head for the relief in headache; flowers are given in blood disease; fruits are useful in pleural and lung diseases.

Mentha arvensis L.

Vernacular Name: Mint, Pudina (Burmese); Field mint, Japanese pepperming (English); Podina, Pudina (Hindi).

Description: Perennial herb, 30-50 cm high; stems quadrangular, erect or prostrate, rooting at the nodes; leaves opposite, ovate, softly tomentose on both sides; margins serrate, inflorescence in axillary capitates whorl; flowers small, white or lilac.

Distribution: Cultivated all over the world.

Parts Used: Whole plant.

Phytochemistry: Plant contains an essential oil (L-menthol, menthhyl acetate, L-menthone, L-α-pinene and limonene).

Uses: Plant is antibacterial and antifibrile; oil is effective in diaphoretic fever, headache, rhinitis, cough sore throat, arthralgia, neuralgia colic, vomiting, dyspepsia, diarrhoea and prurigo.

Tectona grandis L.f.

Vernacular Name: Bur-Kyum(Burmese); Teak (English); Sagaun, Sagvan (Hindi).

Description: A large to very large deciduous tree, 20-35 m in height with light brown or grey bark having shallow longitudinal furrows, fluted and buttressed base and characteristically quadrangular channeled branchlets; leaves simple, opposite, broadly elliptical or obovate, acute or acuminate, coriaceous, rough above, stellatelly-grey tomentose beneath, possessing minute glandular dots, main nerves 8-10pairs; flowers many, white, small, sweet scented, in large erect terminal branched tomentos cymose panicles; fruits hard, bony, irregularly globose drupes enveloped by light brown bladder-like calyx; seeds usually 1-3 ovate, marble white.

Distribution: Throughout India, Pakistan, Mauritius and Bangladesh.

Parts Used: Whole plant.

Phytochemistry: Wood contains calcium, ammonium and magnesium; resins; seeds contain fatty acids.

Uses: Roots are useful in anuria; bark useful in bronchitis, hyperacidity, dysentery, verminosis, burning sensation, diabetes, leprosy and skin diseases; leaves useful in inflammation, leprosy, skin diseases, pruritus, stomatitis, indolent ulcers, haemorrhages, haemoptysis; fruits useful in strangury, vesical calculi, pruritus and stomatitis; flowers used in dipsia, leprosy, skin diseases, strangury and diabetes.

***Vitex negundo* L.**

Vernacular Name: Kyaugng-ban-gyi (Burmese); Five-leaved chaste tree (English); Nirgandi (Hindi/ Sanskrit).

Description: An aromatic large shrub or small tree of about 3 m in height with quadrangular branches; leaves opposite, exstipulate, long-petioled and digitately 3-5 foliolate, all leaflet with petiolules, the middle one longer flowers bluish-purple in panicles up to 30 cm long; fruits globose or ovoid or obovoid, four-seeded drupe, black when ripe.

Distribution: Throughout Asia.

Parts Used: Whole plant.

Phytochemistry: Leaves contain alkaloid nishindine and hydrocotylene; fatty acids; terpenoids.

Uses: Plant is anti-inflammatory, antiseptic, cephalic, alternate, antipyretic, diuretic, emmenagogue, depurative, rejuvenating, ophthalmic, vulnerary and tonic; roots are useful in cephalalgia, otalgia, arthritis, inflammations, dyspepsia, colic, verminosis, flatulence, dysentery, uropathy, wounds ulcers, bronchitis, cough, malarial fever, haemorrhoids, dysmenorrhoea, leprosy, dermopathy, ophthalmopathy and general debility; leaves are useful in sprains, orchitis, gout, splenohepatomegaly, otorrhoea, inflammations and ulcers; bark is useful in verminosis and ophthalmopathy; flowers are useful in diarrhoea, cholera, fever, haemorrhages, hepatopathy and cardiac disorders.

***Vitex trifolia* L.**

Vernacular Name: Kaung-ban (Burmese); Ujjra, Banda (Hindi); Kahbung, Mistletoe (English).

Description: A large, epiphytic, profusely branching shrubby perennial, forming rounded clumps over the host trees; branches dichotomous or whorled; leaves about 5 by 10 cm, sessile, cuneate-oblong, or oblanceolate, with 3-5 longitudinal, basal nerves; flowers are dioecious, sessile; in clusters of 3-5, supported by cocave bracts; perianth-segments 3-4, triangular, deciduous; fruits small, ellipsoid, white, smooth, almost transparent, mucilaginous.

Distribution: Distributed in India, Mauritius, Sri Lanka and Pakistan.

Parts Used: Whole plant.

Phytochemistry: Fatty acids, nicotine alkaloids, amines; flavonoids, chalcones and flavanones; terpenoids (β-amyrin, betalinic acid, lupeol and ester combinations; oleanolic acid, ursolic acid), β-sitosterol, phytosterol glucoside, viscotoxin; mucilage, polyoles; sugars, starch, syringing and tannin, lutein, α- and β-carotene and vitamin C.

Uses: Leaves, stems and berries used in hypertension, arteriosclerosis, tumours, arthrosis, hysteria, epilepsy and ear diseases; also useful in splenic and hepatic enlargements; in lumbago and piles.

Lauraceae

Cinnamomum camphora (L.) J.Presl

Vernacular Name: Payok, Camphor (Burmese); Camphor tree, Formosan wood (English); Kapur, Karpur (Hindi).

Description: Evergreen tree about 15 m tall; trunk bark thick and grooved; leaves alternate, coriaceous, long-petiolate, shining on the upper side, 3-nerved at the base; inflorescence in axillary panicle, shorter than the leaf; flowers small, greenish-yellow; barry globose, black when ripe.

Distribution: Cultivated as a shade-tree in the sub-tropical region including the Indian subcontinent.

Part Used: Roots and wood.

Phytochemistry: Stem and leaves contain essential oil (camphor, D-α pinene, cineol, tempineol, caryophyllin, safrole, limonene, phellandrene, carvacrol, camphorene and azulene).

Uses: Injections of camphor oil and sodium camphosulfonate are prescribed in case of cardiovascular collapse. It is applied externally as an antiseptic, demulcent and anodyne for impetigo, boils, neuralgia and rheumatism.

Cinnamomum verum J.Presl (Syn. C. zeylanicum Blume)

Vernacular Name: Hmanthein-op- (Burmese); Tvak, Darusita (Sanskrit); Cinnamon, Cylon cinnamom (English); Dalcini, Darucini (Hindi).

Description: A moderate-sized evergreen tree of 8-18 m in height and 50 cm in diameter with reddish-brown soft bark, having numerous small water; leaves ovate or elliptic-ovate, shiny above, 3-5 ribbed from a little above the base, the side ribs ending about three-fourths up the base usually round; flowers small in axillary or sub-terminal cymes or panicles; fruits ovoid berry, dark purple in colour having persistent perianth.

Distribution: West coast tropical evergreen and semi-evergreen forests; distributed to India, Bangladesh, China, Mauritius and Indonesia.

Part Used: Bark.

Phytochemistry: Eugenol, O-methyl eugenol, benzaldehyde, cinnamaldehyde, I-α-pinene, α-and β-phellandrene, p-cymene, caryophyllene, benzyl benzoate and linalool; safrole, acetyleugenol, cinnamyl acetate and cinnamyl alcohol.

Uses: Bark is useful in bronchitis, asthma, cephalalgia, odontalgia, cardiac disease, diarrhoea, uropathy, nausea and vomiting, flatulence, fever, halitosis and restoring normal skin colour on the face; cinnamon oil is useful in anorexia, inflammations, stomachalgia, vitiated conditions of odontalgia, vomiting, tuberculosis and ulcers.

Leguminosae

Abrus precatorius L.

Vernacular Name: Ywenge, Ywayglay (Burmese); Gunja (Sanskrit); Jequirity, Crab's eye, love pea, Indian licorice (English); Ghugachi, Chirani, Ratti, Guncaci (Hindi).

Description: A deciduous, wiry climber with tough branches; leaves abruptly pinnate with many pairs of leaflets, the rachis ending in a spine; the leaflets oblong, rounded at both ends, thinly membranous; flowers pink, clustered on tubercles arranged along the rachis of one-sided pedunculate raceme; fruits pods, turgid with a sharp deflexed beak; seeds usually scarlet with a black spot or sometimes pure white.

Distribution: Native to India and grows in tropical and subtropical areas of the world where it has been introduced.

Part Used: Seeds, leaves and root.

Phytochemistry: Roots and leaves contain glycyrrhizin; root contains precol, abrol, abrasine and precasine; seeds contain gallic acid, abrine, hypaphorine, alanine, scrine, valine, choline, trigonelline, precatorine, 5b-cholanic acid, abrin A and abtrin B.

Uses: Seeds used in nervous disorders; abortifacient; paste as a local application for sciatica and paralysis; leaves decoction used for cough, cold colic pain; fresh juice mixed with a bland oil applied to painful swelling; root used for gonorrhoea, jaundice, haemoglobinuric bile.

Acacia concinna (Willd.) DC.

Vernacular Name: KinmuningHpak-ha (Burmese); Saptala, Carmasahva (Sanskrit); Soapnut-acacia (English); Satala, Kochi (Hindi).

Description: A stout prickly climbing shrub with brown branches dotted with white; leaves bipinnate, main rachis bearing sharp hooked prickles and a large gland on the petiole, pinnae 8 pairs or more, leaflets subsessile, sensitive, unequal sided, glabrous; flowers small in globose heads polygamous; fruits short-stalked thin pods, flat, coriaceous, the sutures straight; seeds 6-10 per pod.

Distribution: Throughout India; tropical forests, Mauritius.

Parts Used: Pods, bark.

Phytochemistry: Barkcontains saponin, lupeol, α-spinasterol, acacic acid lactone, glucose, arabinose, rhamnose, hexacosanol, α-spinasterone; pods contain saponins, acacinin-C, -D, and –E; oligosaccharides, sucrose, raffinose, stachyose and verbascose.

Uses: Pods useful in burning sensation, constipation, renal problems, haemorrhoids, erysipelas, leucoderma, leprosy, prurigo, abscesses and eczema.

Acacia farnesiana (L.) Willd.

Vernacular Name: Nanlungyaing (Burmese); Vitkhadira (Sanskrit); Cassie flower (English); Gandhbabul, Vikar (Hindi).

Description: A thorny bush or small tree, attaining a height of 4.5 m. branches spread out from the main trunk; bark with brown streaks; leaves bipinnate with stipular spines; flowers borne in axillary heads; fragrant, bright yellow; pods dull brown, 5.0-7.5 cm long.

Distribution: Tropical part of Indian subcontinent.

Parts Used: Leaves, bark, pods.

Phytochemistry: Eicosane, benzyl alcohol, farnesol, geranyl acetate, linalool, linalyl acetate, nerolided, pinitol, α-terpineol; ripe pods contain gallic acid, ellagic acid, tannin, aromadendrin, kaempterol, naringenin and their glucosides; leaves contain apigenin and rutin glucosides; flower oil contains benzaldehyde, salicylic acid, methyl-salicylate and benzyl-alcohol analdehyde.

Uses: Decoction of the bark together with ginger is useful in the bleeding of the gums; leaves used in gonorrhoea; oil used in perfumery.

Acacia nilotica (L.) Delile (Syn. A. arabica (Lam.) Willd.; Vachellia nilotica (L.) P.J.H.Hurter & Mabb.)

Vernacular Name: Balurl (Burmese); Babbula (Sanskrit); Indian gum, Arabic tree (English); Babul (Hindi).

Description: A Medium sized tree with a short trunk, usually attaining a height of 15 m. bark almost black to dark brown, deeply cracked for longitudinally fissured; leaves bipinnate with spinescent stipules 2.5-5.0 cm long; flowers crowded in long-peduncled globose heads, forming axillary clusters of 2-5 heads, fragrant, golden-yellow; pods white, flat containing 8-12 seeds, 7.5-15 cm long

Distribution: Native to Africa, the Middle East and the Indian subcontinent; also an invasive species in Australia.

Phytochemistry: Bark contains polyphenolic compounds, catechin, epicatechin, quercetin, gallic acid, leucocyanidin gallate, sucrose and tannins, m-Digallic acid and chlorogenic acid; gum contains galactose, L-rhamnose, L-arabinose, four aldobiouronic acids; seeds contain amino acids, fatty acids ascorbic acid and tannins.

Uses: Pods decoction used in urinogenital diseases; leaves infusion used for diarrhoea and dysentery; bark decoction used as a gargle in sore throat and toothache; dry powder applied externally in the ulcer.

Albizia lebbeck (L.) Benth.

Vernacular Name: Kokko, main-Kokko (Burmese); Siris tree (English); Siris (Hindi); Sirisah, Bhandi (Sanskrit).

Description: A medium to large-sized unarmed deciduous tree about 20 m in height with an umbrella-shapedcrown and grey to dark brown rough irregularly cricked bark; leaves abruptly bipinnate, maim rachis with a large gland above the base and one below the upper-most pair of pinnae, pinnae 2-4 pairs, leaflets 5-9 airs with glands between their bases, the lateral ones elliptic-oblong, the two-terminal ovate-oblong, all unequal sided; flowers white, fragrant, in globose umbellate hands; fruits long, characteristic pods, bluntly pointed, tin pale yellow, smooth, shiny, reticulately veined above the seed; seeds 4-12, pale brown, ellipsid, oblong, compressed.

Distribution: Native to Indomalaya, New Guinea and Northern Australia; widely cultivated and naturalised in other tropical and subtropical regions.

Part Used: Bark, flowers, seeds.

Phytochemistry: Bark contains D-catechin, isomers of leucocyanidin and melacacidin, lebbecacidin, friedelin and β-sitosterol; seeds contain protein, calcium, phosphorus, iron, niacin, ascorbic acid, amino acids (arginine, histidine, leucine, isoleucine, lysne, methionine, phenylalanine, threonine, tyrosine, and valine); flowers contain lupeol, α- and β-amyrin.

Uses: Bark is useful in cough, catarrh, asthma, enlarged cervical glands, ophthalmopathy, nyctalopia, strengthening gums, scrofula, skin eruptions, leprosy, leucoderma, sprains, wounds, ulcers, neuralgia, inflammations, erysipelas, diarrhoea and all types of poisoning; flowers are useful in chronic cough and bronchitis; seeds are useful in inflammation, scrofula, skin diseases, leprosy, leucoderma, chronic catarrh, seminal weakness, ophthalmopathy and poisoning.

Albizia procera (Roxb.) Benth.

Vernacular Name: Mai-Tawn, Mai-King (Burmese); White siris, Tall albizia (English); Safed siris (Hindi).

Description: A fast-growing deciduous tree that generally reaches 10 to 20 m in height and has a straight to somewhat curved stem, smoothish light brown to light greenish-grey bark, and a spreading thin crown; leaves about 40 cm long; pinnae about 4–12, 15–20 cm long; leaflets 12–20, oblong-elliptic, rounded or retuse, 2–5 cm long, oblique; flowering generally occurs during the rainy season; distinctly fragrant flowers form whitish globose heads 20 to 24 mm in diameter, borne on racemes 8 to 25 cm long near the ends of twigs; fruits are flattened pods 10 to 20 cm long and 1.8 to 2.5 cm broad, changing from green to deep red or reddish-brown on maturity; each contains 6 to 12 seeds; fruits ripen 6 to 9 months after flowering, during the dry season, and usually remain on the tree until the whole twig bearing the pods is shed.

Destribution: Native to moist deciduous and semievergreen hill forests, swamp forests, and lowland savanna woodlands in Asia from northern India through Southeast Asia; Indian subcontinent, Myanmar, Thailand, Mauritius, Malaysia, Laos, Cambodia and Vietnam.

Part Used: Flowers, seeds, leaves, bark.

Phytochemistry: Seeds contain proceranin A; roots contain α-spinasterol and saponin hydrocyanic acid.

Uses: Plant has anti-cancer activity; decoction of bark used for rheumatism and haemorrhage; useful in treating problems of pregnancy and for stomach-ache; leaves are poulticed onto ulcers.

Bauhinia variegata var. candida Voigt (Syn. Bauhinia alba Wall.)

Vernacular Name: Bwegyin (Burmese); Kancnarah, Kovidarah (Sanskrit); Buddhist Bauhinia, Orchid tree (English); Kachnar, Kancanar (Hindi).

Description: A moderate sized deciduous tree with vertically cracked grey bark, wood moderately hard, greyish brown with irregular darker patches; leaves 0 f2 leaflets, connate for about two-thirds up, leaflets ovate, rounded at apex, 10-15cm long, publescent beneath when young, coriaceous; flowers white or pink, the uppermost petal darker and variegated, usually appearing before the leaves in short axillary or terminaracemes, stamens 5, staminodes absent, fruits flat dehiscent pods, seeds 10-15.

Distribution: Throughout Indian subcontinent.

Parts Used: Root, bark.

Phytochemistry: β-Sitosterol, lupeol, kaempferol-3-glucoside, 5,7-dehydroxyflavanone-4-O-α-L-rhamnopyranosyl-β-D-glucopyranosides, and 5,7-dimethoxyflavanone-4-O-α-L-rhamno-pyranosyl-β-D-glucopyranosides.

Uses: Roots and bark are useful in diarrhoea, dysentery, skin diseases, leprosy, intestinal worms, tumours, wounds, ulcers, inflammations, scrofula, proctoptosis, haemorrhoids, haemoptysis, cough, menorrhagia and diabetes.

Bauhinia racemosa Lam.

Vernacular Name: Palan (Burmese); Burmese Silk Orchid, Bidi leaf tree (English); Katmauli (Hindi).

Description: Deciduous trees to 10 m high, bark grey to black, rough, thinly scaly, with numerous vertical cracks; blaze pinkish-red, turning brown on exposure; leaves simple, bilobed, alternate; stipules small, cauducous; petiole 10-33 mm long, slender, pubescent, swollen at base and at tip; leaves 2-5 x 3.5-11 cm, broader than long, ovate-orbicular, base cordate, apex obtuse, bilobed and mucronate at cleft, margin entire, glabrous above, pubescent beneath, coriaceous; 7-9 nerves from the base, palmate, prominent, intercostae reticulate, prominent; flowers bisexual, 10-12 mm across, yellowish-white, in terminal and leaf opposed few-flowered racemes; bracts and bracteoles 2-3 toothed, minute, pubescent; pedicel 1-2 mm, slender; calyx spathaceous, reflexed, 5-toothed at apex, puberulous; petals 5, 1.5 x 0.2 cm, linear-oblong; stamens 10, all perfect; filaments equal, filiform, pilose at base; anthers versatile; ovary half-inferior, stalked, pubescent, ovules many; style filiform; stigma peltate. Fruit a pod 15-22 x 1.5-2 cm, oblong, blackish-brown, turgid, apex horned indehiscent; seeds 10-20, ovoid.

Distribution: Native to tropical Southeast Asia, China, India, Myanmar, Thailand, Mauritius, Cambodia, Vietnam.

Part Used: Leaves, bark, gum.

Phytochemistry: Leaves containβ-sitosterol, β-amyrin, kaempferol, quercetin, scopoletin and scopolin; heartwood contains stilbene (resveratrol), dibenzoxepin derivative, pacharin.

Uses: Sacred plant of the Hindus, worshipped at the Dashera festival; leaves are used to prepare bidis (Indian thin cigarettes); bark is used as medicine for malaria, dysentery, diarrhoea; stem bark is used in the treatment of headache, fever, skin diseases, tumours, blood diseases, dysentery, and diarrhoea.

Butea monosperma (Lam.) Taub.

Vernacular Name: Pauk (Burmese); Flame of the forest, Bastard teak (English); Dhak, Palas (Hindi)

Description: A medium-sized deciduous tree, very conspicuous when in flower, 12-15m in height with gum-containing grew bark exfoliating in irregular pieces, and somewhat crooked trunk; leaves 3-foliate, leaflets coriaceous, obtuse, glabrous above when old, finely silky and conspicuously reticulately veined beneath; flowers bright orange-red, large, in rigid racemes; fruits pods, thickened at the sutures,.

Distribution: Native to tropical and sub-tropical parts of the Indian Subcontinent and Southeast Asia, ranging across India, Bangladesh, Nepal, Sri Lanka, Myanmar, Thailand, Laos, Cambodia, Vietnam, Malaysia, and western Indonesia.

Part Used: All parts.

Phytochemistry: Root bark contains β-sitosterol, leucoanthocyanidin, amyrin, betulinic acid, stigmasterol, palasonin; gum and bark contain tetramers of leucocyanidin; flowers contain butrin, isobutrin, coreopsin, isocoreopsin and sulphurein, monospermoside and isomonospermoside; leaves contain glucosides; seeds contain proteolytic and lypolytic enzymes, palasonin, monospermoside and somonospermoside.

Uses: Bark is useful in anorexia, dyspepsia, diarrhoea, dysentery, haemorrhoids, intestinal worms, bone fractures, rectal diseases, dysmenorrhoea, gonorrhoea, hepatopathy, ulcer, tumours, hydrocele and diabetes; leaves are useful in pimples, boils, flatulence, worm infestations, inflammations arthralgia and haemorrhoids; flowers are useful in diarrhoea, haemorrhoids, menorrhagia, strangury, fever, leprosy, skin diseases, swelling, hyperdipsia, haemoptysis, arthritis, burning sensation, bone fractures, and are very efficacious in birth control; seed is useful in herpes skin diseases, ringworm, ophthalmopathy, epilepsy, roundworm, arthritis, flatulence, constipation and diabetes.

Butea superba Roxb.

Vernacular Name: Pauk-new (Burmese); Red Kwao Krua (English).

Description: Large woody climber, climbing from left to right, trunk attaining 60-90cm in girth, upto 15m high; bark deep brown; leaves 3foliate; leaflets with stiff short hairs, 32.5-40 x 20 30 cm, base rounded, obovate, the lateral areas asymmetrical, hairy on the nerves beneath; petiole 15-45cm long; petiolule of the lateral leaflets 1.5 cm long, of the terminal upto 9 cm, flowers orange-scarlet, borne in profuse racemes on leafless branches; racemes often 25-30 cm long; calyx campanulate, teeth lanceolate, velvety; corolla much exserted; standard acute, recurved; wings adnate to a curved keel; pod stalked, 12-15 cm long, 1-seeded.

Distribution: Native to Thailand; distributed to Mauritius.

Part Used: Tubers

Phytochemistry: Tubers contain butesuperin A and B, medicarpin, formononetin, prunetin, 3,7,3'-trihydroxy-4'-methoxyflavone and its glycoside, 3,5,7,3',4'-pentahydroxy-8-methoxy-flavonol and its glycoside, daidzein, genistein, coumestrol, procyanidin B2 and B5, (-)-epicatechin, β-sitosterol, campesterol and stigmasterol.

Uses: Used as an aphrodisiac; effective on rejuvenation and sexual vigour.

Caesalpinia crista L.

Vernacular Name: Fever-nut, Kalien-sukyanbo (Burmese); Latakaranjah, Kuberaksi, Kantakikaranjah (Sanskrit); Fever nut, Bonduc nut, physic nut (English); Kantkarej, Kantikaranja (Hindi).

Description: A large straggling, very thorny shrub, branches armed with hooks and straight hard yellow prickles; leaves bipinnate, large, stipules foliaceous, pinnae 7 pairs, leaflets 3-8 pairs with 1-2 small recurved prickles between them on the underside; flowers yellow, in dense long peduncled supra-axillary racemes at the top; fruits inflated pods, covered with wiry prickles, seeds 1-2 per pod, oblong or globular, hard, grey with a smooth shiny surface.

Distribution: Throughout India and nearby countries in the plains on wastelands and coastal areas.

Part Used: Root bark, leaves, seeds.

Phytochemistry: Kernels contain α-, β-, δ-and ε- caesalpins, caesalpin F and homoisoflavone, bonducelline.

Uses: Root bark is useful in amenorrhoea, dysmenorrhoea, fever, cough, asthma, intestinal worms, colic, flatulence and dyspepsia; leaves are useful in elephantiasis, intestinal worms, splenomegaly, hepatomegaly, amenorrhoea, dysmenorrhoea, fever and pharyngodynia; seeds are useful in arthralgia, inflammations, hydrocele, cough, asthma, leucoderma, leprosy, skin diseases, dyspepsia, dysentery, colic, haemorrhoids, intestinal worms, hepatopathy, splenopathy, diabetes and intermittent fevers.

Caesalpinia pulcherrima (L.) Sw.

Vernacular Name: Daungzok (Burmese); Poinciana, Peacock flower, Red bird of paradise, Mexican bird of paradise, Dwarf Poinciana, Pride of Barbados (English); Guletura (Hindi).

Description: It is a shrub growing to 3 m tall; leaves are bipinnate, 20–40 cm long, bearing 3–10 pairs of pinnae, each with 6–10 pairs of leaflets 15–25 mm long and 10–15 mm broad; flowers are borne in racemes up to 20 cm long, each flower with five yellow, orange or red petals; fruit is a pod 6–12 cm long.

Distribution: Native to the tropics and subtropics of the Americas, the West Indies; widely cultivated.

Part Used: Roots, fruits, leaves, flowers.

Phytochemistry: Aerial parts contain (3E)-2,3-dihydro-6,7-dimethoxy-3-[(3-hydroxy-4-methoxyphenyl) methylene]-4H-1-benzopyran-4-one, dihydrobonducellin and 2-methoxydihydrobonducellin; stem contains pulcherimine A,B, ceasalpin J,P, β-sitosterol, etc.

Uses: Root used to induce abortion in the first trimester of pregnancy; leaves used in gastritis and intestinal inflammation, diarrhoea and dysentery, flatulence, ulcers and hepatitis; bark and flowers used in uterine dysfunction.

Cajanus cajan (L.) Millsp.

Vernacular Name: Pesion-ngon (Burmese); Adhaki, Tuvari (Sanskrit); Red gram, Pigeon pea, Congo pea (English); Tuvar, Arhar (Hindi).

Description: An erect shrub, 1.5-3 m in height with many branches provided with skin hair; leaves compound, pulvinate, leaflets oblong-lanceolate, entire, densely silky beneath; flowers yellow in terminal panicles of corymbose racemes; fruits pods, tipped with the persistent lower half of the style, seeds vary in colour from yellow and red to brown or black.

Distribution: Throughout Asia, Africa, and Latin America.

Part Used: Leaves, seeds.

Phytochemistry: Seeds contain amino acids (arginine, alanine, asparagines, cystine glutamic acid, glycine, histidine, leucine, isoleucine, lysine, methionine, phenylalanine, proline, serine, threonine, tyrosine, tryptophan, and valine), α-galactosidase, β-galactosidase isoenzymes, β-L-arabinosidase and nicotinamide deamidase.

Uses: Leaves used for oral ulcers, odontalgia, gingivitis, strangury and inflammations; seeds useful for intestinal worms, oral ulcers, bronchitis, cough, vomiting, haemorrhoids, fever and cardiac diseases; leaves and seeds when applied as a poultice over the breast to induce lactation.

Caesalpinia sappan L.

Vernacular Name: Teinnyet (Burmese); Pattaranjaka (Sanskrit); Sappan wood, Brazil-wood, Bukkum wood (English); Bakam, Patang (Hindi).

Description: Small tree, 5-7 m in height, with hardwood, brownish-red; stems prickly; young shoots tomentose, branches glabrous covered with short spines; leaves alternate, pinnate leaflets trapeziform, glabrous above, tomentose beneath; inflorescence in terminal raceme; flowers yellow peduncle clothed with a ferruginous tomentum; pod compressed, with hard shell and sharp horn; seeds yellowish-brown.

Distribution: Grows wild, in mountains and is cultivated in many places as a hedge plant.

Part Used: Trunk wood.

Phytochemistry: Woody part contains brazilin, brasilein and essential oil (D-α-phellandrene, ocimene, tannin gallic acid and saponin).

Uses: Trunk wood possesses antibacterial, demulcent and haemostatic properties; used in dysentery, diarrhoea, intestinal and uterine haemorrhages, post-partum haematometra, contusions, wounds, dysmenorrhoea, colic furunculosis, impetigo, leucorrhoea and anaemia.

Cassia fistula L.

Vernacular Name: Ngu-Ngu-Shwe (Burmese); Argvadhah, Krtamalah (Sanskrit); Indian laburnum, Purging fistula, Golden-shower (English); Amaltas, Girimala (Hindi).

Description: A moderate-sized handsome deciduous tree, 8-15m in height with greenish-grey smooth bark when young, and rough when old, exfoliating in hard scales; leaves pinnately compound, leaflets 4-8 pairs, acute, bright green, glabrous above, paler and silver-pubescent beneath when young, main nerves numerous; flowers bright yellow in lax pendulous racemes; fruits cylindric pods, 30-60cm long, shortly stipitate, nearly straight, smooth shiny, brownish-black; seeds broadly ovate, horizontally immersed in dark coloured sweetish pulp.

Distribution: Native to the Indian subcontinent and adjacent regions of Southeast Asia; ranges from southern Pakistan eastward throughout India to Myanmar and Thailand, Mauritius and south to Sri Lanka.

Part Used: Whole plant.

Phytochemistry: Plant contains sennosides A and B, rhein and its glucoside, barbaloin, aloin, formic acid, butyric acid, their ethyl esters and oxalic acid, sap, acetyl acid, thiocyanogen and unsapon matter, tannins, phlobaphenes, reducing sugars and oxyanthraquinones.

Uses: Roots are useful in skin disease, tuberculous glands, syphilis and burning sensation; bark is useful in boils, pustules, leprosy, ringworm, colic, dyspepsia, constipation, fever, diabetes, strangury and cardiopathy; leaves used in leprosy, ulcer and intermittent fevers; flowers are useful in skin diseases, pruritus, burning sensation, dry cough and bronchitis; fruits are useful in burning sensation, leprosy, skin diseases, pruritus, flatulence, colic, inflammations, rheumatism, gout, anorexia, hepatomegaly, jaundice, disorders, intermittent fever, strangury, ophthalmopathy and general debility.

Desmodium gangeticum (L.) DC.

Vernacular Name: Prsniparni, Prthakparni, Shalparni (Sanskrit); Salpan, Salvan (Hindi).

Description: An erect, diffusely branched undershrub, 90-120 cm in height with a short woody stem and numerous prostrate branches provide with soft grey hairs; leaves unifoliate, ovate to ovate-lanceolate, membranous and mottled with grey patches; flowers white, purple or lilac in elongate lax, terminal or axillary racemes; fruits moniliform, 6-8 jointed glabrescent pods, joints of sparsely pubescent with hooked hairs, joints separating when ripe into indehiscent one-seeded segments; seeds compressed reniform.

Distribution: Tropical Africa, E. Asia - China, Japan, Indian subcontinent, Myanmar, Mauritius, Thailand, Cambodia, Laos, Vietnam, Malaysia, Indonesia, Philippines, Australia.

Parts Used: Roots, leaves.

Phytochemistry: Roots contain flavonoids, alkaloids, pterocarpanoids (gangetic, gangetinin and desmodin); aerial parts contain indol-3-alkylamines and their derivatives.

Uses: A decoction of the leaves is used against stones in the gall bladder and kidneys; leaves are applied as a poultice to the head as a treatment for headache; roots are considered to be alternative, astringent, bitter tonic, diuretic, expectorant and febrifuge; decoction of the root is employed to treat kidney problems, oedema, swellings, chronic fever, coughs, biliousness, diarrhoea and dysentery; or as a sedative for children; roots are applied to the gums as a treatment for toothache; decoction is used externally to clean wounds and ulcers.

Dalbergia lanceolaria L.f.

Vernacular Name: Bur-Thitpagan (Burmese); Bastard rosewood (English); hardi, takoli (Hindi).

Description: Small trees, branchlets pubescent; leaflets alternate, 5 or 6 pairs, 3-4 x 2.5 mm, oblong-obovate, apex obtuse, base cuneate; petiole ca 4 mm; panicles terminal or axillary on tender shoot, pubescent; flowers to 1 cm across; calyx tube ca 4 mm, rusty tomentose; corolla white, standard 1 x 0.5 mm, ovate, oblong, wings and keels to 8 x 3 mm; stamens 10, diadelphous; ovary ca 3 mm; pods to 7 x 11.5 cm, lanceolate, acute; seeds 1 or 2.

Distribution: Native to India, Sri Lanka, Nepal, and Burma; distributed to Mauritius and Thailand.

Part Used: Root, stem bark.

Phytochemistry: Root bark contains apiose, lanceolarin; tannins

Uses: Bark is used as an analgesic and anti-diarrhoeal; seed oil is used in relieving pain and inflammation; useful in rheumatoid arthritis and osteoarthritis.

Erythrina suberosa Roxb.

Vernacular Name: Kathit (Burmese); Corky coral tree, Indian coral tree (English); Dhaul dhak (Hindi).

Description: Deciduous trees, to 10 m high, bark grey, corky, deeply cracked; branchlets tomentose, armed`; leaves trifoliate, alternate; stipules about 5 mm long, lateral, lanceolate; rachis 7.5-12.5 cm long, stout, puberulent, pulvinate; petiolule upto 10 mm; stipels gland like, leaflets 15-10 x 5.5-12 cm, rhomboid-ovate, base deltoid or truncate, apex acute or obtuse, margin entire or sinuate, glabrous above and wooly pubescent beneath, coriaceous; 3-ribbed from the base, lateral nerves 4-5 pairs, pinnate, prominent, intercostae reticulate, faint; flowers bisexual, about 4 cm long, bright scarlet, in axillary and terminal racemes; bracts lanceolate, cauducous; calyx tube about 5 mm long, campanulate, splitting to become bilabiate, glabrous; corolla exserted; petals 5, standard oblong, 3.8 x 2 cm, sessile, the wings minute, keels about 1.8 cm long, connate; stamens 10, monadelphous, the vexillary filament free in the upper two thirds; filaments 6 and 8 mm; anthers uniform; ovary inferior, oblong, downy-pubescent, stipitate, 1-celled, ovules many; style to 1 cm, curved, subulate at apex, not bearded, stigma capitate; fruit a pod, to 15 cm long, linear-falcate, torulose, follicular, with spongy packing between seeds; seeds 2-5, dark reddish-brown, subreniform.

Distribution: Native to Indian subcontinent.

Part Used: Stem bark, leaves.

Phytochemistry: Flowers contain alkaloids (erysodine and erysothrine); flavonoids.

Uses: Bark used in dysentery; leaves used in the ulcer.

Indigofera tinctoria L.

Vernacular Name: Me-me-nai (Burmese); Nilini (Sankrit); Indian indigo (English); Nil (Hindi).

Description: A branching shrub uoto 2 m high, leaves with 7-13 leaflets, green when fresh and greyish back on drying, tender branches bluish red in colour, flowers many in nearly sessile lax spicate racemes which are much shorter the leaves, red or pink; fruits cylindric pods, pale greenish grey when young and dark brown on ripening with 10-12 seeds.

Distribution: Throughout Indian subcontinent.

Parts Used: Whole plant.

Phytochemistry: Rotenoids, deguelin, dehydrodeguelin, rotenol, rotenone, tephrosin and sumatrol.

Uses: Roots, stems and leaves are useful for promoting the growth of hair and in gastropathy, splenomegaly, cephalalgia, cardiopathy, chronic bronchitis, asthma, ulcer and skin diseases; juice expressed from the leaves is useful in the treatment of hydrophobia; extract of the plants is good for epilepsy and neuropathy.

Mimosa pudica L.

Vernacular Name: Hti-gayon-Kon (Burmese); Lajja (Sanskrit); Sensitive plant (English); Lajwanti, Chui mui (Hindi)

Description: Semi-prostrate, prickly course herb or subshurb up to 0.5 m tall; leaves alternate, bipinnately compound which fold up when disturbed; flowers pink with several stamens up to 8 mm long, borne on globose heads; fruit a flat, hairy legume (pod) breaking into 2-4 one-seeded segments.

Distribution: Native to South America and Central America; distributed in Asian countries such as Thailand, Mauritius, Indonesia, Malaysia, Philippines, Japan, and Jamaica.

Part Used: Whole plant.

Phytochemistry: Alkaloid (mimosine), flavonoid C-glycosides, sterols, terenoids, tannins, fatty acid, norepinephrine, gentisic acid, jasmonic acid and D-panitol.

Uses: Leaves used in haemorrhoids and urinary infections; decoction of the roots is also used in urinary infection, dysentery, fever, syphilis, leprosy, stomach worms, veneral disease, insect bite, insomnia, nervousness and piles.

Pterocarpus marsupium Roxb.

Vernacular Name: Padauk, Malabar, Kino, (Burmese); Indian kino tree, Malabar kino tree (English); Bijasal, Vijayasar, Bija (Hindi).

Description: A medium-sized to large tree 15-30 m in height with dark brown of grey bark having shallow cracks, exfoliating in thin flakes and exuding a red gummy substance (Gum kino) on injury; leaves compound, imparipinnate, leaflets 5-7 coriaceous, oblong, obtuse, emarginated or even bilobed at apex, glabrous on both surfaces mina nerves numerous, prominent; flowers yellow in terminal panicles, corolla with crisped margins; fruits nearly circular, glabrous, flat winged pods, convexly curved between stipe and style, wings veined, seeds, 1-2, convex bony.

Distribution: Native to India, Nepal, and Sri Lanka; distributed to Mauritius.

Parts Used: Heartwood, leaves, flowers, gum.

Phytochemistry: Heartwood contains liquiritigenin, isoliquiritigenin; alkaloids and resin; non-glucosidal tannin kinotannic acid, kinoin, catechol, protocatechuic acid, pectin and gallic acid.

Uses: Heartwood is anthelmintic, constipating, anodyne, alternate and elephantiasis; used in inflammations, fractures, bruises, leprosy, skin diseases, leucoderma, erysipelas, urethrorrhea, diabetes, rectalgia, rectitis, ophthalmopathy, haemorrhages, verminosis, diarrhoea, dysentery, odontalgia, gout, rheumatoid arthritis, cough, asthma, bronchitis and greyness of hairs; leaves are useful in boils, sores and skin diseases; flowers are useful in anorexia and leavers; gum is useful in urethrorrhea, odontalgia, diarrhoea, psoriasis, wounds and ulcer, helminthiasis, intermittent fevers, hepatopathy and ophthalmia.

***Senna alata* (L.) Roxb. (Syn. *Cassia alata* L.)**

Vernacular Name: Thinbaw-mezali (Burmese); Emperor's candlesticks, Candle bush, Candelabra bush, Christmas candles, Empress candle plant, Ringworm shrub, Candletree (English); Dadmurdan (Hindi).

Description: A shrub stands 3–4 m tall, with leaves 50–80 cm long; leaves close in the dark; inflorescence looks like a yellow candle; fruit, shaped like a straight pod, is up to 25 cm long; seeds are distributed by water or animals; seed pods are nearly straight, dark brown or nearly black, about 15 cm long, and 15 mm wide; on both sides of the pods is a wing that runs the length of the pod; pods contain 50 to 60 flattened, triangular seeds.

Distribution: Northern S. America - northern Brazil, Colombia, Venezuela, the Guyanas; Indian subcontinent.

Part Used: Whole plant.

Phytochemistry: Leaves contain anthraquinones (aloe-emodin, chrysophanol, isochrysophanol and rhein), alkaloids (tyramine), β-sitosterol, chrysophanic acid; bark contains tannins; petals contain anthraquinones, glycosides, steroids, tannins and volatile oil; stem contains chrysophanol, emodin, rhein and aloe emodin.

Uses: Herb has effective fungicidal properties, for treating ringworm and other fungal infections of the skin; the plant is laxative, antibacterial, antitumor, anti-inflammatory, diuretic, analgesic, vulnerary, hypoglycaemic, and antispasmodic.

***Senna occidentalis* (L.) Link (Syn. *Cassia occidentalis* L.)**

Vernacular Name: Dangywe (Burmese); Septicweed, Coffee senna, Coffeeweed, Mogdad coffee, Negro-coffee, Senna coffee, Stephanie coffee, Stinkingweed, Styptic weed (English) Kasunda (Hindi).

Description: Annual undershrub, subglabrous, foetid, few feet high; leaves alternate, compound, paripinnate; rachis channelled, presence of a gland at the base of the rachis; stipulate, stipules obliquely cordate, acuminate; leaflets 4–5 pairs, size (3.7 cm x 2 cm - 7 cm x 3.5 cm), obate to oblong- lanceolate; acuminate, margin ciliate, glabrous or pubescence; inflorescence axillary corymb and terminal panicle; flowers complete, bisexual, slightly irregular, zygomorphic, pentamerous, hypogynous, pedicelate; bractate, bracts white with pinkish tinge, thin, ovate- acuminate, caducous; yellow; sepals 5, gamosepalous, tube short, 5 lobed, obtuse, glabrous, imbricate, odd sepal is anterior; petals 5, polypetalous, alternisepalous, sub-equal, with distinct claw, conspicuously veined, ascending imbricate, posterior petal is the innermost; stamens 10, free, unequal in size, 7 perfect and 3 reduced to staminode, filaments unequal, anther dithecous, basifixed, introrse and dehiscing by terminal pores; carpel 1, ovary superiour, unilocular, many ovuled, marginal placentation; style simple; stigma terminate, capitate; fruit pod, dehiscent, woody, 12.5 cm X 0.7 cm, glabrous, recurved, subcompressed, distinctly torulose, 23-30 seeded.

Distribution: S. America - Argentina, Uruguay, Paraguay, Brazil, Peru, Ecuador, Colombia, Venezuela, the Guyanas; C. America - Panama to Mexico; Caribbean; Indian subcontinent.

Part Used: Whole plant.

Phytochemistry: Plant contains anthraquinones; roots contain emodin; seeds contain chrysarobin (1,8-dihydroxy-3-methyl-9-anthrone) and N-methylmorpholine.

Uses: Seeds used to treat diuretic, haemorrhoids, gout, laxative, rheumatism, diabetes, and rheumatism; the plant is diuretic, febrifuge, stomachic and tonic; used in the treatment of hypertension, dropsy, fevers, biliousness, ringworm and eczema.

Tamarindus indica L.

Vernacular Name: Bur-magyi (Burmese); Tamarind tree (English); Imli, Amli, Ampli (Hindi).

Description: A large to very large evergreen tree up to 30 m in height with dark grey bark having longitudinal fissures and deep cracks; leaves paripinnate up to 15 cm long. Rachis slender, channeled, leaflets 10-12 pairs, subsessil, oblong; flowers yellow, striped with red in laxz, few-flowered racemes at the end of the branchlets; fruits pods, brownish ash-coloured, slightly curved, subcompressed, with a shallow pit on each side of the flat faces; seeds enveloped by a toughy leathery membrane (the endocarp) and pulpy mesocarp, tests shining, hard.

Distribution: Throughout Indian subcontinent.

Parts Used: Roots, Fruits, Seeds.

Phytochemistry: Fruit contains tartaric acid, citric maleic acid and potassium bitartrate and traces of oxalic acid, kernel, and polysaccharides; leaves contain flavonoid glycosides saponaretin, vitexin, orientin and homoorientin; leaves, bark and flower contain hordenine.

Uses: Root bark is useful in diarrhoea, asthma, amenorrhoea, singivitis and ulcer; leaves are useful in swellings, fever, scalding urine, gastropathy, helminthiasis, wounds, ulcer, jaundice, scabies, tumours, ringworm, boils, smallpox, otalgia and conjunctivitis; fruits are useful in gastropathy, bilious vomiting, datura poisoning, alcoholic intoxication dipsia, scabies, pharyngitis, pharyngodynia, stomatitis, constipation, haemorrhoids and ophthalmopathy; seeds are useful in stomachalgia, diarrhoea, dysentery, dipsia, burning sensation, haematuria, giddiness, vertigo, hepatopathy, inflammations, chronic ulcer, abscess, haemorrhoids, vaginopathy, metroptosis, diabetes and general debility.

Tephrosia purpurea (L.) Pers.

Vernacular Name: Bur-me-yaing (Burmese); Wild indigo, Purple tephrosia (English); Sarphomka (Hindi).

Description: A much-branched suberect herbaceous perennial, 30-60 cm in height with spreading branches; leaves imparipinnate, leaflets 11-21, narrow, oblancloate; flowers red or purple in extra axillary racemes; fruits slightly curved pods, 3-4.5 cm, long; seeds 5-10 per pod, grey, smooth.

Distribution: Distributed in Indian subcontinent.

Parts Used: Whole plant.

Phytochemistry: β-Sitosterol, lupeol, retin, delphinidin chloride, cyaniding chloride, isolonchocarpin, lanceolatins A and B, pongamol, karangin, kangone, 5, 7-demethoxy-8,flavanone, 2-methoxy-3-9-dihydroxycoumestone, flevichapparins B and C, methyl karanjic acid and purpurin.

Uses: Roots are useful in inflammations, skin diseases, scrofula, elephantiasis, dyspepsia, stomachalgia, flatulence, haemorrhoids, asthma, bronchitis, anaemia, hepatosplenomegaly, verminosis, strangury, dysmenorrhoea, chronic fever, boils pimples, odontalgia and gingivitis; leaves are useful in dyspepsia, pectrol diseases haemorrhoids, syphilis, gonorrhoea and bruises; seeds are useful in skin disease and rat poisoning.

Trigonella foenum-graecum L.

Vernacular Name: Pe-matha (Burmese); Fenugreek, Greek hayes (English); Methi, Mutti (Hindi); Methika, Methi, Kalanusari (Sanskrit).

Description: An aromatic, erect annual, 30-60 cm in height; leaves pinnate, 3-foliate, leaflets toothed; flowers white or yellowish-white, axillary; fruits pods, 5-7.5 cm long with long persistent beak; seeds 10-20 per pod, greenish-brown along with a deep groove across one corner.

Distribution: Cultivated throughout the Indian subcontinent.

Parts Used: Leaves and seeds.

Phytochemistry: Seed contains galactomannan, carbohydrates, sugar, amino acids, fatty acids, vitamins, saponins, diosgenin, gitogenin, neogitogenin, homoorientin, saponaretin, neogigogenin and tigogenin.

Uses: Leaves are used for swellings and burns; seeds are good for fever vomiting, anorexia cough, bronchitis, and colonitis; extermally in the form of poultics they are used for boils, abscess and ulcers; infusion of the seed is good cool drink for smallpox patients; powdered seeds find application in veterinary medicine; seeds possess antibacterial property.

Linaceae

Linum usitatissimum L.

Vernacular Name: Bur-hnan-Kyat (Burmese); Atasi (Sanskrit); Linsed, Flax plant (English); Alsi, Tisi (Hindi).

Description: An annual herb, 60-120 cm high; stems solitary or few, corymbosey branched; branches ascending towards the apex; leaves upto 3.8 cm long, linear-lanceolate, flowers small, blue, bluish violet or white in terminal panicles; fruits capsular with 5-cells; seeds compressed, ellipsoid, smooth, dark brown, shiny.

Distribution: Cultivated throughout the world.

Parts Used: Flowers, seeds oil.

Phytochemistry: Seeds contain protein, fatty acids, carbohydrates, carotene, thiamin, riboflavin, niacin, pantothenic acid, choline and vitamin E.

Uses: Flowers are considered as cardiotonic; seeds used in coughs, catarrh and bronchial affections, ureteritis, gonorrhoea and diarrhoea; crushed linseed is applied in the form of a poultice for the relief of local inflammations, ulcer, boils and carbuncles.

Lythraceae

Lawsonia inermis L.

Vernacular Name: Bur-dan, Henna-Plant (Burmese); Medhini, Madayantika, Nakharanjani (Sanskrit); Henna, Egyptian priven, Cypress shrub (English); Mehandi (Hindi).

Description: A glabrous much-branched deciduous shrub with 4-gonous lateral branches often ending in spines; leaves simple, opposite, entire, lanceolate, petioles very short or absent; flowers white, or rose-coloured, fragrant, in large terminal pyramidal panicled cymes, stamens 8, in 4 pair inserted on the calyx tube; fruits globose capsules, tipped with the style and supported by the persistent calyx, seeds numerous, smooth, pyramidal.

Distribution: Cultivated in throughout Asia.

Parts Used: Roots, leaves, flowers, seeds.

Phytochemistry: Plant contains lawsone, behenic, arachidic, stearic, palmitic and oleic acids.

Uses: Roots are useful in burning sensation, dipsia, leprosy, skin disease, strangury, amenorrhoea, dysmenorrhoea, and premature graying of hair; leaves are useful in wounds, ulcers strangury cough, bronchitis, cephalalgia, hemicranias, lumbago, rheumatalgia, inflammations, diarrhoea, dysentery, leprosy, leucoderma, scabies, boils, hepatopathy, splenopathy, anaemia, haemorrhages, haemoptysis, fever, ophthalmia, amenorrhoea, dysmenorrhoea, falling of hair, greyness of hair and jaundice; flowers are useful in cephalalgia, burning sensation, sardiopathy, amentia, insomnia, and fever.

Punica granatum L.

Vernacular Name: Bur-thale (Burmese); Dadima (Sanskrit); Pomegranate (English); Anar (Hindi).

Description: Shrub to small tree to up to 6 m high; leaves mostly opposite, short-petiolate, blades oblong-elliptical up to 8 cm long; flowers showy and up to 6 cm board, bisexual, 5-8 petals, reddish and up to 2.5 cm long, numerous stamens surrounding a conspicuous hypanthial tube, the flowers usually occurring terminally or in axils; fruit a red spherical berry; flowers available during the summer.

Distribution: Native to Iran; widely cultivated throughout the Middle East and Caucasus region, north Africa and tropical Africa, the Indian subcontinent, Central Asia, the drier parts of southeast Asia, and parts of the Mediterranean Basin; introduced into Spanish America and California

Part Used: Seeds, flowers.

Phytochemistry: Apigenin glucoside, betulinic acid, callistephin, tannins, chrysanthemin, coniine, coumestrol, cyaniding and its diglucoside, cyanin, diadzein, diadzin, delphin, delphinidin and its glucosides, ellagic acid and its derivative, estrone, freidelin, gallic acid, genistein, genistin, lipids, hygrine and norhygrine, luteolin glycosides, mannitol, pelargonin, pelletierine and its derivative, piperidine derivatives, polyphenols, sedridine, xanthoxylin and estradiol.

Uses: Decoction of the seed is used to treat syphilis; juice of the fruit is used to treat jaundice and diarrhoea; rind of the fruit is ground in water and drunk every morning by diabetics; fruit for common cold; juice of the flowers is used to treat nose bleeds; fruit pulp and seed are a stomachic; root and stem bark have astringent and anthelmintic properties.

Magnoliaceae

Magnolia champaca (L.) Baill. ex Pierre. (Syn. Michelia champaca L.)

Vernacular Name: Bur-Saga, Sagawa (Burmese); Champak, Golden champa (English); Champa (Hindi).

Description: Tree grows to 50 m or taller; trunk can be up to 1.9 m in diameter; flowers are fragrant, tepals 15-20, yellow, inverted-lanceshaped, 2-4 x 0.4-0.5 cm; staminal connective is protruding and forming a long tip. Buds, young twigs, young petioles, and young leaf blades are pale yellow velvet-hairy; twigs are ascending and forming a narrow umbelliform crown; stipular scar 0.3-1 x as long as petiole; leaf-stalks are 2-4 cm, leaves elliptic or ovate, 10-20 x 4.5-10 cm, slightly puberulous below, base broadly wedge-shaped or rounded, tip long-pointed tp falling off; fruit is 7-15 cm; mature carpels obovoid-ellipsoid, 1-1.5 cm, tuberculate; seeds 2-4 per carpel, rugose.

Distribution: Native to Bangladesh, China, India, Indonesia, Malaysia, Mauritius, Myanmar, Nepal, Thailand, and Vietnam.

Part Used: Flower, fruits, bark.

Phytochemistry: Stem bark contains iiriodenine, macheline and lanuginosine; leaves contain polyisoprenoid, β- sitosterol and liridenine, roots contain champakin, liriodenine and micheliolide; flowers contain essential oil (champacene and linalool)

Uses: Flowers and stem bark are useful in diabetes, quick wound healing, cardiac disorders, gout and dysuria; oil obtained from flowers used in perfumery.

Malpighiaceae

Hiptage benghalensis (L.) Kurz

Vernacular Name: Bur-Bewe (Burmese); Madhavi, Atimukha (Sanskrit); Clustered hiptage (English); Madhavilata, Madhumalati (Hindi).

Description: A large, woody much-branched evergreen climbing shrub with silky young parts; leaves simple, opposite, coriaceous, elliptic-oblong, ovate lanccolate, 10-23 cm long; flowers yellowish-white, fragrant, in erect finely tomentose recemes, petals twice as long as the sepals, the uppermost petal broader than the rest, fringed on the margin; fruit 2-3.

Distribution: Throughout the Indian subcontinent.

Parts Used: Bark, leaves, flowers.

Phytochemistry: Bark contains hiptagin and tannin.

Uses: Bark, leaves and flowers are useful in wounds ulcer, cough, asthma, cardiac debility, inflammations, skin diseases leprosy, rheumatism and hyperdipsia.

Malvaceae

Abutilon indicum (L.) Sweet

Vernacular Name: Bauk-Khway-galay (Burmese); Atibala (Sanskrit); Country mallow (English); Kanghi (Hindi).

Description: A Perennial, softly tomentose, up to 3 m in height; stems round, frequently tinged with purple; leaves ovate to orbicular-cordate, 1.9-2.5 cm, long; flowers solitary on jointed peduncles, orange-yellow or yellow; capsules hispid, hardly larger thin the calyx, awns erect; seeds 3-5, reniform, tubercled or minutely stellate-hairy, black or dark brown; flowers and fruits almost throughout the year.

Distribution: Indian subcontinent.

Part Used: Seeds, leaves, bark, root.

Phytochemistry: Leaves contain mucilage, tannin, organic acid, Asparagine; root contains asparagine.

Uses: Seeds used in piles; leaves locally applied to boils and ulcer; decoction used in toothache and tender gums; given internally for inflammation of bladder; roots used in fever, chest infection and arthritis.

***Bombax ceiba* L. (Syn. *Salmalia malabarica* (DC.) Schott & Endl)**

Vernacular Name: Letpan (Burmese); Salmali, Moca (Sanskrit), Red silk cotton tree (English); Semal (Hindi).

Description: A tall deciduous tree with a straight buttressed trunk having a clear bole of 24-30 m and spreading branches, bark grey or brown covered with hard, sharp, conical prickles; leaves large, spreading, glabrous, digitate, leaflets lanceolate, 3-7, entire; flowers red, numerous, appearing when the tree is bare of leaves, stamens many arranged in five bundles of 9-12 each and an inner bundle of 15; fruits capsules; dehiscing by 5 leathery or woody valves; seeds smooth, black or grey embedded in long white wool; dried gum light brown in colour resembling the galls, and gradually become opaque and dark brown.

Distribution: Distributed throughout the Asian subcontinent.

Part Used: Roots, gum, bark, leaves, flowers, young fruits, seeds.

Phytochemistry: Bark contains tannins, lupeol, β-sitosterol and its D-glucoside.

Uses: Roots are used in dysentery; gum is useful in dysentery, haemoptysis of pulmonary tuberculosis, influenza, menorrhagia, burning sensation, strangury, haemorrhoids, and blood impurities; bark is used for fomenting and healing wounds; paste of it is good for skin eruption; leaves are good for strangury and skin eruptions; flowers are good for skin troubles, splenomegaly and haemorrhoids; young fruits are useful in calculus affections, chronic inflammations and ulceration of the bladder and kidney; seeds are useful in treating gonorrhoea, and chronic cystitis.

Hibiscus rosa-sinensis L.

Vernacular Name: Bur-Kaung-Yan, Pan-swe-lwe (Burmese); Japa, Java, Rudra pushpam (Sanskrit); Red hibiscus, Rose of China, Chinese hibiscus (English); Jasut, jasum, Gudhal (Hindi).

Description: An evergreen woody, glabrous, showy shrub, 5-8 ft. high; leaves bright green, ovate, entire below, coarsely toothed above; flowers solitary, axillary, bellshaped, with pistil and stamens projecting from the center; capsules roundish, many seeds.

Distribution: Cultivated throughout the Indian subcontinent.

Parts Used: Leaves and flowers.

Phytochemistry: Taraxeryl acetate, β-sitosterol, campesterol, stigmasterol, cholesterol, erogosterol, lipids, citric, tartaric and oxalic acids, fructose, sucrose, flavonoids, falconoid glycosides, hibiscetin, cyanidine, cyanin glucosides and alkanes.

Uses: Plant used to induce abortion, ease menstrual cramps and to help in childbirth; treat headaches; leaves are used to treat postpartum relapse sickness, to treat boils, sores and inflammation.

Pterygota alata (Roxb.) R.Br. (Syn. Sterculia alata Roxb.)

Vernacular Name: Bur-sinkadet, letkok (Burmese); Buddha coconut (English).

Description: A tall tree; leaves are carried on 3-10 cm long stalks, crowded towards the ends of branches; leaves are blade broadly ovate-heartshaped, 10-25 cm long, 7-15 cm broad, wavy, smooth, pointed or tapering; flowers are borne in small, few-flowered racemes; flowers are 1-1.5 cm across, on 2-3 mm long stalks; flowers have no petals, sepals are 5, nearly free, linear-lance- shaped or elliptic, 1.2-1.5 cm long, 3-4 mm broad, fleshy, densely ferruginous pubescent outside, sparsely pubescent and purple with red streaks within; anthers in male flowers are united into 1-2 mm broadhead on 4-6 mm long staminal column; in bisexual flowers, sessile anthers are arranged in clusters of 4 or 5 in the sinuses formed by the carpels; carpels are 5; ovaries sessile, 2-3 mm long, pubescent; style recurved; fruit is large, woody, 7-12 cm in diameter, obliquely round; seeds are about 40 per follicle, oblong, compressed, in 2 rows, winged.

Distribution: Buddha coconut is native to India, and found variously in South East Asia; occur in tropical Asia and Africa; distributes naturally in China, Vietnam, India, Mauritius and Philippines.

Part Used: Stem, leaves.

Phytochemistry: Stem and leaves contain phenylpropanoid galactoside, triterpenes and flavonoids.

Uses: Stems and leaves used to treat skin diseases.

***Sida cordifolia* L.**

Vernacular Name: Kat-si-ne "Bala" (Burmese); Bala (Sanskrit); Kungyi (Hindi).

Description: An undershrub, branched, softly hairy and with stellate hairs, nearly in all the parts of the plant; leaves are 2.5-5 cm long, cordate, ovate-oblong, crenate, obtuse or subacute and not acuminate and the petioles are 1.2-3.8 cm long; pedicels are solitary and few together, short, some upto 1.2-2 cm long bearing yellow flowers; calyx is 6-8 mm long with ovate and acute lobes; corolla exceed the calyx slightly; fruit is 6-8 mm in diameter; strongly reticulated with two spiny projections on the top.

Distribution: Distributed throughout the tropical and sub-tropical Indian subcontinent.

Part Used: Whole plant.

Phytochemistry: Stem, roots and seeds contain alkaloid (ephedrine), fatty acids, phytosterol, mucin, and resin; rutin.

Uses: Plant has tonic and aphrodisiac properties; used for the fever, fits, leucorrhoea, micturition, gonorrhoea, colic, nervous disorders, general debility and heart irregularity; root juice is used for the healing of the wound and the juice of the whole plant is given for spermatorrhoea; bark of the plant is effective in curing facial paralysis and sciatica and the leaves are used for curing the bloody flux; useful in ophthalmia, rheumatism; shivering fits and improves sexual strength.

Menispermaceae

***Cissampelos pareira* L.**

Vernacular Name: Kywet-nabaung (Burmese); Akaudi, Paari (Hindi); Ambashtha, Patha, Laghu patha (Sanskrit).

Description: Slender, twining, perennial herb; leaves 2.5-6.0 x2.0-4.5 cm, ovate or orbicular, peltate, truncate, cordate at base, softly pubescent on both the surfaces, palmately 5-7 nerved; petioles up to 8 cm long; flowers small, greenish-yellow, male flowers greenish-yellow, in axillary, fascicled, pilose cymes or panicles; female flowers in pendulous racemes 6.5-10.5 cm long; drupes ovoid; endocarp reniform, compressed with hollowed tubercled sides on the back; seeds 0.05-0.06 cm long, curved.

Distribution: Common throughout in lower valleys between 600-1800 m.

Part Used: Leaves and roots.

Phytochemistry: Tubers contain berberine, palmatine, and other bisbenzylisoquinoline alkaloids.

Medicinal uses: Roots are antidiarrhoeal, antipyretic, diuretic and laxative; used in leucorrhoea, gonorrhoea and chronic inflammation of the bladder; decoction of roots is given in enlargement of spleen; pounded roots are used internally in indigestion and headache; roots are also used to lower the blood pressure; leaf juice is used in eye problems; externally the plaster of crushed leaves is applied on various skin ailments, burns, wounds and also applied on forehead in fever.

Tinospora sinensis (Lour.) Merr. (Syn. T. cordifolia (Willd.) Miers)

Vernacular Name: Bur-sindon-ma-new, (Burmese); Vatsadani, Sudarsana, Amrta (Sanskrit); Giloy, Gruch (Hindi).

Description: Deciduous, climbing shrubs; stem with prominent, scattered lenticels; bark papery, greyish-white; latex watery; tender branches with distinct white hairs; leaves alternate, broadly egg-shaped, 7-17 x 4-13 cm, base heart-shaped, apex acuminate, margin entire, sparingly pubescent in upper surface and densely pubescent in lower surface, basal nerves 5-7; leaf-stalk up to 110cm long, hairy; flowers are formed in racemes from the axils of fallen leaves and also on the old stem, greenish-yellow, many, 5 mm across; male and female flowers are seen separately in the same plant; drupes 1-3 in cluster, spherical-shaped, about 1.5 cm across, turning orange-red when ripe.

Distribution: Indigenous to the tropical areas of India, Myanmar, Mauritius and Sri Lanka.

Parts Used: Stems.

Phytochemistry: Plant contains columbin, tinosporaside, jatrorhizine, palmatine, berberine, tembeterine, tinocordifolioside, phenylpropene disaccharides, choline, tinosporic acid, tinosporal, and tinosporon.

Uses: Stems are used for treating piles, ulcerated wounds, liver complaints, chronic rheumatism and also as a muscle relaxant.

Moraceae

Ficus benghalensis L.

Vernacular Name: Bur-Pyi-nyaung (Burmese); Nyagrdhan, Vatah (Sanskrit); Banyan tree (English); Bat, Bargad (Hindi).

Description: A very large tree upto 30 m in height with widely spreading branches bearing many aerial roots functioning as prop roots, bark greenish-white, leaves simple, alternate often in clusters at ends of branches, stipulate, 10-20 cm long and 5-12.5 cm broad, broadly elliptic to ovate, entire, coriaceous, strongly 3-7 ribbed from the base; the fruit receptacles are axillary, sessile, in pairs, globose, brick-red when ripe, enclosing male, female and fall flowers; fruits small, crustaceous achenes, enclosed in the common fleshy receptacles.

Distribution: Native to the Indian Subcontinent.

Parts Used: Aerial root, bark, leaves, buds, fruits.

Phytochemistry: Bangalenoside, leucocyanidin 3-O-β-D-galactosyl cellobioside 5,3'-dimethyl ether and leucopelargonidin-3-O-α-L-rhamnoside 5,7-dimethyl ether, leucopelargonidin glycoside leucocyanidin glycoside.

Uses: Aerial roots are useful in obstinate vomiting, leucorrhoea and osteomalacia of the limbs; bark is useful in burning sensation, haemoptysis, haemorrhages, diarrhoea, dysentery, diabetes, enuresis, ulcer skin diseases, gonorrhoea, leucorrhoea and hyperdipsia; leaves are good for ulcer, leprosy, allergic conditions of skin, burning sensation and abscesses; latex is useful in neuralgia, rheumatism, lumbago, bruises, mastitis, ulorrhagia, ulitis, odontopathy, haemorrhoids, gonorrhoea, inflammations, crack of the sole and skin diseases.

Ficus religiosa L.

Vernacular Name: Bur-nyaung-bawdi (Burmese); Pippalah, Asvatthah (Sanskrit); Peepal tree, Sacred fig (English); Pippal, Pipli (Hindi).

Description: A large deciduous tree with few or no aerial root, often epiphytic, the dropping branchlets bear long-petioled ovate, cordate, shiny leaves which produce a characteristic rustling sound when the wind blows; leaves bright green, the apex produced into a linear-lanceolate tail about half long as the main portion of the blade; the receptacles occurring in pairs, axillary, depressed globose, smooth, purplish when ripe.

Distribution: Native to the Indian Subcontinent.

Parts Used: Bark, leaves, tender shoots, fruits, seeds.

Phytochemistry: Arabinose, mannose, glucose, phenolic glucoside ester, flacourtin; steroids, ramontoside, β-sitosterol and its β-D-glucopyranoside.

Uses: Bark has antibacterial activity; powdered bark is used for inflammation and burns; leaves and tender shoots have purgative properties and recommended for wounds and skin disease; fruit is laxative and digestive; the dried fruit pulverized and taken in water cures asthma; seeds refrigerant and laxative; latex is good for neuralgia, inflammations and haemorrhages.

Ficus elastica Roxb. ex Hornem.

Vernacular Name: Bur-nyaung-Kyot-paung (Burmese); Rubber fig, Rubber bush, Rubber tree, Indian rubber bush (English).

Description: a large tree in the banyan group of figs, growing to 30–40 m tall, with a stout trunk up to 2 m in diameter; trunk develops aerial and buttressing roots to anchor it in the soil and help support heavy branches; broad shiny oval leaves 10–35 cm long and

5–15 cm broad; leaf size is largest on young plants, much smaller on old trees; leaves develop inside a sheath at the apical meristem, which grows larger as the new leaf develops; when it is mature, it unfurls and the sheath drops off the plant; inside the new leaf, another immature leaf is waiting to develop.

Distribution: Native to northeast India, Nepal, Bhutan, Burma, China, Malaysia, Mauritius, and Indonesia; naturalized in Sri Lanka, the West Indies, and the USA.

Part Used: Whole plant.

Phytochemistry: Root contains chlorogenic, p-coumaric, ferulic and syringic acids; stem contains chlorogenic p-coumaric and ferulic acids; leaves contain caffeic acid; latex contains polyisoprenes and resin.

Uses: Decoction of the aerial roots is used as a vulnerary; latex is effective for trichuriasis.

Musaceae

Musa paradisiaca L.

Vernacular Name: Hnget-pyaw (Burmese); Kadali (Sankrit), Banana (English); Kela (Hindi).

Description: A tall herb with aerial pseudo stem dying after flowering, leaves oblong, narrowed to base; flowers unisexual in spikes, dropping, females at the bottom and males at the top, bracts conspicuous, dull brown, falling off in succession; fruits berries in several clusters, golden yellow colour on ripening.

Distribution: Cultivated throughout Asian countries.

Part Used: Roots, leaves, fruits, stem.

Phytochemistry: Fruits contain carbohydrates, catecholamines (norepinephrine, serotonin, dopamine), tryptophan, indole compounds, pectin, leucocyanidin, quercetin and its 3-O-galactoside, 3-O-glucoside, and 3-O-rhamnosyl glucoside.

Uses: Roots are useful in venereal diseases, helminthiasis, scabies, leprosy, skin diseases and debility; leaves are useful in scabies, inflammations, ophthalmopathy, blisters and burns; fruit is useful in dipsia, helminthiasis, scabies, pruritus, bronchitis, pharyngodynia, nephropathy, strangury and general debility; flowers are good for dysentery, diabetes, ascites and dropsy.

Myristicaceae

Myristica fragrans Houtt.

Vernacular Name: Bur-zadeik-pho (Burmese); Jatiphala (Sanskrit); Nutmeg tree (English); Jaiphal, Japatri (Hindi).

Description: A lofty tree with branches slender; leaves 7-10 cm, coriaceous, dark green, entire, elliptic-oblong or lanceolate, acuminate both at apex and base, glauxous beneath, nerves about 8 pairs, slender; petiole 6-12 m. flowers dioecious; male flowers 3-5 in racemes, 2.5-5 cm, flowers 6 mm long ellipsoid nodding, with a scale-like bracteole under the glabrate perianth; anthers 9-12, connate in a cylindric stipulate column; pistillate flowers, yellow, small axillary, often solitary; fruit ovoid, sub-globose or pyriform, 4-5 cm. long, valved; seed one, almost completely surrounded by a fleshy, irregularly slit or holed arillode.

Distribution: Indigenous to Indonesia; widely grown across the tropics including China, Taiwan, Indonesia, Malaysia, Mauritius, Caribbean, India, Sri Lanka and South America.

Part Used: Fruit.

Phytochemistry: Fatty acids (β-pinene, myristicin, linalool, safrol, dl-α-pinene, dipentene, and linalyl acetate).

Uses: Oil is a potent brain booster; relieves stress; liver tonic; prevents infection; the herb is also beneficial for the heart as it stimulates blood circulation; it has anti-inflammatory properties and used to treat joint and muscle aches.

Myrtaceae

Syzygium cumini (L.) Skeels (Syn. Eugenia cumini (L.) Druce)

Vernacular Name: Thabyebyu, sapace-sb-bye (Burmese); Jaman, Jambolan, Black plum (English); Jamun, Jambhal, Jaman (Hindi).

Description: A medium-sized to large tree, 15-30 m in height with smooth light grey bark having dark patches; leaves simple, opposite, variable in shape, about 2.5 cm broad and 7.5-15 cm long, acuminate, nerves joining in a distinct inframarginal nerve, gland-dotted, smooth and shiny; flowers greenish-white in trichotomous panicls; fruits oblong or ovoid-oblong, dark purple with pinkish juicy pulp; one-seeded.

Distribution: Throughout Asian countries.

Parts Used: Bark, leaves, fruits.

Phytochemistry: Seeds contain jambolin, ellagic acid, tannin, gallic acid, fatty acids, starch and resin; flowers contain acetyl oleanolic acid, ellagic acid, isoquercitrin, quercetin, kaempferol and myrcetin.

Uses: Bark is useful in fever, gastropathy, strangury and dermatopathy; leaves are antibacterial and are used for strengthening the teeth and gums; leaves are used for vomiting; fruit and seeds are used in diabetes, diarrhoea, pharyngitis, splenopathy, urethrorrhea and ringworm.

Nelumbonaceae

Nelumbo nucifera Gaertn.

Vernacular Name: Padung ma (Burmese); Ambuja, Padma, Pankaja, Kamala (Sanskrit); Kanwal, Kamal (Hindi); Lotus (English).

Description: Perennial aquatic herb; root-stock stout, cylindrical, embedded in the mud; leaves peltate, rediately nerved; margins wavy; petiole long, aculeate, inserted in the middle of the leaf; flowers large, solitary, handsome and fragrant, rosy or white, carpels numerous, ovoid, fleshy, sunk separately in cavities of receptacle, maturing into nut-like achenes; skin hard and blackish-brown when ripe.

Distribution: Cultivated in Asian and African countries.

Part Used: Leaves, seed, receptacles, filament, plumules.

Phytochemistry: Leaves contain alkaloids (nuciferine, roemerine, nor-nuciferine), flavonoid (quercetin); plumules contain proteins, sugars and vitamins; receptacles contain quercetin.

Uses: Ripe seeds produce a wholesome effect in cases of neurasthenia, spermatorrhea and metrorrhoea; leaves and seed cores in decoction are effective for insomnia, haemorrhage and haematemesis; filaments or the receptacles in the form of a decoction are used in treating bloody stools, uterine haemorrhage and haematemesis.

Nyctaginaceae

Boerhavia diffusa L.

Vernacular Name: Payan-nawa (Burmese); Punarnava, Sophaghin (Sanskrit), Hogweed, Pigweed (English); Sant, Bishkhapra, Gadahpurna (Hindi).

Description: A perennial diffuse herb with stout rootstock and many procumbent branches; leaves simple, opposite, short-petioled in unequal pairs, ovate-oblong, acute or obtuse, rounded or subcordate at base, glabrous above, and whitish beneath; flowers pale rose coloured, small, short-stalked, in irregular of terminal panicles at the ends of branches; fruits highly viscid, easily detachable, one-seeded, indehiscent with a thin pericarp.

Distribution: Distributed throughout the world.

Part Used: Whole plant.

Phytochemistry: Alkaloids, triacontanol hentriacontane, β-sitosterol, ursolic acid, 5, 7-dihydrixy-3, 4-dimethoxy-6, 8-dimeyhyl flavones, glucose, fructose, sucrose, hypoxanthine-9-L-arabinoaide, moulding hormone, and β-ecdysone.

Uses: Plant is useful in all types of inflammations, strangury, leucorrhoea, ophthalmia, lumbago, myalgia, scabies, cardiac disorders, jaundice, anaemia, dyspepsia, constipation, cough, bronchitis and general debility.

Nymphaeaceae

Nymphaea alba L.

Vernacular Name: Kya-byu (Burmese); White water lily (English); Lily (Hindi).

Description: A perennial aquatic plant with deep roots, plate-shaped leaves on long cylindrical stems, and large-petalled white flowers occasionally tinged pink.

Distribution: Native to Europe; distributed to the Asian subcontinent.

Part Used: Rhizomes and flowers.

Phytochemistry: Rhizome contains alkaloids, resin, glycosides and tannins.

Uses: Rhizome used to treat bronchial catarrh and kidney pain, and taken as a gargle for sore throats; flowers have long been reputed to reduce sexual drive; useful in the treatment of insomnia, anxiety and similar disorder.

Oleaceae

Jasminum grandiflorum L.

Vernacular Name: Bur-myatlay (Burmese); Jati, Malati (Sanskrit); Spanish jasmine, Common jasmine, Catalonian jasmine (English); Jati, Chameli (Hindi).

Description: A large nearly glabrous, suberect, twining evergreen shrub; leaves imparipinnately compound, opposite, terminal leaflets larger than the laterals, ovate, lanceolate, acute or acuminate, lateral ovate; flowers very fragrant, white tinged with pink outside in lax terminal and axillary cymes, calyx lobes long and linear; fruits elliptic, globose berries, black when ripe.

Distribution: Cultivated throughout the world.

Parts Used: Whole plant.

Phytochemistry: Linalyl acetate, benzyl benzoate, benzyl alcohol, geraniol, nerol, I-α-terpineol, d-and di-linalool, and alcohol with an odour reminiscent of β, γ-hexenol, farnesol nerolidol, eugenol, p-creosol, lactones, benzaldehyde, jasmone, methyl anthranilate and indole.

Uses: Plant is bitter, astringent, acrid, thermogenic, aphrodisiac, antiseptic, anodyne, depurative, emmenagogue, emollient, diuretic, anthelmintic, deobstruent, dentifrice, suppurative and tonic; roots are useful in cephalalgia, paralysis, facial paralysis, mental debility, chronic constipation, flatulence, strangury sterility, dysmenorrhoea, amenorrhoea, ringworm, leprosy, skin diseases and giddiness; leaves are useful in odontalgia, fixing loose, teeth, ulcerative stomatitis, leprosy, skin diseases, otorrhoea, otalgia, strangury, dysmenorrhoea, ulcers, wounds and corns; flowers are useful in stomatopathy, cephalopathy, odontopathy, ophthalmopathy, leprosy, skin diseases, pruritus, strangury, dysmenorrhoea, ulcer.

Phyllanthaceae

Phyllanthus emblica L. (Syn. Emblica officinalis Gaertn.)

Vernacular Name: Bur, zibyu, Tasha (Burmese); Amalaki, Dhatri (Sanskrit); Emblic myrobalan, India gooseberry (English); Amlika Amalak, Amvala (Hindi).

Description: A small to medium-sized deciduous tree, 8-18 m in height with thin light grey bark exfoliating in small thin irregular flakes; leaves simple, very many, subsessile, closely set along the branchlets, distichous, light green having the appearance of pinnate leaves; flowers greenish-yellow, in axillary fascicles, unisexual, males numerous on short slender pedicels, females few, subsessile, ovary 3-called; fruits globes, fleshy, pale tallow with six obscure vertical furrows enclosing 6 trigonous seeds in 2-seeded 3crustaceous cocci.

Distribution: Throughout Indian subcontinent.

Parts Used: Root bark, leaves, fruits.

Phytochemistry: Amino acids (alanine, aspartic acid, glutamic acid lysine, and proline); fruit pulp contains protein, fat, carbohydrates, fibre, minerals, niacin, and vitamin.

Uses: Root bark is useful in gonorrhoea, jaundice, diarrhoea and myalgia; leaves are useful in conjunctivitis, inflammation, dyspepsia, diarrhoea and dysentery; fruits are useful in diabetes, cough, asthma, bronchitis, cephalalgia, ophthalmopathy, dyspepsia, colic, flatulence, hyperacidity, peptic ulcer, erysipelas, skin diseases, leprosy, haematemesis, inflammations, anaemia, emaciation, hepatopathy, jaundice, strangury, diarrhoea, dysentery, haemorrhages, leucorrhoea, menorrhagia, cardiac disorders, intermittent fevers and greyness of hair.

Piperaceae

Piper betle L.

Vernacular Name: Bur-Kun, pu-betel (Burmese); Nagavalli, tambula (Sanskrit); Betel leaf vine, Betel leaf pepper, Betel pepper (English); Pan, Tambuli (Hindi).

Description: A slender creeper with adventitious roots; stems glabrous, sulcate, thickened at the nodes,; leaves alternate, heart-shaped, palmately nerved, glabrous, and shining on both side; inflorescence in drooping, dense axillary spike, consisting of male and female flowers berry globose, hairy at the apex (rarely produced); all parts of the plant have a special aroma.

Distribution: Cultivated throughout the Indian subcontinent.

Parts Used: Leaves and roots.

Phytochemistry: Plant contains an essential oil (eugenol, carvacrol, chavicol, allyl catechol, chavibetol, cineol, estragole, methyl eugenol, p-cymene, caryophyllene and cadinene); tannin; sugars; carotene, thiamine, riboflavin, nicotinic acid, vitamin C and amino acids.

Uses: Leaves possess antibacterial properties and are beneficial in the treatment of purulent periodontosis in the form of a collutory made of the juice or extract; poultice of the leaves and a wash with the decoction are used in treating wounds, burns, impetigo, furunculosis, eczema and lymphangitis; leaves if topically applied to the chest cure cough and asthma, and if applied to the breast arrest lactation; roots are used in treating rheumatism.

Piper nigrum L.

Vernacular Name: Bur-nagayokkanug, Peikclin (Burmese); Maricha, Ushana, Hapusha (Sanskrit); Black pepper, Common pepper, White pepper (English); Kalimirch, Kalamorich, Golmorich (Hindi).

Description: Climbing perennial plant, stems glabrous, rooting at the nodes; leaves alternate, coriaceous, rounded at the base; apex pointed; recurved nerves prominent beneath; inflorescence in drooping spike of dioecious flowers, opposite to and shorter than the leaf; berry globular, red when ripe, turning black after drying, 3-4 mm in diameter, strongly scented and bitter to the taste.

Distribution: Native to India; extensively cultivated in tropical regions.

Part Used: Fruit, seeds.

Phytochemistry: Seeds contain alkaloids (piperine, piperetin, chavicine); essential oil (terpene, phellandrene, caryophyllene, piperonal-dihyrocarbeol and caryophyllene oxide).

Uses: Berries are used for treating dyspepsia, vomiting, diarrhoea and colic; powdered berries applied topically to cure toothache.

Plumbaginaceae

Plumbago zeylanica L.

Vernacular Name: Kant-gyok-phyu (Burmese); Ceylong Leadwort (English); Chita, Chitarak, Chitra (Hindi).

Description: A perennial, sub-scandent shrubs; glabrous; flowers white in elongated spikes; capsule oblong, pointed contained in viscid glandular persistent calyx.

Distribution: Native to southern India and Malaysia; naturalized in much of South-East Asia and in Africa.

Parts Used: Leaves, root.

Phytochemistry: Plumbagin, phenolics and tannin.

Uses: Root and leaves used to treat infections and digestive problems such as dysentery; externally, leaves paste is applied to painful rheumatic areas or to chronic and itchy skin problems; paste acts as a counter-irritant; raising blisters and increasing circulation, it speeds the clearing of toxins from the affected area.

Poaceae

Cynodon dactylon (L.) Pers.

Vernacular Name: Myin-sa-myet (Burmese); Niladurva, Durva (Sanskrit), Dhub grass, Barmuda, Bahama grass, Conch grass (English); Dub, Durba (Hindi).

Description: A prostrate extensively creeping glabrous, highly branched perennial grass, rooting at very nods, forming matted tufts; leaves narrow, linear, soft smooth, distichous at the base, ligule a very fine ciliate rim; inflorescence terminal spikes. Green or purplish, rachis slender, involucral glumes acute to subulate-mucronulate, floral glume obliquely oblong to semi-ovate; fruits grins, oblong, laterally compressed about1 mm long.

Distribution: Throughout the world.

Parts Used: Whole plant.

Phytochemistry: β-Inone, 2-propionic, 4-hydroxybenzoic, 2-propionic and 3-methoxy-4ydroxybenzoic acids, phytol, β-sitosterol-D-glucoside, stigmasterol acetate, phagostimulant, phytone, glycosides, saponins, tannins, flavonoids and carbohydrates.

Uses: Plant is useful in hyperdipsia, burning sensation, haemoptysis, haematuria, haemorrhages, wounds, haemorrhoids, conjunctivitis, cephalalgia, erysipelas, leprosy, skin diseases, vomiting, diarrhoea, dysentery, strangury, colporrhagia, abortion and general debility.

Saccharum officinarum L.

Vernacular Name: Kyan (Sugarcane) (Burmese); Ikshu, Khanda, Sarkara (Sanskrit); Sugar cane (English); Ganna (Hindi).

Description: Erect, perennial grass with stout culms and solid, sweet, juicy, green to purplish stems 3-4 cm in diameter and up to or exceeding 3 m in height. Leaves sheathing and overlapping, lance-shaped, upto to 2 m long and 6 cm broad, mature plants bear erect, dense clusters of small wind-pollinated flowers.

Distribution: Widely cultivated.

Part Used: Whole plant.

Phytochemistry: Abscisic acid, apigenin and its glycoside, 5-O-methylapigenin, arabinose, arunodin, para-hydroxybenzoic acid, calcium, campesterol, coumarin, cylindrin, orient and its derivatives, fructose, galactose, gibberellins, glucose, O-methyllupeol, luteolin, phytosterol, potassium, saccharans, schaftoside and isoschaftoside, β-sitosterol, sucrose, invert sugar, swertiajaponin, swertisn, tarasxerol methyl ether, tricin and tricin glycosides, vicenin and xylose.

Uses: Leaf ash is used to treat sore eyes the stem juice is used to treat sore throats; non-sugar part is used as antidiabetic.

***Saccharum spontaneum* L.**

Vernacular Name: Bur-Kaing (Burmese); Thatch Grass, Wild sugarcane (English); Kas, Kus (Hindi).

Description: A tall, erect, perennial grass upto 6 m in height with slender culms and aggressive rhizomes bearing extensive roots; leaves simple, sheathing, narrowly linear, finely acumination, rigid coriaceous, ligule ovate, membranous; spikelets in pairs, on silky hairy panicles, pale or greyish white to purplish-grey in colour; in each spikelet pair one is pedicelled and the other sessile, glumes 4, coriaceous and brown below, obscurely keeled, one-nerved; grains oblong to subglobose.

Distribution: Throughout Indian subcontinent.

Part Used: Whole plant.

Phytochemistry: Stem contains quinones, terpenes, alkaloids, flavonoids, saponins, tannins, carbohydrates, protein, coumarin, phenol, steroids and glycosides; root contains terpenoids, steroids, glycosides, tannins, flavonoids and carbohydrates.

Uses: Roots are useful in burning sensation, strangury, renal and vesical calculi, dyspepsia, haemorrhoids, menorrhagia, dysentery, agalactia, phthisis and general debility.

***Zea mays* L.**

Vernacular Name: Pyaung-bu, Phu-sa-pyaung (Burmese); Maize, Indian corn (English); Makai, makka, Bhutta (Hindi).

Description: Robust annual herbaceous plant, 1-2 m tall; leaves strap-shaped, distichous; margins ciliate, wavy; flowers unisexual, both sexes on the same plant; the male in a terminal spike, the female closely crowded in the axil of the leaves, enclosed in numerous large foliaceous bracts; style silky, elongate, the fruit is a caryopsis.

Distribution: Cultivated everywhere in the world.

Part Used: Seed grains.

Phytochemistry: Grains contain starch, glucose, fructose, sucrose, vitamins E, C. K, β-carotene and phenolics.

Uses: Maize is used as a diuretic in the treatment of heart disease, hypertension, cystitis, urethritis, urinary lithiasis, cholecystitis, hepatitis, rheumatism and diabetes mellitus.

Portulacaceae

Portulaca oleracea L.

Vernacular Name: Mye-byit (Burmese); Brihalloni, Lonika, Lonamla (Sanskrit); Garden purslane, Common purslane (English); Khursa, Baralaniya, Kulfa (Hindi).

Description: Prostrate annual herb; stems succulent, green or reddish; leaves alternate, fleshy, shinning glabrous; base attenuate, apex truncatel nerves inconspicuous; flowers bright yellow in the terminal cluster, without stalks; capsule globose ovoid, opening transversely; seeds numerous, shining black.

Distribution: Distributed Indian subcontinent.

Part Used: Whole plant.

Phytochemistry: Plant contains carotene, vitamins C, B_1, B_2, organic acids, nicotinic and oxalic, noradrenalin, and biflavonoid liquiritin.

Uses: Plant is used as an antibacterial, anti-inflammatory and anthelminthic; used in treating bacillary dysentery and dysuria; fresh plant used as an anthelminthic against oxyuriasis and ascariasis; poultices of fresh leaves are used to treat mastitis, boils and impetigo.

Rhamnaceae

Ziziphus jujuba Mill.

Vernacular Name: Zee-bin-Zee-daw (Burmese); Common jujube; Date (English); Khajoor (Hindi).

Description: A spiny deciduous shrub or s small tree up to 10 m high; leaves alternate, ovate to ovoid-lanceolate, 3-7 cm long by 2-3.5 cm wide; apex slightly obtuse; base oblique; margin closely serrulate, 3-veined, inflorescence an axillary cyme; April-May; flowers perfect, 7-8 flowers in each cluster, greenish-yellow; drupe 1.5-5 cm long, ovate to oblong, dark reddish-brown when ripe, containing a single stone surrounded by fleshy pup; September-October.

Distribution: Cultivated in India, Japan, China, Myanmar, Mauritius, Africa, Malaysia, Afghanistan, and Australia.

Part Used: Fruits.

Phytochemistry: Carbohydrate, fat, protein, amino acids, anthocyanins; bark contains leucocyanidin; stem contains leucopelargonidin, betulinic and ceabothic acids; leaves contain rutin, mauritines A, B, C, D, E, and F, farngufoline and amphbines B, D and F.

Uses: Poor apetiate, general fatigue; loss bowels, palpitation insomnia; night sweats and hysteria.

Rosaceae

Prunus persica (L.) Batsch

Vernacular Name: Bur-hmit-na (Burmese); Peach (English); Aru, Shaftalu (Hindi).

Description: A large deciduous shrub or small tree up to 8 m high; stem erect, branched twigs glabrous; special dwarf shoots bear flowers or else the flowers arise from long branches; leaves exstipulate, subsessile with many pairs of glands at the base of each leaf-blade; alternate, simple, elliptic-lanceolate, 8-15 cm long; apex acute; margin serrulate, glabrous; flowers, cup-shaped, appearing before or with the leaves; calyx a campanulate tube, 5-lobed; petals, pink, stamens many, as long as the petals; carpel one; fruit a flashy and downy drupe enclosing a hard furrowed one-seeded stone.

Distribution: Cultivated in China, India, Mauritius and Japan.

Parts Used: Seeds, fruits.

Phytochemistry: Sugars, thiamine, ascorbic acid, vitamin A; fatty acids.

Uses: Fruits edible; infusion of leaves or bark give for whooping cough; flowers anthelmintic; root-bark used in dysmenorrhoea; amenorrhoea, contusions, strains, and constipation.

Rubiaceae

Neolamarckia cadamba (Roxb.) Bosser (Syn. Anthocephalus cadamba (Roxb.) Miq.)

Vernacular Name: Ma-U (Burmese); Kadambah, Kutsitangah (Sanskrit); Wild chinchona (English); Kadamb (Hindi).

Description: A medium to large sized deciduous tree with clean cylindrical stem horizontal branches and rounded crown, bark dark grey with longitudinal fissures peeling off in thin scales; leaves elliptic-oblong or ovate, 30 cm long and 15 cm broad pubescent beneath; flowers yellow of orange in contemporary herbal medicine.

Distribution: Native to South and Southeast Asia.

Part Used: Bark, leaves, fruits.

Phytochemistry: Leaves contain cadambine, cadamine, isocadamin, 3α-dihydrocadambine, 3α-isodihydro-cadambine, 3β-dihydrocadambine, 3β-isodihydro-cadambine, hentriacontanol and β-sitosterol; stem bark contains sapogenins, cadambagenic acid, quinovic acid, β-sitosterol, saponins A,B, C and D; heartwood contains alkaloids, cadambine, 3α-dihydrocadambine, and isodihydrocadambine.

Uses: Bark is useful in inflammations, haemoptysis, gastropathy, fever, strangury, cough, diarrhoea, vomiting, burning sensation dipsia, ophthalmopathy, wounds, ulcer and debility; decoction of the leaves is good for ulcer, wounds and metrorrhea; fruits are useful in gastric irritability and fever.

Rutaceae

Aegle marmelos (L.) Corrêa

Vernacular Name: Okshit (Burmese); Bael tree, Holy fruit tree (English); Bel (Hindi); Vilvah, Sivadrumah, Sriphalah (Sanskrit).

Description: A medium-sized armed deciduous tree upto 8.0 m high with straight, sharp, axillary thorn and yellowish-brown shallowly furrowed corky bark; leaves trifoliate, aromatic, alternate, leaflets ovate or ovate-lanceolate, crenate pellucid punctate, the lateral subsessile and the terminal long-pertioled; flowers greenish-white, sweet-scented in axillary panicles; fruits globose, woody berry with yellowish ring; seeds numerous oblong, compressed, embedded in orange-brown sweet gummy pulp.

Distribution: Native to Bangladesh and India; present throughout Southeast Asia.

Part Used: Roots, leaves, fruits.

Phytochemistry: Fruit contains marmalosin; bark contains coumarin, fragrine, marmesin, sterols, triterpenoids and umbelliferone; leaves contain α and β-phellandrene, d-limonene.

Uses: Roots are useful in diarrhoea, dysentery, dyspepsia, stomachalgia, cardiopalmus, seminal weakness, uropathy, vomiting, intermittent fever, swellings and gastric irritability in infants; leaves are useful in ophthalmia, deafness, inflammations, catarrh, diabetes and asthmatic, complaints; unrips fruits are useful in diarrhoea, dysentery and stomachalgia; ripe fruits are good for the heart and brain and in dyspepsia.

***Citrus medica* L.**

Vernacular Name: Thanbaya (Burmese); Matulungah (Sanskrit); Citron (English); Baranimbu, Bijaura (Hindi).

Description: An evergreen armed shrub with straggling thorny branches and smooth yellowish-brown bark; leaves oblong or elliptic with acute or rounded apex, coriaceous, glabrous, pellucid-punctate, dull dark green above; flowers white tinged with pink, scented, in axillary cymes; fruits large berries oblong or globose, fleshy, rind thick, rough, irregular or warted, yellow when ripe; seeds few, smooth.

Distribution: Cultivated various Asian countries including India, Mauritius, Pakistan, Bangladesh and Thailand.

Parts Used: Fruits, roots.

Phytochemistry: Campesterol, stigmasterol, β-sitosterol, cholesterol.

Uses: Roots are useful in constipation, colic, flatulence, strangury, renal and vesical calculi, tumours, vomiting, helminthiasis, and dental caries; buds and flowers are useful in anorexia, vomiting tumourus abdominal disorders, asthma, cough, and hiccough; fruits are useful in pharyngodynia, cough, asthma, hiccough, hyperdipsia, otalgia, anorexia, vomiting, hepatopathy, flatulence, haemorrhoids, haemoptysis, amenorrhoea, dysmenorrhoea, dysentery, leprosy and skin disease; seeds are useful in inflammations, skin disease, amenorrhoea, dysmenorrhoea and haemorrhoids.

***Citrus aurantium* L.**

Vernacular Name: Kabala, mak-sawak (Burmese); Seville, Sour orange (English); Khatta nimbu (Hindi).

Description: Small tree to 10m tall with large spines on younger branches; leaves alternate with winged-petioles, the blade ovate, bluntly toothed, emitting strong citrus odour due to the presence of copious oil glands; flowers axillary, borne singly or a few, white and very fragrant; fruit a yellow-orange hesperidium, several to numerous seed within; flowers usually during warmer month and fruit available later in the year.

Distribution: Cultivated and naturalized in almost all Asian countries.

Parts Used: Leaves, bark, fruits.

Phytochemistry: Fruits contain4-methulacetophenone, carotenoids, essential oil (monoterpenses and sesquiterpenes), coumarins, caffeine, isoquinoline alkaloids, synerphrine, 5-methul and N-methul tyramines; flavonoids, triterpenoids, steroids (desmosterol, ergosterol, β-sitosterol and stigmasterol).

Uses: Leaves and bark used for headaches; juice of the fruit is mixed with coconut oil or castor oil is given as a laxative; leaves are used treat abdominal pain; bark or leaves are boiled and taken to treat urinary tract infection.

Murraya koenigii (L.) Spreng.

Vernacular Name: Bur-Kyuangthwe (Burmese); Surabhininiba (Sanskrit); Curry leaf tree (English); Kathnim, Mitha neem, Kurry patta, Gandhela, Barsanga (Hindi).

Description: An aromatic shrub or small tree; leaves pinnate, leaflets mostly ovate, crenate- dentate; flowers white in corymbose cymes; berries purplish black.

Distribution: Throughout tropical and sub-tropical areas; cultivated in India, Sri Lanka, Mauritius and neighbouring countries.

Part Used: Leaves.

Phytochemistry: Essential oil (β-caryophyllene, β-gurjunene, β-elemene, and β-phellandrene); alkaloids (mahanimbine, girinimbine, koenimbine and mahanine).

Uses: Leaves are extensively used for flavouring curries, chutneys, soup, pickles; tea made by leaves used for cough and cold.

Toddalia asiatica (L.) Lam.

Vernacular Name: Bur-Kyan-za (Burmese); Forest pepper, Lopez-root tree Wild orange tree, (English); Jangli-Kalimirc, Kanc, (Hindi); Tiksnaksah, Kancanah (Sanskrit).

Description: A variable prickle sarmentose rambler up to 15 m in height, circular lenticels and hooked prickles; leaves digitately trifoliate, leaflets oblong, elliptic or obovate, crenulate, glabrous and gland-dotted; flowers white or greenish-yellow in axillary and terminal pubescent panicles, polygamous, fruits, subglobose, fleshy, orange coloured; seeds reniform, surrounded by colourless mucilage.

Distribution: Native to many countries in Africa and Asia.

Parts Used: Roots, leaves, flowers, fruits.

Phytochemistry: Essential oil, berberine; roots contain resin; leaves contain toddaline, toddalolactone, pimpinellin, isopimpinellin, resins and glycosides; coumarins-norbraylin, 5,7,8-trimethoxycoumarin.

Uses: Roots are useful in odontalgia, paralysis, malarial and intermittent fever, dyspepsia, tonic, flatulence, diarrhoea, cough, bronchitis, nausea, wound, filthy ulcers, epilepsy, gonorrhoea, and general debility; flowers used externally in wasp-stings.

Zanthoxylum armatum DC. (Syn. Z. alatum Roxb.)

Vernacular Name: Tezaba (Burmese); Toothache tree (English); Tumru, Tejphal (Hindi); Tumburuh (Sanskrit).

Description: An armed shrub or a small tree, upto 6 m in height with dense glabrous foliage and straight pickles on stems, blaze yellowish-brown rapidly darkening on exposure; leaves compound, imparipinnate, rachis winged, often bearing pink prickles, leaflets 5-11, lanceolate, more or less serrate, each serrature with a pellucid gland, glabrous, dark flossy green above, pale beneath, the terminal larger than the laterals; flowers yellow or green in dense terminal or axillary panicles, polygamous; fruits reddish subglobose glabrous follicles; seeds solitary, globose, shining, black.

Distribution: Native to warm temperate and subtropical areas worldwide.

Part Used: Bark, fruits.

Phytochemistry: Plant contains pipevine, linalool, linalyl acetate, citral, cinnamatic, limonene, sabinene.

Uses: Bark and fruits are useful in tumours, odontalgia, cephalalgia, otopathy, ophthalmopathy, dyspepsia, colic, flatulence, helminthiasis, diarrhoea, hepatopathy, splenopathy, strangury, fever, leucoderma leprosy, skin disease cough, asthma, paralysis, gout, convulsions, diabetes, wounds ulcers, stomatopathy, pharyngopathy, cardiac debility and general debility.

Santalaceae

Santalum album L.

Vernacular Name: Santagu-byu (Burmese); Sandal tree, White-sandal tree (English); Safed candan (Hindi); Candanah, Srikhandam (Sanskrit).

Description: A medium-sized evergreen, semiparasitic, glabrous tree with slender drooping branches, reaching up to 18 cm in height; bark grey or brownish-black; rough with short vertical cracks; leaves simple, opposite, elliptic-lanceolate glabrous, entire; flowers brownish purple, reddish-purple or violet in terminal and axillary paniculate cymes; fruits globose drupes, purple-black with ribbed endocarp; seeds hard, globose or obovoid; heartwood is light yellowish brown when fresh, turning dark brown to dark reddish-brown on exposure; wood is highly scented.

Distribution: Native to semi-arid areas of the Indian subcontinent; planted in India, China, Sri Lanka, Mauritius, Indonesia, Malaysia, the Philippines and Northern Australia.

Part Used: Heartwood.

Phytochemistry: Leaves, fruits and seeds contain santalbic acid, palmitic acid, oleic acid, linoleic acid and sugar; fruits contain betulic acid and β-sitosterol and fatty acids; leaves contain betulinic acid, β-sitosterol, glucose, fructose and sucrose; heartwood contains an essential oil (α- and β-santalenes, santenol and teresantanol).

Uses: Heartwood is useful in dipsia, cephalalgia, foul odour due to hyperhidrosis, skin diseases, leprosy, forgetfulness, amentia, psychopathy, cardiac debility, haemoptysis, hyperacidity, jaundice, strangury, cough, bronchitis, inflammations, cystitis, dysentery, gastric irritability, menorrhagia, leucorrhoea, spermatorrhoea, intermittent fever and general debility.

Solanaceae

Datura stramonium L.

Vernacular Name: Padaing-Khatta (Burmese); Thornapple (English); Datura (Hindi).

Description: Robust annual growing to 1 m; has lobed oval leaves, long white or violet trumpet-shaped flowers, and spiny fruit capsules similar to those of the horse chestnut.

Distribution: Grows in the Americas, Europe, Asia and North Africa.

Parts Used: Leaves, flowers, seeds.

Phytochemistry: Tropane alkaloids, flavonoids, withanolides, coumarins and tannins.

Uses: Common remedy for asthma, whooping cough, muscle spasm and the symptoms of Parkinsonism; relaxes the muscles of the gastrointestinal, bronchial and urinary tracts, and reduces digestive and mucous secretions.

Nicotiana tabacum L.

Vernacular Name: Say, (Tabacco) (Burmese); Tamakhuh (Sanskrit); Tabacco (English); Tamaku (Hindi).

Description: A glandular-pubescent erect herb with few branches; leaves simple, ovate, oblong-lanceolate or elliptic upto 100 cm or more length, the lower ones semi-amplexicaul and decurrent; flowers light red, white or pink in many-flowered panicled racemes; fruits narrowly elliptic, ovoid capsule, seeds many, brown, with fluted ridges.

Distribution: Cultivated in the hotter parts of the Indian subcontinent; tropical and sub-tropical areas.

Parts Used: Leaves.

Phytochemistry: Nicotine, nornicotine, anabasine, l-nicotine, nicotyrine, nicotimine, l-noricotine, d-nornicotine, piperidine, pyrrolidine, N-methylpyrroline, 2, 3'-dipyridyl, l-Anabasine, N-methyl-L-anabasine, l-anatabine, N-methyl-l-anatabine, nicotonic, necotelline and Nyosmine.

Uses: Leaves are useful in odontalgia, dental caries, inflammations, helminthiasis, dyspepsia, flatulence, haemorrhages, bronchitis, asthma, scabies skin diseases, ulcer, painful timers, strangulated hernia, purulent discharge from the nose, tubercular glands of the neck, and otalgia.

Solanum anguivi Lam. (Syn. S. indicum L.)

Vernacular Name: Bur-Khayan-Kazaw (Burmese); Brhati, Simhi (Sanskrit); Poison berry (English); Barhanta, Birhatta (Hindi).

Description: A much branched, common, very prickly undershrub 0.3-1.5 m in height; leaves simple, large, ovate, subentire, sinuate or lobed, sparsely prickly on both sides, base cordate, often unequal; flowers blue in extra-axillary cymes, peduncles stellately hairy and prickly; fruits globose, berries, reddish of dark yellow; seeds smooth or minutely pitted.

Distribution: Throughout India and nearby countries including Mauritius.

Parts Used: Root.

Phytochemistry: Fruits contain enzyme and alkaloids; roots and leaves contain solanine and solanidine.

Uses: Roots are useful in odontalgia, dyspepsia, flatulence, colic, verminosis, diarrhoea, pruritus, leprosy, skin diseases, strangury, cough, asthma, bronchitis, amenorrhoea, dysmenorrhoea, fever, cardiac disorders and vomiting; leaves and fruits are digestive, laxative and antibacterial, and are useful in odontalgia, pruritus and ringworm.

Withania somnifera (L.) Dunal

Vernacular Name: Dahat-tha-gaing (Burmese); Winter cherry (English); Asgandh (Hindi); Asvagandha, Varahakarni (Sanskrit).

Description: An erect branching undershrub reaching about 150 cm in height usually clothed with minutely stellate tomentum; leaves ovate up to 10 cm long; flowers greenish or lurid yellow in axillary fascicles; fruits globose berries which are orange coloured when mature enclosed in a persistent calyx; fleshy roots when dry are cylindrical, gradually tapering down with a brownish white surface and pure white inside when broken.

Distribution: Cultivated in India, Nepal, China, Yaman and Mauritius.

Part Used: Roots, leaves.

Phytochemistry: Roots contain pyrazole alkaloids; withasomnine, steroldal lactones, withaferin A and withanolides; starch, reducing sugars, hentriacontane, glycosides, dulcitol and withaniol.

Uses: Tubers are useful in leucoderma, constipation, insomnia, tissue-building and nervous breakdown; leaves are bitter and are recommended in fever, painful swellings and ophthalmitis; paste of the roots and bruised leaves are applied to carbuncles, ulcer and painful swelling.

Vitaceae

Vitis vinifera L.

Vernacular Name: Sapyit (Burmese); Common grape-vine, Wine grape, European grape (English); Angoor (Hindi); Draksa, Mrdvika (Sanskrit).

Description: A large, deciduous tendril climber, tendrils leaf-opposed, often bifid; leaves simple, more or less deeply 3-5 lobed, orbicular-cordate, irregularly toothed, glabrescent above, grey tomentose beneath, thin, membranous; flowers small green in leaf-opposed panicled cymes; fruits bluish-black, purplish or greenish berries, ovoid to globose; seeds 2-4 with a discoidal tubercle on the back.

Distribution: Cultivated throughout the world.

Parts Used: Fruits, leaves, stems, flowers.

Phytochemistry: Leaves contain thiamine, niacin, biotin, tocopherol, hulme, hexokinase, catalase, ascorbic acid oxidase, perosidase, polyphenol oxidase, ascorbic acid oxidase, peroxidase and polyphenol oxidase.

Uses: Fruits are useful in dipsia, constipation, amentia, cardiac debility, haemoptysis, haemorrhages, anaemia, atrangury, emaciation, fever, leprosy, skin diseases, dyspepsia, colic, flatulence, cough, asthma, bronchitis; leaves are useful in cephalalgia, strangury, scabies, skin diseases, syphilis, haemorrhoids, splenomegaly, vomiting and stomatorrhagia; flowers are useful in bronchitis, hepatopathy, anaemia, amenorrhoea and dysmenorrhoea.

Xanthorrhoeaceae

Aloe vera (L.) Burm. f. (Syn. A. barbadensis Mill.)

Vernacular Name: Sha-zaung-Let-Pat (Burmese); Barbados aloe, Jaffarabad aloe (English); Ghikuamr, Ghikumari (Hindi); Ghartkumari (Sanskrit).

Description: A coarse perennial with a short stem and shallow root system; leaves fleshy in rosettes, sessile, often crowded with horny prickles on the margins convex below, 45-60 cm long tapering to a blunt point, surface pale green with irregular white blotches; flowers yellow or orange in racemes; fruits loculicidal capsule.

Distribution: Cultivated all over the world.

Part Used: Leaves, leaf-juice.

Phytochemistry: Leaves contain barbaloin, chrysophanol glycoside, aloe-emodin; mucilage of leaves contains glucose, galactose, mannose, galacturnic acid, aloesone and aloesin.

Uses: Plant is bitter, sweet, cooling, anthelmintic, aperients, carminative, deobstruent, depurative, diuretic, stomachic, emmenagogue, ophthalmic and alexeteric; juice is used in dyspepsia, amenorrhoea, burns, colic, hyperadenosis, hepatopathy, splenopathy, skin diseases, constipation, spanomenorrhea, abdominal tumours, dropsy, carbuncles, sciatica, lumbago and flatulence; used for local application in painful inflammations, chronic ulcers and catarrhal and purulent ophthalmia.

Zingiberaceae

Curcuma longa L.

Vernacular Name: Sa-nwin (Tunneie) (Burmese); Turmeric (English); Haldi (Hindi).

Description: A perennial herbaceous plant that reaches up to 1 m tall; highly branched, yellow to orange, cylindrical, aromatic rhizomes are found; leaves are alternate and arranged in two rows; divided into leaf sheath, petiole, and leaf blade; from the leaf sheaths, a false stem is formed; petiole is 50 to 115 cm long; simple leaf blades are usually 76 to 115 cm long and rarely up to 230 cm, width of 38 to 45 cm and are oblong to elliptic, narrowing at the tip.

Distribution: Native to southern Asia; cultivated throughout the world.

Part Used: Rhizomes.

Phytochemistry: Curcuminoids (curcumin, demethoxycurcumin, and bisdemethoxycurcumin); volatile oils (turmerone, atlantone, and zingiberene); sugars, proteins, and resins.

Uses: Rhizomes are used to treat a variety of internal disorders, such as indigestion, throat infections, common colds, or liver ailments, as well as topically to cleanse wounds or treat skin sores.

Zingiber officinale Roscoe

Vernacular Name: Gyin-gyin-sein (Burmese); Ginger (English); Adarak (Adrak); Ardrakam (Sanskrit).

Description: A slender, perennial rhizomatous herb; leaves linear, sessile, glabrous; flowers yellowish-green in oblong, cylindric spikes, ensheathed in a few scarious, glabrous bracts; fruits oblong capsules. The rhizomes are white to yellowish brown in colour, irregularly branched, somewhat annulated and laterally flattened. The growing tips are covered over by a few scales. The surface of the rhizome is smooth and if broken a few fibrous elements of the vascular bundles project out from the cut ends.

Distribution: Cultivated throughout the world.

Part Used: Rhizomes.

Phytochemistry: α-Curcumene, α-curcumene, β-D-curcumene; α-bergamotene, β-camphene, γ-bisabolene, β-bourbornene, d-borneol and its acetate, calamine, d-camphene, car-3-ene, α-cedrol, citral, citronellol.

Uses: Raw ginger is useful in anorexia, dyspepsia, pharyngopathy and inflammations; dry ginger is useful in dropsy, otalgia, cephalagia, asthma, cough, colic, diarrhoea, flatulence, anorexia, cardiopathy, pharyngopathy, cholera, nausea, vomiting, elephantiasis and inflammations.

Zygophyllaceae

Tribulus terrestris L.

Vernacular Name: Su-le-thein (Burmese); Land-caltrops, Puncture-vine (English); Gokharu (Hindi); Goksurah, Svadamstra (Sanskrit).

Description: An annual or perennial, prostrate herb with many slender, spreading branches and silky-villous young parts; leaves abruptly simple, pinnate, opposite, leaflets almost sessile, rounded or oblique at the base, mucronate at the apex; flowers bright yellow, solitary, pseudo-axillary or leaf-opposed; fruits, a 5 angled or winged spinous tuberculate woody schizocarp, separating into five cocci, each coccus having two long, stiff, sharp divaricate spines towards the distal half and two shorter ones nearer the base; seeds one or more in each coccus.

Distribution: Throughout Indian subcontinent including Mauritius and Myanmar.

Parts Used: Whole plant.

Phytochemistry: Saponins, diosgenin, gitogenin, chlorogenin, ruscogenin, 25-D-spirosta-3.5-diene, kaempreroI, kaempferol-3-glucoside, kaemperol-3-rutinoisd and falconoid tribuloside.

Uses: Roots and fruits are useful in strangury, dysuria, renal and vesical calculi, anorexia, dyspepsia, helminthiasis, cough, asthma, consumptions, cardiopathy, haemoptysis, spermatorrhoea, anaemia, scabies, ophthalmia, ulocace and general weakness; leaves are useful in gonorrhoea, gleet, ulitis, inflammation, menorrhagia, strangury, leprosy, skin diseases, verminosis and general weakness; seed is useful in epistaxis, haemorrhages and ulcerative stomatitis.

Colour Plates of Important Medicinal Plants

Paederia foetida

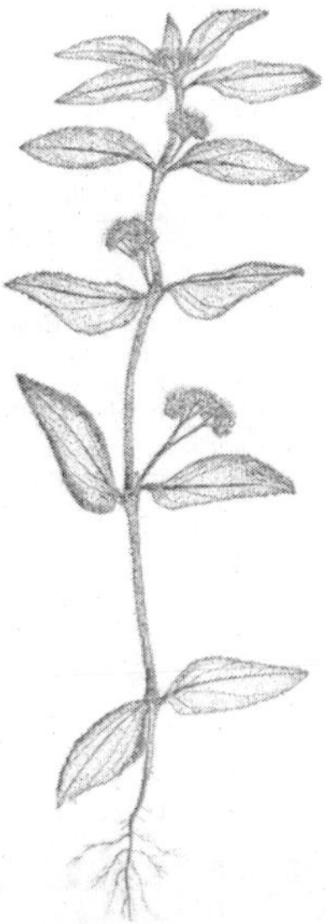

Euphorbia hirta

Cissampelos pareira

Drimia indica

Cassia occidentalis

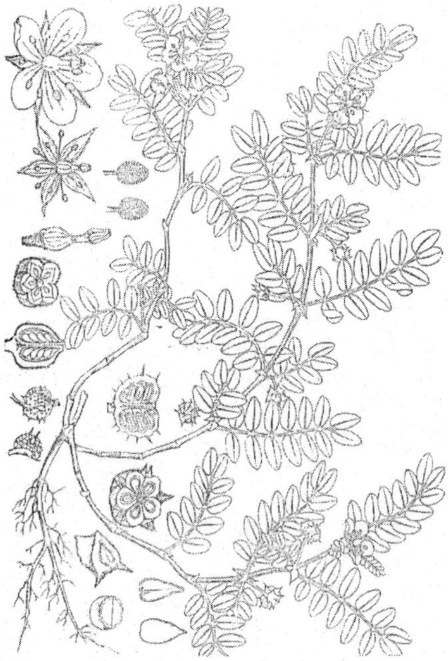

Tribulus terrestris

Phyllanthus emblica

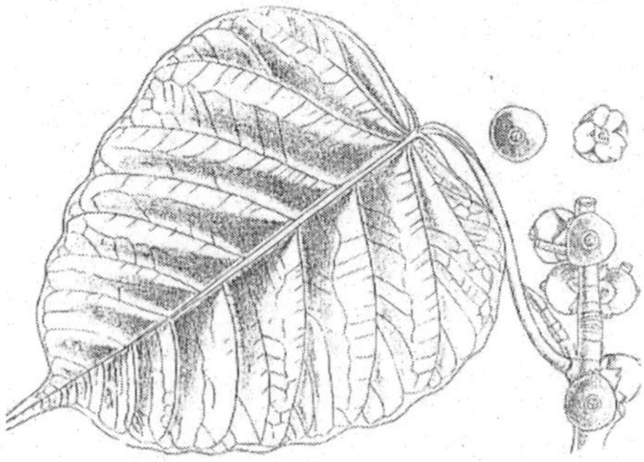

Ficus religiosa

Centella asiatica

Anacyclus pyrethrum

Artemisia vulgaris

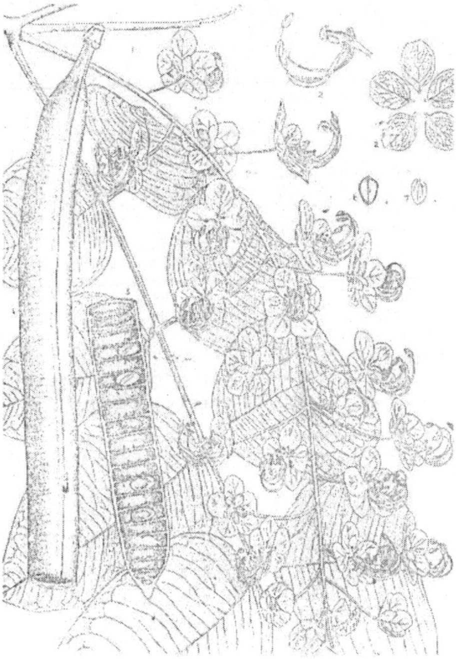

Cassia fistula

REFERENCES

Bouton, Louis (1864). *Plants medicinales,* Maurice.

Chopra, R.N., Nayar, S.L. and Chopra I.C. (1956). *Glossary of Indian medicinal Plants.* C.S.I.R New Delhi.

CRC World Dictionary of Medicinal and Poisonous Plants Common Names, Scientific Names, Eponyms, Synonyms, and Etymology.

Dhiman A.K. (2005). *Medicinal Plants of Uttaranchal State.* Chowkhamba Sanskrit Series Office, Varanasi, India

Edible Plants of North West Himalaya (Uttarakhand).

Gurib-Kakim, Ameenah (1995-96). *Plants Medicinales de Maurice Tome I Universite de Maurice [Medicinal Plants of Mauritius Tome I University of Mauritius].* Mauritius Sugar Industry Research Institute.

Kurian, J.C., Mutatkar R.K., Samraj E. (1995). *Plants that heal.* P.H. Lall, Pune-India.

Malcy-de-Chazal 1803-1880. *The medicinal Plant of Mauritius.*

Owadaly, A.W. (1976). *A guide to the Royal Botanic Garden,* Pamplemousses.

Prajapati, N.D., Purohit, S.S., Sharma, A.K., Kumar, T. (2007). A hand book of Medicinal Plants. *A Complete Source Book.*

Robbins, Christopher (1995). *Herbalism,* 77-85, Fulcham Place Road, London W685B.

Rodger, Alex (1922). *LACE- List of trees, shrubs, & principal climbers, etc., recorded from Burma, with vernacular names*; Rangoon Superintend, Government Printing, Burma, 1922.

Sawyer, A.M. and Nyan, Daw (1927). *A classified list of the plants of Burma.* Rangoon Superintend, Government Printing and Stationary, Burma, 1927.

Shah R. (2004). *Nature's Medicinal Plants of Uttaranchal* (Trees, Shrubs & Climbers). Gyanodaya Prakashan, India

Shah, N.C, (1993). *Ayurveadic medicine in Mauritius,* India View.

Shanker, V., Prakash S. and Uniyal, M.R. (1972). Studies in Ashtavarg-chromatographic examination of sugars in the tubers of Harbenaria intermedia and Mycrostylis Wallichii. *Jour. Res. Ind. Med.* 7 (1), 91-92, 1972.

Sussman, Linda K. (1979). Herbal medicine in Mauritius. *Journal of Ethno-Pharmacology,* Netherland.

The Wealth of India. National Institute of Science Communication and Information Resources (NISCAIR), India

Uniyal, M.R. 1996. *Cultivation of Medicinsl Plants needed for Agriculture of Agro Forestry system in India.*

Uniyal, M.R. 1997. *Mauritius dweep me uplabdh Ayurvedic Jadi bootiya se upchar [Treatment with Ayurvedic herbs available in Mauritius Island].*

Uniyal, M.R., Semwal, D.K., Raj, K., Semwal, R.B. *Indigenous plants of Myanmar used in traditional medicine.* ISBN: 978-81-934455-2-5; AYUSH Health India, New Delhi, 2017.

Uniyal, M.R., Semwal, D.K., Raj, K., Semwal, R.B., Chauhan, A. *Traditional Medicinal flora of Mauritius* (Indigenous and Endemic). ISBN 978-3-16-148410-0; Pravek Kalp Private Limited, Noida, India, 2018.

W.H.O. 1978. *Primary Health Care Reports of International Conference on Primary Health care Alma-Ata,* Germany.

W.H.O. 1989. *Medicinal Plants in China.*

W.H.O. 1990. *The use of Traditional medicine in Primary health Care.*

W.H.O. 1991. *Guidelines for the Assessment of herbal medicine.*

W.H.O. 1992. *Natural Resources and Human Health and Plants of Medicinal and Nutritional Value.*

Watt, G. (1972). High Commission of India Mauritius. *Dictionary of Economic Products of India,* New Delhi.

Wiart, Christophe. *Medicinal plants of the Asia-Pacific Drugs for the future.*

ABOUT THE AUTHORS

Dr. Mayaram Uniyal

Dr. Mayaram Uniyal is one of the most experienced and prominently known experts of Ayurvedic Herbs in India. He earned A.M.B.S. (Ayurvedacharya); Ayurved Bhaskar Charak Samhita and Netra Vigyan with a specialty in Dravyaguna. He served as Director, CCRAS, Ministry of AYUSH, Govt. of India as well as a Senior Consultant, Medicinal and Aromatic Plants of Govt. of Uttarakhand. Besides, he is a member of Ayurvedic Pharmacopeia and Fellow of National Ayurveda Vidhyapeeth, AYUSH Ministry, Govt. of India. Formerly, he served as Technical Consultant, Ayurveda in Govt. of Himachal Pradesh and Member of National Medicinal Plants Board, Indian System of Medicine and Homeopathy, etc. Dr. Uniyal is an author of more than 50 useful and informative books of Medicinal and Aromatic Plants. Among them, few books are Medicinal flora of Garhwal Himalaya; Ashtang Samgrahki Vanoshadiyan evam Vargikaran; Laddakh ki Sanskriti evam Paramparagat Tibeti Chikitsa Pranali; Bihar ki Adivashi Jadibootiyan; Sanskriti Sahitya me Paryavaran evam Kalidaski Vanaspatiyan; Prayogatmak Abhinav Dravyaguna Vigyan; Bharat ki Jadibootiyon ka Krishikaran; Uttarakhand Himalaya ki Vanoshadhiyan evam Khanij; Ayurved Anusandhan Darshika; Uttarakhand him shikharon par divyasanjeevani aushadhiyan; Uttarakhand vanoshadhi darshika; Himalaya luptprayah vanyajantu kastoori mrig evam mahoshadi kastoori, etc.

Dr. D. K. Semwal

Dr. D. K. Semwal is currently serving as a faculty member at the Department Phytochemistry, Uttarakhand Ayurved University, India. He has been a Postdoctoral Research Fellow at Tshwane University of Technology, South Africa. He received his MSc and PhD (Phytochemistry) from HNB Garhwal University, India. He served as a Lecturer at HNB Garhwal University, India, a R&D scientist at an India based pharmaceutical industry. He received prestigious Dr DS Kothari Postdoctoral Fellowship from University Grants Commission, Govt. of India and worked at Panjab University, Chandigarh. He has been a PI of six R&D projects funded by different agencies. Presently, he is serving as an editor and a reviewer for many international journals. He has been published about 100 research papers in various journals of international repute and many books are already in his account. His current research interests are phytochemistry, phytomedicine, clinical pharmacology, drug discovery and Ayurveda.

Dr. Ruchi B. Semwal

Dr. Ruchi B. Semwal is presently working as Assistant Professor at Govt. Postgraduate College, Rishikesh, India. She earned her PhD in the area of Natural Products having medicinal and dyeing properties. She screened more than 50 plants of medicinal use in various laboratories in India and overseas for their scientific authentication. She developed various methods to isolate bioactive molecules from the plants. Besides, she optimized some traditional dyeing methods in the laboratories for industrial use. Around 70 research papers and 5 books are already in her credit. She has been a recipient of many prestigious awards including NRF Postdoctoral Fellowship, UGC-Dr DS Kothari Postdoctoral Fellowship and UGC-Research Fellowship for PhD. Her research interests are natural products chemistry and bioactivity of natural molecules.

INDEX

D

E

F

G

H

I

J

P

Q

R

S

T

U

V

W

X

Y

Z